CHEMICAL MUTAGENS

Principles and Methods for Their Detection

Volume 1

Sponsored by the Environmental Mutagen Society

CHEMICAL MUTAGENS

Principles and Methods for Their Detection
Volume 1

Edited by Alexander Hollaender

Division of Biology
Oak Ridge National Laboratory
Oak Ridge, Tennessee

with the cooperation of
Ernst Freese, Kurt Hirschhorn, and Marvin Legator

℗ PLENUM PRESS • NEW YORK-LONDON • 1971

ISBN-13: 978-1-4615-8968-6   e-ISBN-13: 978-1-4615-8966-2

DOI: 10.1007/ 978-1-4615-8966-2

Library of Congress Catalog Card Number 73-128505
SBN (Vol. 1) 306-37101-4
SBN (2-Volume set) 306-37100-6

© 1971 Plenum Press, New York
Softcover reprint of the hardcover 1st edition 1971
A Division of Plenum Publishing Corporation
227 West 17th Street, New York, N.Y. 10011

United Kingdom edition published by Plenum Press, London
A Division of Plenum Publishing Company, Ltd.
Davis House (4th Floor), 8 Scrubs Lane, Harlesden, NW10 6SE, England

Contributors to Volume 1

Bruce N. Ames
Biochemistry Department
University of California
Berkeley, California

P. Brookes
Chester Beatty Research Institute
Institute of Cancer Research:
Royal Cancer Hospital
London, England

John W. Drake
Department of Microbiology
University of Illinois
Urbana, Illinois

Ernst Freese
Laboratory of Molecular Biology
NINDS-NIH
Bethesda, Maryland

Bernard Heinemann
Research Division, Bristol Laboratories
Division of Bristol-Myers Company
Syracuse, New York

Roger M. Herriott
Department of Biochemistry
School of Hygiene and Public Health
The Johns Hopkins University
Baltimore, Maryland

Harold Kalter
Children's Hospital Research Foundation
    and Department of Pediatrics
Univ. of Cincinnati College of Medicine
Cincinnati, Ohio

P. D. Lawley
Chester Beatty Research Institute
Institute of Cancer Research:
Royal Cancer Hospital
London, England

Elizabeth C. Miller
McArdle Laboratory for Cancer Research
University of Wisconsin Medical Center
Madison, Wisconsin

James A. Miller
McArdle Laboratory for Cancer Research
University of Wisconsin Medical Center
Madison, Wisconsin

T. R. Manney
Department of Microbiology
Case Western Reserve University
Cleveland, Ohio

R. K. Mortimer
Donner Laboratory
University of California
Berkeley, California

Bernard S. Strauss
Department of Microbiology
The University of Chicago
Chicago, Illinois

C. Yanofsky
Department of Biological Sciences
Stanford University
Stanford, California

Foreword[*]

These volumes on methods for detecting chemical mutagens have been sponsored by the Environmental Mutagen Society. Immediately after the Society was organized in March 1969, certain urgent needs became obvious. One was the need for a registry of compounds that have been tested for mutagenesis. Such a registry has been organized under the title of Environmental Mutagen Information Center (EMIC). Another essential item was an organized description of the methods now available for detecting chemical mutagens and for determining the relation between them and teratogens and carcinogens. These volumes are being published to fill this urgent need.

A good part of the study of chemical mutagens is based on experience with radiation as a mutagenic agent. The broad experience obtained in that area over the last 40 years is being put to good use in the chemical field, and although it should be remembered that there are many more complications connected with chemical mutagenesis than with radiation mutagenesis, these complications create the probability that new information about the structure and function of the gene will grow out of studies on chemical mutagenesis.

It is most urgent that new compounds with which people come in contact be tested by some of the methods described in these volumes, and that the mutagenicity of these compounds be evaluated as soon as possible. There seems to be little question that carcinogenesis, teratogenesis, and mutagenesis are related for many compounds, but for other compounds this relation does not appear to exist.

The field of chemical mutagen studies is very active, and it is expected that new information will become available in the near future which will be incorporated into later editions of these volumes.

[*]Research sponsored by the U.S. Atomic Energy Commission under contract with Union Carbide Corporation.

As editor I want especially to thank Dr. Ernst Freese for helpful co-operation in preparing these volumes, and to express my appreciation to Drs. Kurt Hirschhorn and Marvin Legator, the other members of the editorial board.

Alexander Hollaender

January 1971

Preface

The purpose of these volumes is to encourage the development and application of testing and monitoring procedures to avert significant human exposure to mutagenic agents. The need for protection against exposure to possibly mutagenic chemicals is only now coming to be generally realized. The recently issued Report of the Secretary's Commission on Pesticides and Their Possible Effects on Health (the Mrak Report—U.S. Department of Health, Education and Welfare, December 1969) has made an important start. Its Panel on Mutagenicity recommends that all currently used pesticides be tested for mutagenicity in several recently developed and relatively simple systems. Whether recommendations such as these are actually put into effect will depend on convincing government, industry, and the public that the problem is important, that the proposed tests would be effective, and that they can be conducted at a cost that is not prohibitive.

Why is it important to screen environmental agents for mutagenic activity? To those who will read this book, the answer is self-evident. The *sine qua non* of all that we value and all that we are is our genetic heritage. Our knowledge of genetics and molecular biology clearly establishes the possibility that exposure of human germ tissue to certain exogenous agents can cause genetic damage. Although some mutagenic agents are quite toxic in other ways and therefore may be detected before causing major damage, other mutagens may be far more mutagenic than toxic and thus could escape notice until very serious genetic damage had already been done. We must also remember that, by its nature, this genetic damage can be cumulative over generations while even the most insidious nongenetic poison cannot accumulate in the body beyond the lifetime of an individual. Thus, the safeguards we need must be specific and they must be sensitive.

Among the particular tests already developed to the point of routine applicability, the three recommended by the Mrak Commission panel, used in conjunction, seem the most suitable at present. These are the dominant lethal test, the host-mediated microorganism assay, and direct cytogenetic observation, all performed in mammals. These tests give reproducible results and are relatively cheap and simple to do—simpler and cheaper, for example, than currently standard carcinogenicity tests.

In addition to these mammalian and mammalian host-mediated systems, there are several well-developed microbiological test systems which are so sensitive, and so simple, that they are clearly worthwhile as an adjunct. The Mrak Panel recommends they be so used. Drosophila is also a sensitive and fairly convenient organism for mutagenicity testing, but for routine screening the use of microorganisms is simpler.

The greatest additional need is for a direct screening test against agents that mainly cause "point" mutations in mammals. These mutations are generally recessive. Although several methods can detect point mutations in other organisms, testing for single-locus mutations in laboratory rodents is the only experimental method described in this collection that is designed to detect recessive mutations induced in mammalian germ tissue. The method was developed in order to study radiation mutagenesis. Its exploitation has led to discoveries of great relevance to man that could not have been made by experimentation with microorganisms or *Drosophila*. However, its utility for the screening of possible chemical mutagens is limited by its cost, by the smallness of the number of investigators and facilities able to conduct such work, and by problems of interpretation that are just beginning to be explored. For example, we need to know what kinds of recessive mutations are picked up by testing against the particular recessive alleles that are chosen as testers. It may be that a new choice of test loci will be needed in order to achieve efficient detection of recessive mutations corresponding to single amino acid substitutions.

The application of the single-locus method to the study of radiation mutagenesis in mice led to the discovery of large differences in sensitivity, depending on the sex and on the stage of gametogenesis and suggesting the operation of repair mechanisms. The study of chemical mutagenesis is sure to reveal even greater complexities than these, for there are several quite different classes of mutagenic agents. Furthermore, there may be considerable variation in metabolic reactions that inactivate mutagens or that produce mutagens from other compounds. Though it detects only some of the mutagenic effects, a workable mammalian single-locus system that could detect single amino acid substitutions would be worthwhile to test compounds known to be mutagenic in other systems, so as to build up a better understanding of the complexities that may be relevant to human exposure. However, for a long time to come, the simpler tests in mice or other laboratory animals outlined in the Mrak Report may be the only effective experimental method for testing the mutational sensitivity of the germ tissue itself.

Quite recently, systems of great sensitivity have been developed for detecting the induction of recessive mutations in cultured mammalian somatic cells, including human cells. Chromosome aberrations can also be readily detected. Cell cultures can be used to test agents directly and also to detect mutagenic products of metabolism formed after administration of agents to living animals. So far, these cultures respond to more agents than intact animals do. But it is to be expected that these techniques will develop into practical routine screening programs of great power and relevance to human hazards.

Finally, no laboratory screening program can completely substitute for actual monitoring of the mutation rate in the human population. Our rapidly growing knowledge of the basic mechanisms of mutation can help to direct suspicion at certain agents. Laboratory screening systems can certainly reduce the chance of serious human exposure. But now that man has created for himself an unnatural and rapidly changing environment, he ought to have the wisdom to institute a continuing scan of his own muation rate. Otherwise, some chemical or combination of chemicals or perhaps a virus that no one knew or thought to test or some mechanism not detectable in laboratory systems may inflict serious damage before it is detected and before its genetic nature is realized.

The monitoring of certain appropriately chosen dominant and sex-linked recessive "sentinel phenotypes" is practical and could be started now on a trial basis, to be expanded as experience is gained. Monitoring point mutations that change the electrophoretic behavior of human enzymes and other proteins could be started on the same basis. Increased knowledge of fundamental human genetics would accrue as a side benefit. When the subject is first broached, the cost of any significant program for monitoring the human mutation rate may seem excessive. Let us assume that the cost for a system based on protein electrophoresis able to detect a doubling of the point-mutation rate in one year is some tens of millions of dollars, not an unreasonable estimate. Is this too much to spend on population monitoring? It would represent a few percent of the cost of childbirth in the United States, a few tenths of a percent of the Department of Health, Education and Welfare budget, and a few hundredths of a percent of the Federal budget.

Just as in the case of carcinogenicity testing, there will be important differences in response between different systems for mutagenicity. However, this is no excuse for doing nothing. Available testing methods are vastly better than no testing at all. And any intelligent testing program is sure to generate information and interest that will lead to a better understanding of the complexities caused by metabolic reactions, repair mechanisms, and other factors. Indeed, as compared with cancer testing, we are already in a better position to conduct meaningful tests since we have a fairly good understanding of the molecular basis of mutation but no comparable insight into carcinogenesis.

Given the importance of averting serious mutational damage and considering that practical means are now available for screening against environmental mutagens, we have a responsibility to implement at least a pilot screening program without further delay.

Matthew Meselson

The Biological Laboratories
Harvard University
Cambridge, Massachusetts

Contents of Volume 1

Chapter 2

**Correlation Between Teratogenic and Mutagenic Effects of
Chemicals in Mammals** 57
by Harold Kalter

Chapter 3

**The Mutagenicity of Chemical Carcinogens: Correlations,
Problems, and Interpretations** 83
by Elizabeth C. Miller and James A. Miller

Chapter 4

Effects on DNA: Chemical Methods 121
by P. Brookes and P. D. Lawley

Chapter 5

Physical–Chemical Methods of the Detection of the Effect of Mutagens on DNA 145
by Bernard S. Strauss

Chapter 9

The Detection of Chemical Mutagens with Enteric Bacteria 267
by Bruce N. Ames

Contents of Volume 2

Chapter 14

Chapter 15

Chapter 18

Root Tips for Studying the Effects of Chemicals on
Chromosomes 489
by B. A. Kihlman

Chapter 19

Cytogenetic Studies in Animals 515
by Maimon M. Cohen and Kurt Hirschhorn

Chapter 20

Specific Locus Mutation in Mice 535
by B. M. Cattanach

Chapter 21

Dominant Lethal Mutations in Mammals 541
by A. J. Bateman and S. S. Epstein

Chapter 22

The Host-Mediated Assay, a Practical Procedure for Evaluating Potential Mutagenic Agents in Mammals 569
by M. S. Legator and H. V. Malling

Chapter 23

Human Population Monitoring 591
by James F. Crow

CHAPTER 1

Molecular Mechanisms of Mutations

Ernst Freese

Laboratory of Molecular Biology
*NINDS-NIH**
Bethesda, Maryland

I. INTRODUCTION

Most hereditary information of cells apparently is carried by DNA. In growing or adult cells, most DNA resides in chromosomes inside the cell nucleus, whereas a smaller fraction (0.1—10%, depending on organism and cell type) is found in plastids (mitochondria, chloroplasts, etc.). In egg cells, up to 90% of DNA can be contained in extrachromosomal material (mitochondria). The amount of chromosomal DNA per cell is usually constant, but it differs greatly for different organisms (see Table 1). Whereas some viruses contain only 5600 nucleotide pairs of DNA, bacteria contain about 5×10^6 and human cells about 6×10^9 nucleotide pairs.

Just before DNA duplication starts, each type of information usually occurs once within a chromosome, in one DNA double strand (in chromosomes of some higher organisms two or more identical DNA strands may be present). Just after DNA duplication, each type of information is present twice, once in each *chromatid* of a chromosome. *Haploid* organisms have only one chromosome of each kind (per nucleus or nuclear body), whereas *diploid* organisms have two non-sex chromosomes (*autosomes*) of each kind

* National Institute of Neurological Diseases and Stroke, National Institutes of Health, Public Health Service, U.S. Department of Health, Education and Welfare.

TABLE 1. Genetic Properties of Organisms Used for Cytological or Genetic Studies

Organism	Cell (particle) duplication time (hr)	Time needed for genetic crosses (days from germ to germ)	Number of chromosomes per cell (or virus)[a]	Amount of DNA per cell (or virus) (10^{-12} g)[b,c]
Unicellular				
Polyoma virus			1	0.000006
Bacteriophage T4	0.05	0.5	1	0.0002
Yeast, Saccharomyces cerevisiae (d)[a]		4	34	0.046
Bacillus subtilis (h)[e]	0.5	0.5	1	0.005
Salmonella typhimurium (h)	0.5	0.5	1	0.011
Bread mold, Neurospora crassa (h)		14	7	0.017
Aspergillus nidulans (h)		8	8	0.05
Multicellular				
Fruit fly, Drosophila melanogaster		10	8	0.2
Parasitic wasp, Habrobracon juglandis,				
female (d)		10	20	
male (h)		10	10	

Horse bean, *Vicia faba* (d)	20	180	12	40
Tradescantia paludosa (d)		180	12	60
Onion, *Allium cepa* (d)	20	180	16	50
Maize, *Zea mays*		360	20	15
Sea urchin, *Arbacia* (d)		360	40	1.4
Frog, *Rana pipiens*		1000	26	15
Woolly opossum	≤ 24	70	14	6
Amphiuma (d)		1000	30	170
Chicken		30	78	2.6
Mouse (d)	≤ 24	60	40	6
Rat (d)	≤ 24	70	42	6
Chinese hamster, *Cricetulus griseus* (d)	12–22	60	22	
Rat kangaroo (cell line), *Potorus tridactylis*	96		13	
Man (d)	12–22	7000	46	6

a Altman and Dittmer (1964).
b Sober (1968).
c 1 g DNA means ≈ 0.093 g DNA phosphorus or 1×10^{21} nucleotide pairs.
d d means diploid (rat kangaroo cell lines have an aneuploid chromosome set).
e h means haploid (for microorganisms per nucleus or nuclear body).

and, depending on the organism and the sex, one or two (different or identical) sex chromosomes. Equivalent portions of two autosomes are called *homozygous* if they have the same and *heterozygous* if they carry different information. In the heterozygous case, phenotypically expressed information is *dominant*, unexpressed information *recessive*. Dominant information usually gives rise to functional (e.g., enzymic) protein, recessive information to nonfunctional or no protein. Sometimes information is expressed in some tissues and not in others (of the same state of differentiation); the probability of expression is called *penetrance*. When the genetic change of one biochemical function (enzyme) has several phenotypic effects, it is called *pleiotrophic*.

A knowledge of the DNA structure and its duplication in conjunction with the chemistry of DNA alterations makes it possible to explain many mutagenic effects in molecular terms. The DNA double helix consists of two strands of polydeoxynucleotides which are held together by hydrogen bonds between the complementary bases A–T and G–C and by van der Waals forces between adjacent bases that are stacked parallel along the fiber axis (see Fig. 1). During duplication, DNA must rotate to unwind the strands. To maintain a reasonable rate of rotation in the viscous cellular milieu, swivel points are introduced by a DNAse that cuts one of the two strands at certain (specific?) places. Each of the separated strands is copied,

MOLECULAR BIOLOGY OF THE GENE

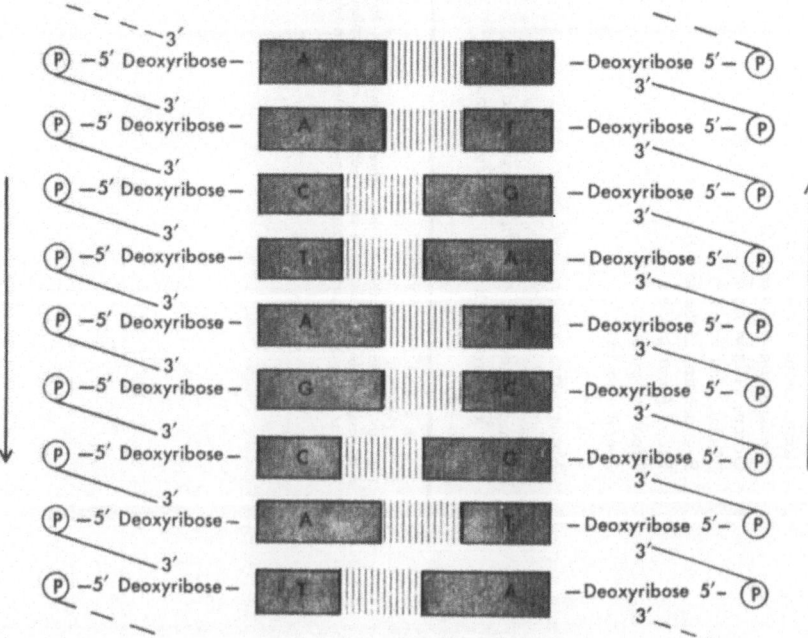

FIGURE 1A. Chemical structure of double-stranded DNA. (From Hayes, 1964.)

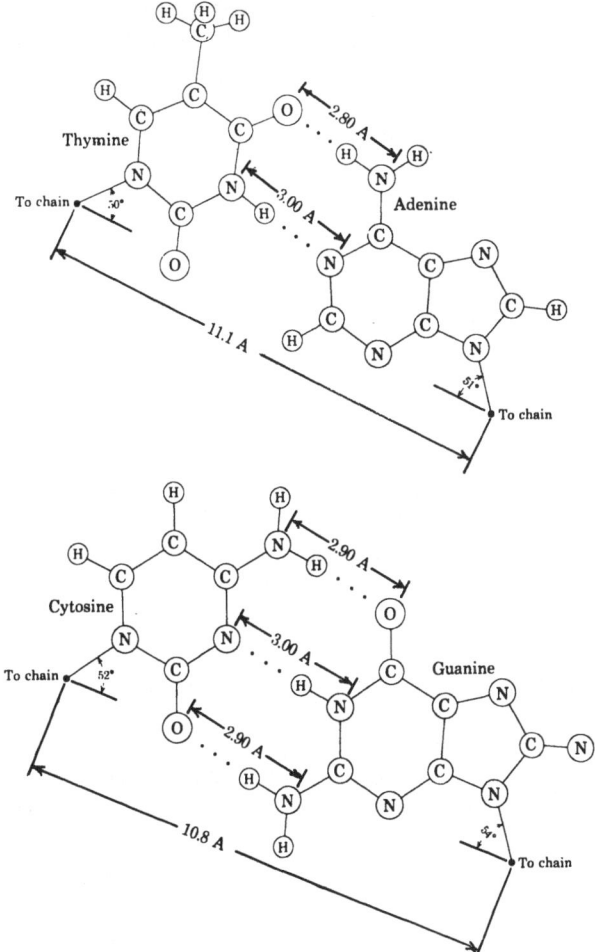

FIGURE 1B.   Base pairing of double-stranded DNA. (From Hayes, 1964.)

which apparently requires the pairing—by hydrogen bonding—of complementary bases; two new double strands are then produced, each containing one old and one new strand. At some time later the swivel gaps are closed again by a ligase.

It is known from genetic studies that all information in a chromosome is arranged one-dimensionally, which implies that it could all be carried by one long DNA double strand. But it is not known whether this strand is interrupted by small peptide links and how the DNA is condensed into a chromosome. Chromosomes also contain RNA, histones, and other proteins;

some of these components are essential for the developmental state of different DNA regions.

The information contained in DNA is determined by the precise sequence of bases and is preserved by its precise duplication. One of the two DNA strands can be transcribed into RNA by complementary base pairing. rRNA (made from certain areas of DNA) and 5sRNA end up in *ribosomes*, which are the factories for protein synthesis. tRNA is used for the *transfer* of amino acids in the cytoplasm to the ordered structure of proteins. The information for such proteins (enzymes) is transcribed from DNA into *messenger* = mRNA and then translated on the ribosomal surface by means of a triplet code into the corresponding sequence of amino acids. The concept *gene* denotes a functional unit on DNA, by which is meant that stretch of DNA which contains the information for one peptide or enzyme, or one tRNA, etc. Although the operational definition of this term is not precise, the concept of functional units, making up the total information on DNA, is useful. To obtain some idea about the length of a DNA stretch corresponding to a gene, one may regard a typical polypeptide of molecular weight 30,000. It contains about 300 amino acids for which one needs about 900 nucleotides in mRNA. Hence, one gene contains roughly 1000 nucleotides, a bacterial chromosome could contain about 5000 genes, and a human chromosome on the average about 100,000 genes.

From the number of known enzymes one can estimate that in bacteria about one half of the DNA information is used for vegetative growth (cell duplication), and relatively few genes are used for differentiation (such as sporulation). For the duplication of cells in higher organisms, roughly the same amount of vegetative information is needed (some of which is present several times in different DNA regions); the much larger amount of DNA in these organisms suggests that most of it is available for differentiation. However, much information may not be used at all, as is indicated by the minimal phenotypic effect of certain chromosome deletions (and perhaps the extensive heterochromatic regions).

II. CAUSE AND TYPES OF ALTERATIONS OF THE HEREDITARY MATERIAL AND PROTECTION AGAINST THEM

A. General Causes of Genetic Alterations

Chromosomes are large molecular structures which undergo complex reactions during their duplication, segregation, and differentiation. It is not surprising that these structures can be altered by many physical and chemical reactions (summarized in Table 2). External agents can react with

TABLE 2. Causes of Mutations

Physical:	Mechanical tearing apart of DNA
	Cutting by ionizing radiation or ^{32}P decay
	Nondisjunction of chromosomes
	High temperature
Chemical:	Alteration or removal of DNA bases
	Incorporation of altered bases
	Intercalation of oligocyclic aromatic compounds
	Alteration of DNA backbone
Enzymic:	Production of chemicals affecting DNA
	Mistakes of the DNA replicating system
	Alteration of the DNA replicating system
	Mistakes in recombination or repair

chromosomes either directly or after enzymic activation, or they can interfere with enzymes needed for chromosome duplication or segregation. Since the information strand in a chromosome consists mostly (if not exclusively) of DNA, all reagents which can react with DNA will most likely cause an alteration in the information by that reaction. A few agents (e.g., colchicine) act differently because they interfere with spindle formation and chromosome segregation.

Since cell membranes contain phospholipids, charged compounds do not enter cells at a significant rate unless they are transported in by specific proteins. In contrast, lipophilic groups generally increase the probability of passive uptake into the membrane, whereas the hydrophilic portion of such a molecule enables its release into the cell.

B. Protective Mechanisms

Considering the huge number of DNA nucleotides per chromosome, it is surprising that not more alterations in the nucleotide sequence occur during the functioning and duplication of DNA. Actually, the maintenance of the correct genetic information in a cell is so important that many mechanisms have evolved to protect chromosomes against damage and to repair damage once it has been inflicted. Some of these mechanisms are summarized in Table 3. The knowledge of their existence and their specificity is extremely important for the understanding of the effects of environmental mutagens. Some mutagenic agents which are very effective on isolated DNA may not exert much of a mutagenic effect in cells, because either they cannot enter the cell or the nucleus (e.g., negatively charged compounds), or they are inactivated by enzymes (H_2O_2), or their effect is repaired (UV). In contrast, other compounds which affect isolated DNA little or not at all

TABLE 3. Cellular Protection Against DNA Alterations

Structural:	Cell membrane, only certain molecules can enter
	Nuclear membrane, protects during DNA replication
	Condensation of DNA into chromosomes, avoids tearing of DNA during segregation
	Precision of chromosomal segregation
Enzymic:	Destruction of dangerous chemicals
	Specificity of nucleotide kinases and replicases, avoids incorporation of wrong nucleotides
	Excision of wrong bases + repair
	Repair of single-strand lesions by copying of complementary strand and sealing of gap
	Repair of double-strand breaks by stickiness and joining of broken ends
	Maintenance of pH, ion concentration, etc.

may frequently alter the hereditary material if they are activated inside the cell (urethan), or if their effect on DNA cannot be repaired.

Alterations known to be repaired in certain organisms are thymine- (or generally pyrimidine-) dimers induced by UV (Smith, 1966; Witkin, 1966), one of the methylation reactions of DNA (Strauss and Robbins, 1968) and the restitution of chromosome breakage by X-rays due to "stickiness" of the broken ends (Swanson, 1957). The repair reactions of DNA involve excision of the altered nucleotides, repair synthesis in which the complementary strand is copied, and subsequent joining of the open ends by ligase (Howard-Flanders, 1968). Different organisms differ in the types of available repair enzymes; mutants with repair deficiencies have been isolated in bacteria (Strauss, 1968; Witkin, 1969). In the human hereditary disease xeroderma pigmentosum, thymine dimers can no longer be excised; people afflicted with this disease are extremely sensitive to UV, which produces skin cancer (Cleaver, 1969).

Since mutagens can be activated by some enzymes and inactivated by others, it is impossible to predict whether a compound that is strongly mutagenic in one organism will be mutagenic in another. Nor is it possible to assure that a compound which is not mutagenic in one organism will not be mutagenic in another. Nevertheless, owing to a common evolution of vegetative cell properties, many biochemical reactions are similar in different organisms, and the existence of similar protective mechanisms can be expected. This similarity will be closest for organisms that are most closely related by evolution, an important consideration for selecting test systems relevant to man. With respect to some mutagens that have been added to our environment during the technical development of the last 100 years, protective mechanisms could not evolve: an organism will be partially protected against such new effects only if some protective mechanism evolved with respect to other genetic alterations is effective also against the new ones.

C. Types of Genetic Alterations

An alteration in the genetic information may be lethal to the cell or all of its early progeny or it may produce any one of three major types of hereditary alterations: a change in *ploidy* (number of chromosomes per cell), a *recombination* of existent information, or a *mutation* of such information (Srb *et al.*, 1965) (see Fig. 2). Theoretically, the different types of hereditary alterations are well defined, but experimentally their distinction is sometimes difficult.

Polyploidy implies that a nucleus (or nuclear body) contains, in place of the normal set of $2n$ chromosomes, $3n$, $4n$, etc., chromosomes. In organisms with well-visible chromosomes, polyploidy can be detected cytologically, but in some organisms it can be detected only indirectly (by a low frequency of mutation induction or by abnormal segregation in genetic crosses). Polyploidy can be caused by colchicine or other alkaloids, which prevent spindle and cell plate formation and thus chromosome segregation; occasionally it results from treatment by methylated purines (Kihlman, 1966). Suboptimal concentrations of colchicine sometimes produce *aneuploid* cells in which the chromosome number is increased (or decreased) by less

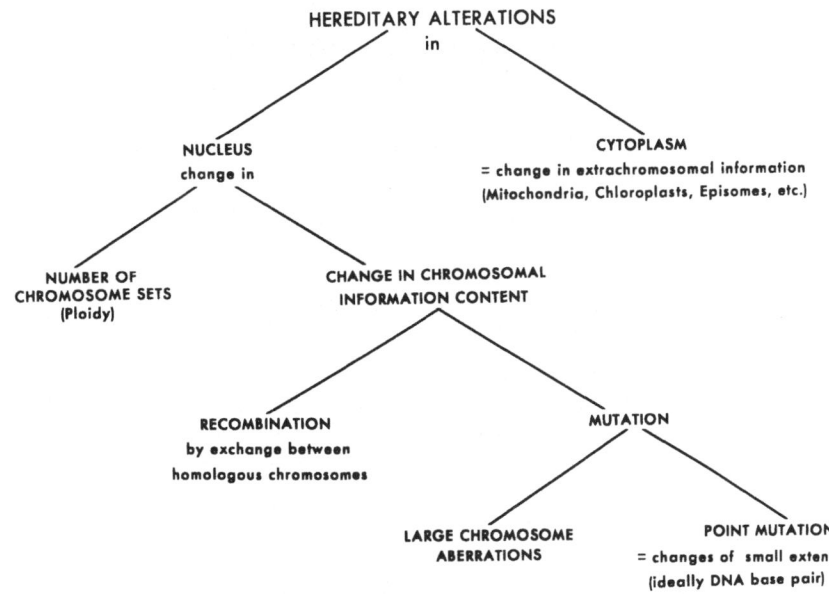

FIGURE 2. Subdivisions of hereditary alterations.

than a whole chromosome set, n. Aneuploid cells are here considered mutants of the normal cell. Whereas polyploidy usually increases the cell size, aneuploidy can have drastic consequences for differentiation, because it changes the ratio of metabolites in the cell. The change from a diploid $(2n)$ to a haploid (n) chromosome set occurs naturally during the reduction division of meiosis (in animals in the gonads) and diploid cells are reformed upon mating of germ cells (fertilization). Occasionally, two or more chromosomes do not separate properly (owing to asynapsis or nondisjunction) during meiosis of oogonia; in the subsequent mating an aneuploid chromosome set is formed which is either lethal (triploidy $= 3n$) or causes abnormal development (as in mongoloids with one extra chromosome). The fact that mongoloids are delivered more frequently by very young or by old mothers suggests a hormonal deficiency as the cause of nondisjunction. Agents altering the normal hormonal balance (estrogen, birth control pills?) may occasionally have the same effect.

Recombination is the (almost) reciprocal exchange of information between two homologous DNA molecules or chromosomes. It usually occurs during meiosis or during the mating, transduction, or transformation exchange of information in microorganisms. In certain microorganisms (aspergillus, yeast) recombination has also been observed during mitosis. Recombination can be induced by agents which induce inactivating DNA alterations (for references see Freese and Freese, 1966), presumably because they invite repair enzymes to cut out the altered DNA and thus increase the probability of mating between two DNA molecules. A reciprocal translocation between homologous sites of two *homologous* chromosomes may be called recombination. But we regard as mutation, rather than recombination, the more easily detectable reciprocal translocation between nonhomologous sites of any two chromosomes. Recombination can lead to greatly altered or even nonviable progeny (although the latter may be relatively rare, becuase lethal gene combinations are automatically eliminated in the course of evolution).

A *mutation* is any hereditary alteration in the information content or in the distribution of the hereditary material in an organism, a cell, or a virus which cannot be attributed to polyploidy or to recombination. The word "hereditary" implies that the mutation can be inherited indefinitely and without further change in some suitable environment. It is important to specify in which biological system the mutation has occurred. The cells of a higher organism can be divided into *germ cells* and their stemlines (*germinal cells*), which are used for the propagation of the organism, and *somatic* (body) cells, which comprise the majority of cells and which do not contribute to the progeny of the organism but die with it. Consequently, one distinguishes somatic and germinal mutations.

A mutation may lead to morphological changes or to biochemical requirements. In a diploid organism a dominant mutation is readily expressed, whereas a recessive mutation is immediately expressed only if the dominant gene is located in a functionally inactive chromosome: otherwise, a recessive mutation becomes expressed only one or many generations later if it occurs in homozygous form. It may then either produce an altered phenotype or a lethal effect (recessive lethal mutation). Dominant lethal mutations can be distinguished from other lethal effects only in multicellular organisms, where cells can multiply many times before a developmental crisis occurs as the result of a developmental mutation. It requires elaborate cytological or genetic studies before dominant lethal mutations can be distinguished from the lethal effect of chromosome breaks. Chromosome breaks per se do not constitute mutations because they usually lead to the death of the cell itself or all of its early progeny, owing to the loss of some vital information. However, in most cells, broken chromosomes can heal either by restitution or by the exchange of two broken chromosome portions that happen to be close enough in space and time. As a result of this exchange, mutations can arise; indeed, all known chromosome-breaking agents also produce mutations. The frequency of these mutations is relatively small compared to the lethal effect, but, as we shall see later, mutations are potentially much more harmful to a higher organism than is cell lethality.

Very large chromosome mutations can be observed cytologically in appropriate organisms. They consist of deletions, insertions, or inversions of chromosome pieces, translocation of material between two chromosomes or aneuploid chromosome sets (see also Fig. 4). The more frequently occurring smaller mutations can be detected only by a changed phenotype and their extent can be determined only by genetic mapping. For this mapping, the mutant is crossed against a set of other mutants with different phenotypes, to determine on which chromosome (*linkage map*) and approximately where on it the mutation is located, and against mutants with the same phenotype, to determine the fine-structure location of the mutation. The exact location is pinpointed by crosses against double or deletion mutants. One can further measure the frequency of reverse mutation to the original phenotype or analyze the amino acid alterations in the altered protein. When the mutant cannot recombine with at least two other mutants, which in turn can recombine with each other, it contains a large genetic alteration (usually deletion). Such mutants do not revert to the original genotype and usually not to the original phenotype. If they can revert to a phenotype similar to the original one they do this by a *suppressor mutation*. If a mutant behaves as if only one biochemical function has been altered and if it can recombine with all (but one) other mutants it is called a *point mutation*. Since the proven extent of this "point" depends on the number

and type of mutants used in this test, a point mutation is not precisely defined in terms of nucleotide pairs altered; nevertheless, the term is generally
quite useful. Point mutations usually can revert to the original phenotype,
although sometimes with a low frequency (10^{-8}). Ideally, one would like
to identify a point mutation as an alteration of a single nucleotide pair in
DNA, but since a sequence analysis of DNA is at present impossible, the
identification of a point mutation as the change of a single nucleotide pair

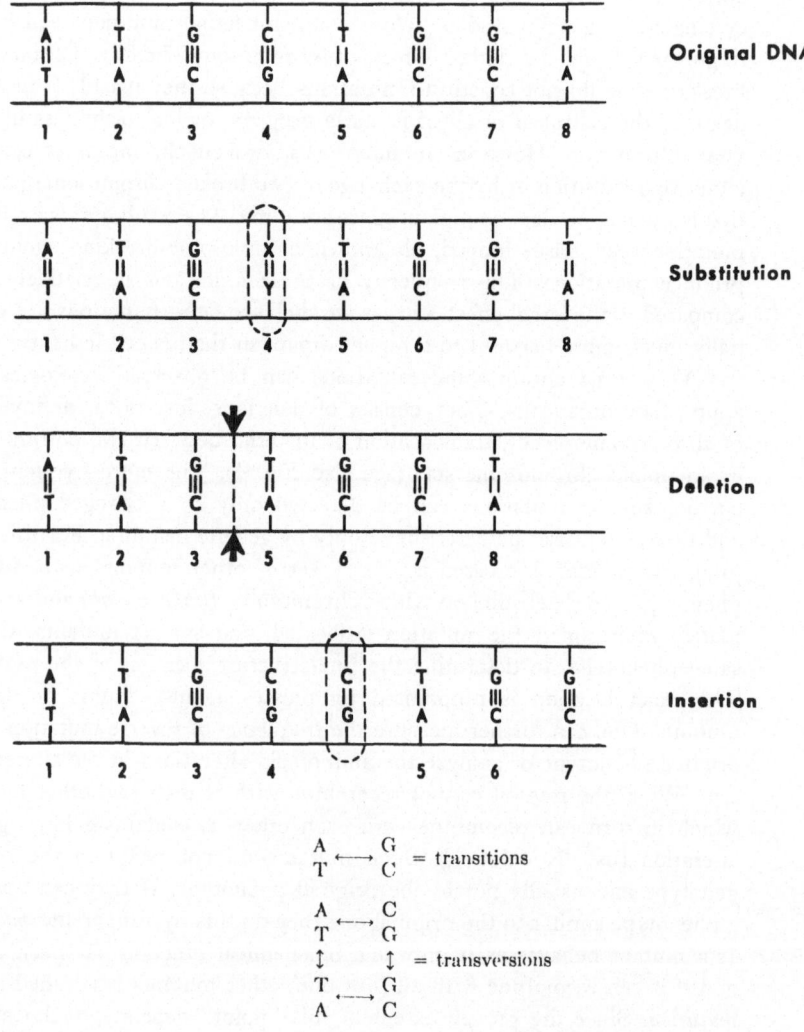

FIGURE 3. Upper: Types of point mutations. Lower: Types of base-pair substitutions.

can be obtained only more or less indirectly, although in some cases quite convincingly (utilizing reverse mutations and amino aicd changes in the corresponding protein). However, such a fine-structure analysis of point mutations is feasible only in microorganisms. In more complex organisms, the extent of point mutations can only be guessed from the revertibility (hard to measure), the number of functions or proteins affected (if any protein change can be observed), or the number of altered amino acids in the mutated protein (e.g., hemoglobin).

Point mutations can be subdivided into transitions, transversions, deletions, and insertions, as shown in Fig. 4. Base-pair substitutions (transitions and transversions) are the most frequent spontaneous mutations, but they can often not be recognized phenotypically for two reasons: (a) for many coding triplets, the third base can be replaced by another base without changing the coding properties of the triplet, i.e., the amino acid remains the same; (b) many transitions cause amino acid interchanges, e.g., among nonpolar amino acids (Ala, Gly, Ile, Leu, Thr, Val), which do not significantly alter the tertiary folding of the corresponding enzyme (Margoliash and Smith, 1965).

D. Consequences of Genetic Alterations in Higher Organisms

The consequences of treatment by a mutagenic agent depend on the developmental state of the organism, the type of cell that is affected, and the type of genetic alteration that has been produced. As Table 4 illustrates, the more drastic effect of cell death is usually least harmful to the organism, since a dead cell can be easily replaced by one of the many live cells both before determination (to develop specialized organs) has occurred (i.e., before gastrula) and after many equally differentiated cells have been formed. Only early in differentiation (gastrula and later), when just a few cells have been determined (the anlage has been formed), can the death of a few cells be disastrous for the development of a whole organ and cause a teratogenic effect (see Kalter, 1968 and this volume). The death of many cells in an adult organism undoubtedly reduces its vitality and may contribute to early aging.

In contrast, mutations can have serious effects at any time, because the mutated cell duplicates and can thus greatly affect the normal course of development. Mutations in germ cells may lead to an abnormal development of the fetus, resulting in early abortions, fetal death, stillbirth, or abnormally developed offspring. Even more insidiously, germinal mutations may affect development at a later stage or only in certain tissues, resulting in abnormal puberty, neurological diseases, and early death, or they may give rise to the many known inborn errors of metabolism.

During the development of the organism, somatic mutations may lead

TABLE 4. Phenotypic Effect of Lethal or Mutagenic Alterations of Individual Cells on a Multicellular Organism

Cell type, developmental stage	Dead cell	Mutated cell	
		Dominant, e.g., loss of repressor by deletion, effect of extra chromosome, translocation in germinal cells	Recessive, e.g., point mutation, deletion, or cryptic aberration (translocation)
Somatic cells			
Very early development (to blastula)		Abnormal development (teratogenic); fetal death	—
Early development (from gastrula on)	Abnormal development (teratogenic)	Same	—
Late development to adult	—	Neoplastic	—
Germinal cells	—	Lethal (fetal death), or other phenotypic effects such as malformation, metabolic or neurological disease, mongolism, Turner's syndrome; expressed in next and all following generations	Lethal or other phenotypic effects; expressed in next or any later generation when in homozygous form when or dominant information is in functionally inactive chromosome region
Germ cells	—	Lethal, malformation, metabolic or neurological disease; expressed in next and all following surviving generations	Same

to abnormal differentiation (teratogenic changes). In the adult organism, most somatic mutations are harmless if they occur in resting cells which do not significantly affect the rest of the organism. But even in the adult organism, somatic mutations can affect duplicating cells, leading to abnormal cell types, or they may occur in cells whose duplication is reinitiated; in both cases neoplasms (cancer) may result.

In man the frequency at which abnormal development occurs is quite high. At least 15% of conceptions are aborted (Warburton and Fraser, 1964) and 7% of full-term children exhibit congenital malformations at birth or develop them later (Mellin, 1963). The frequency of abortions very early in development is not even known. Most remarkable is the fact that 20% of the aborted babies have visible chromosome anomalies (Carr, 1967; World Health Organization Bulletin, 1966). Since the frequency of cytologically unobservable genetic changes is usually higher than that of observable ones, it is possible that practically all abortions are due to genetic alteration. How many of these alterations are truly spontaneous and how many are induced by environmental factors is not known. The abortions are one way of continuing natural selection in man. But unfortunately a significant number of mutations are not as deleterious; they express themselves as diseases or merely as a reduction in vitality. Since mankind has decided to eliminate natural selection among human beings, it is also responsible to reduce human misery by curtailing the propagation of detrimental traits and by eliminating environmental mutagens which can increase the frequency of such traits.

III. DETECTION OF CHROMOSOME ABERRATIONS AND SYSTEMS OF GENETIC ANALYSIS

A. Cytological Examination of Large Chromosome Aberrations

Chromosome aberrations can be seen under the light microscope only in organisms whose chromosomes are sufficiently large that structural details can be carefully examined. The classical objects of investigation have been mostly plant cells such as the root tips of the broad bean (*Vicia faba*), the onion (*Allium cepa*), *Tradescantia paludosa*, or *Zea mays* (for chromosome numbers see Table 1) (see Kihlman, 1966). The ultimate consequences of mutations could be observed by an alteration in the band pattern of functional activity in the polytenic salivary gland chromosomes of *Drosophila melanogaster* (2000 homologous chromatids stacked in parallel). More recently, investigations in animal cells have become useful, in particular when cells with small chromosome numbers are used (woolly opossum and a rat kangaroo cell line). Large alterations have been observed even in human chromosomes (see Cohen and Hirschhorn in Volume 2). In the near future it may become feasible to separate chromosomes by physical techniques and to automate their analysis by flying-spot scanning and computer evaluation. But even with the greatest accuracy, the loss, addition, or exchange of only a rather large number of nucleotide pairs of DNA (probably larger than 10^7) would be cytologically detectable.

Chromosomes become sufficiently visible only after they have finished their duplication and have condensed in the equatorial plate at metaphase (the nuclear membrane has dissolved by then). They remain visible through anaphase and into telophase, at which time the chromosomes have separated completely and begin to form their new nuclear envelope. Following radiation or chemical treatment, two principal types of chromosome aberrations have been observed: chromosome and chromatid aberrations (see Fig. 4). In *chromosome aberrations* both chromosome strands (i.e., chromatids) seen in metaphase show breakage or exchange reactions at homologous sites. A breakage reaction that had occurred in the G1 phase, before DNA duplication, apparently could not be healed in time; during S-phase (DNA synthesis) the two separated chromosome portions duplicated and consequently both daughter chromatids show the break (Taylor, 1953; Hsu *et al.*, 1962). In *chromatid aberrations* only one chromatid or daughter chromosome shows a break at one site, suggesting that the DNA alteration occurred after the DNA of this chromosome area had been duplicated (see Fig. 3). Most induced aberrations apparently are repaired (restitution) before they can show up as breaks in anaphase. The frequently (in metaphase) observed achromatic gaps, which do not produce breaks later (Revell, 1959; Evans, 1962), may represent single-stranded DNA regions in which the other broken strand is being repaired. The reactions which lead to stable broken chromosomes or to exchange figures in anaphase (Fig. 3) seem to result from exchange reactions between two chemically altered homologous or nonhomologous sites in sister chromatids or in chromatids of different chromosomes; the alterations must occur close enough in time and space. Since agents usually affect only one DNA strand at a site, these exchange reactions may come about when two DNA double helices encounter each other while they are being repaired. An exchange similar to recombination may then occur between the two (nonhomologous) regions. This consideration shows that repair and recombination enzymes may play an important role in the production of chromosome aberration.

It can be seen from Fig. 4 that many chromosome aberrations lead to the loss of a chromosome segment without attached centromere. If this segment, which contains many genes, carries even one gene essential for survival, the cell will die. Even for diploid organisms, which have heterozygous chromosomes, the probability of cell death is appreciable. In addition,

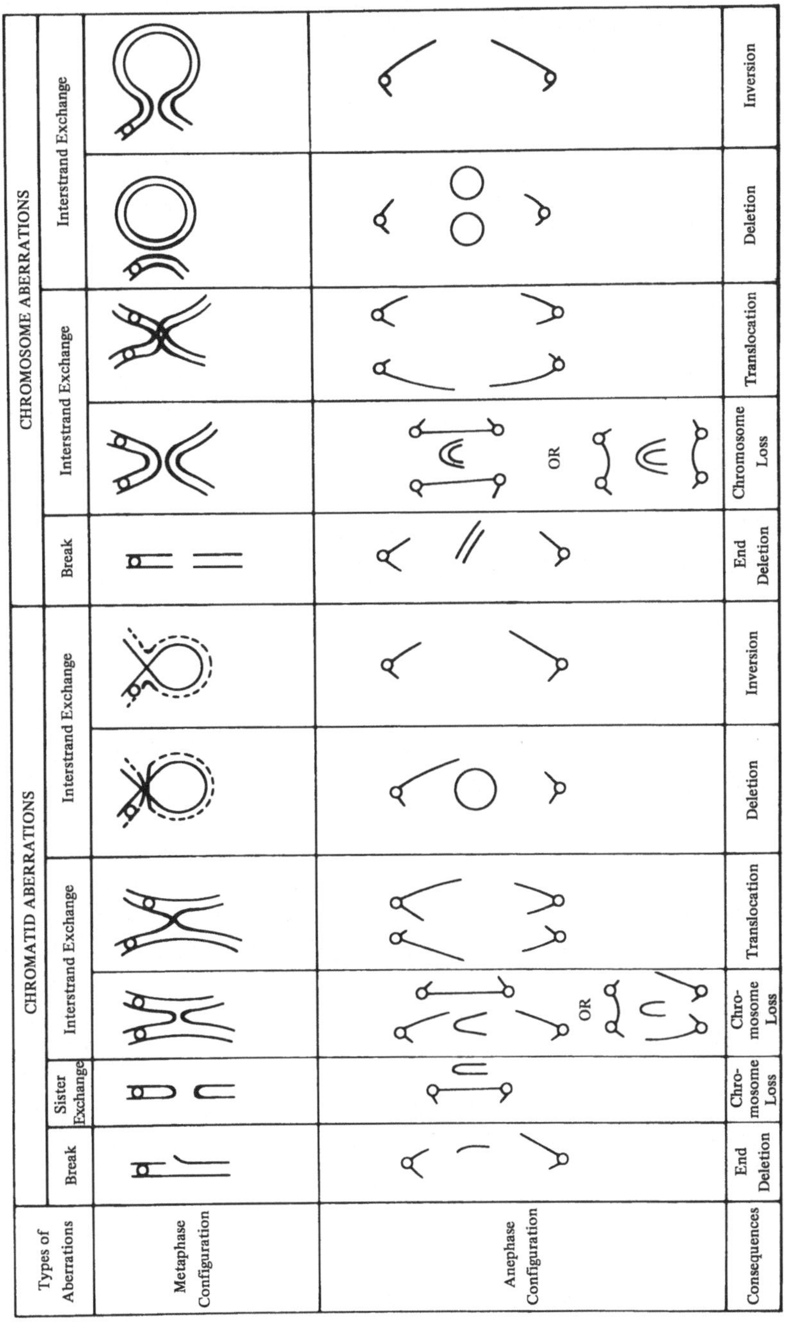

FIGURE 4. Chromosome loss and chromosome mutations resulting from breaks before (chromatid aberrations) and after (chromosome alterations) DNA duplication.

many of the chromatid exchanges produce dicentric chromosomes. Since the two centromeres move (in one half of the divisions) to opposite poles, the chromosome strands tend to break randomly at any place between the centromeres; the sister strands of the broken ends will fuse and produce a new dicentric chromosome in the next division. As a result, the breakage–fusion–bridge cycle will ensue, which in most cases will eventually lead to the death of all progeny cells, due to the random loss of genetic material. But if only a small portion of genetic material was deleted, duplicated, or inverted, or if a translocation between two chromosomes occurred without loss of genetic material, the chances are good that a viable mutant was formed. If such a mutation occurred in germinal cells, it may still be lethal in the next generation, because the new diploid cells may contain a nonviable combination of mutated chromosomes (e.g., a chromosome from a translocation may have lost important information and gained unessential information instead).

For unknown reasons, chromosome breaks seem to occur frequently at specific (heterochromatic) sites of certain chromosomes (see Swanson, 1957). However, it is possible that this specificity results from a selection of breaks which can proceed to metaphase, whereas breaks in important functional areas may lead to immediate cell death.

A cytological investigation is also feasible for very small chromosomes, e.g., virus or bacterial DNA molecules, under the electron microscope. In bacteriophages λ and T4, relatively small deletions have been detected Davis and Davidson, 1968; Bautz and Bujard, 1969).

B. Genetic Examination of Mutations

The types of organisms used for genetic investigations have changed considerably during the last 100 years. This change was necessary because increasingly larger numbers of organisms had to be processed in order to obtain statistically aignificant results for increasingly smaller dimensions of the genetic material. Consequently, smaller organisms with shorter generation times were preferred. Another reason for change was the need for organisms that were haploid and could grow on a defined medium so that mutations with biochemical rquirements could be easily detected. Very recently, having uncovered the different types of mutations and the specificity of different mutagens in microorganisms, the trend has been reversed again to include in mutagenic tests diploid tissue cultures and higher organisms that, although much more difficult to handle, are of more immediate relevance to man. The development of different genetic systems will be outlined in the following, to point out their relative advantages; most of the systems will be described in detail in other chapters of this book. Table 1 gives a list of some of the organisms and the times needed for genetic crosses.

Following the initial discoveries of Mendel's laws in garden peas, the first spontaneous mutation (aneuploidy due to reciprocal translocation) was discovered in *Oenothera* by De Vries, and other spontaneous mutations were quantitated in *Antirrhinum* by Baur. More detailed genetic studies were possible when Morgan used the fruit fly, *Drosophila melanogaster*, in which recombination and a one-dimensional linkage map, as well as its cytological correlation to bands in salivary gland chromosomes were discovered. In this organism Muller proved the induction of recessive lethal mutations by X-rays, using strains with special genetic markers. *Drosophila* posed two inherent difficulties for a further genetic and biochemical analysis (see Abrahamson and Lewis, in Volume 2). It is a diploid organism that grows only on a complex medium. Since most mutations with biochemical deficiencies are recessive, they will not show up in the diploid organisms hatched from mutated germ cells; rather, they can be discovered only by back-crosses yielding organisms homozygous in the mutations. This problem of diploidy is common to most higher organisms or their tissue cultures, and it has rendered an analysis of the induction of point mutations difficult; this applies even to the otherwise biochemically and genetically well-studied diploid yeast, *Saccharomyces cerevisiae*, in contrast to the haploid yeast *Schizosaccharomyces pombe* (see Mortimer and Manney, this volume). In tissue cultures of the Chinese hamster the problems of mutation induction have recently been partially overcome either by measuring the induction of mutants resistant to an antibiotic or by killing off some unwanted information through DNA sensitization to 5-bromouracil (see Chu, in Volume 2). The problem of diploidy can be partially overcome by using the parasitic wasps, *Habrobracon juglandis* or *serinopas*, in which the female is diploid, while the male, having only one chromosome set from an unfertilized egg, is haploid (see Smith and von Borstel, in Volume 2). But even here the problem remains of growth being possible only on a complex medium (larvae); biochemically deficient mutants cannot be easily isolated.

The next important contribution to the evolution of genetics came from the use of haploid microorganisms that could grow on a defined minimal medium. In these systems *forward mutations* (from functionally active to inactive genes) exhibiting some biochemical requirement can be isolated rather easily, especially when strong mutagens are used to increase the mutation frequency. But the major advantage is the selectivity with which the usually infrequent revertants or recombinants of the standard (wild) phenotype can be counted by plating the cells on media lacking a required biochemical. One revertant among 10^8 organisms can thus be scored.

The first biochemical mutants were isolated in the bread mold, *Neurospora crassa* (Beadle and Tatum, 1941), and a genetic analysis of these mutants as well as others in *Aspergillus nidulans* followed promptly (see Roper, in Volume 2). Since these fungi can form heterokaryons, which contain nuclei of two different strains in a common cytoplasm, cross feeding of mutations altered in different genes can be observed, whereas mutations altered in different DNA sites of the same gene usually do not cross feed. The cross feeding test (sometimes called *cis trans* test) can be used in many microorganisms in some form or another; it has been very useful to determine the functional extent of a gene. But certain mutations within one peptide-forming DNA region can nevertheless cross feed to some extent; this intragenic *complementation* apparently results from the interaction of two or more proteins forming a partially active enzyme; it may be restricted to enzymes which normally contain several peptide subunits (Ahmed *et al.*, 1964). *Neurospora* can be used to measure the induction of both point mutations and deletions in the same genetic system of adenine-requiring mutants (see de Serres and Malling, in Volume 2). As will be seen later, this is important because the relative frequency at which point mutations and deletions are induced differs greatly for different mutagens.

The next development, allowing a more rapid analysis of larger populations, came with the use of bacteria. The investigations of biochemical mutants in bacteria have led to the realization that genes for related biochemical functions of a biochemical path are often controlled coordinately as an "operon." The controlling compound (repressor) attaches to the operator region of DNA, whereas the RNA polymerase apparently attaches to a "promoter" region which is adjacent to the operator. Using the tryptophan synthetase gene of *Salmonella*, a correlation between the amino acid alterations and the type and extent of mutations has confirmed and extended the theories of mutagenesis that had been derived from work with phages (Yanofsky, 1965). Certain bacterial mutants today provide an excellent tool to measure the induction of both forward and reverse mutations for different types of base-pair changes (transitions, transversions, deletions, insertions). In addition, suppressor mutations (amber, ochre), which can incorporate an amino acid into the site of a nonsense code triplet, can be utilized to recognize very specific types of base-pair changes (see Ames and also Yanofsky, this volume).

Almost in parallel with bacterial genetics, phage genetics developed (see Drake and also Heinemann, this volume). Bacterial viruses inject almost only their nucleic acid into bacteria so that any mutation induced by treatment of isolated phages can be attributed to a nucleic acid alteration. Particularly useful were the genetic markers of the rII phenotype in phage T4, which allowed a thorough genetic fine-structure analysis by means of many overlapping deletion mutants (Benzer, 1961). In this system base-pair transitions and transversions (Freese, 1959) and small deletions and insertions (Crick *et al.*, 1961) could be characterized. In principle, this system could be used to measure the frequency at which larger chromosome alterations can be induced by different agents. Using the lysozyme gene of phage T4, base-pair deletions and insertions have been correlated to shifts in the reading frame by which triplets in DNA are translated into amino acids in proteins (via messenger RNA) (Terzaghi *et al.*, 1966).

Many bacteria carry phage DNA in a hidden (latent) form inserted into the bacterial DNA but functionally inactive or only partially active. These bacteria are called "lysogenic" because the phage DNA can be induced by certain agents to liberate from the bacterial chromosome, to multiply, produce phages, and consequently to lyse the induced bacteria. All agents that induce inactivating DNA alterations also induce this phage development, providing a convenient test system for inactivating DNA alterations (see Heinemann, this volume).

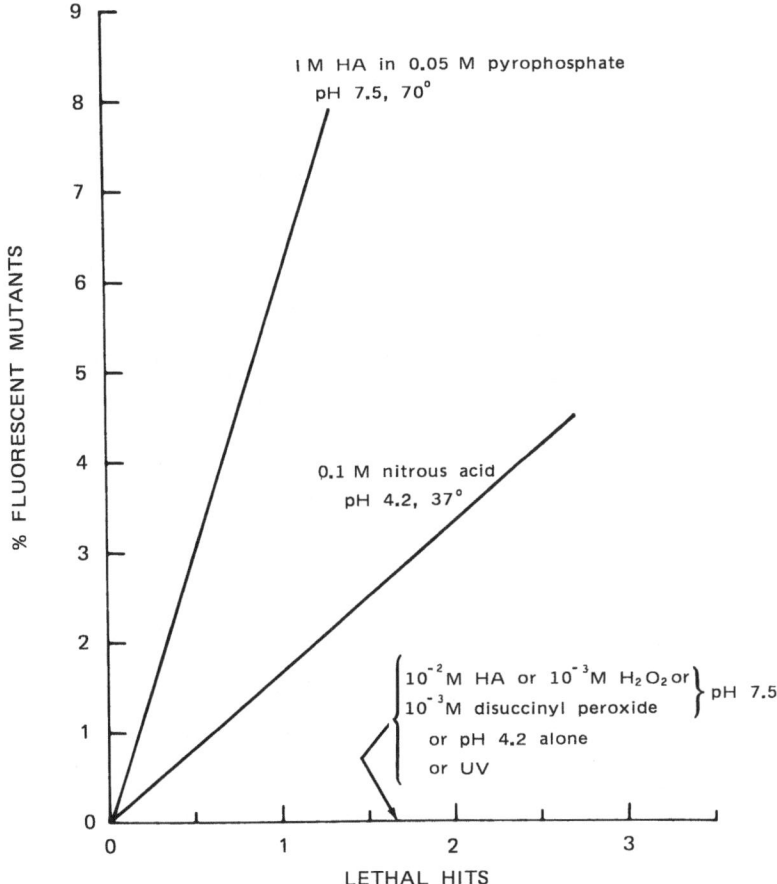

FIGURE 5. Increase of point mutations (fluorescent mutations) versus lethal hits after treatment of transforming DNA by various agents.

The extensive use of microorganisms as a genetic tool had created a dichotomy between microbial geneticists who investigated only point mutations and cytologists who could detect only large chromosome alterations. Since one agent (e.g., alkylating agents) induced both genetic effects, it was tacitly assumed that these effects arose by the same reaction with DNA. This fallacy was not easily disproved. The fact that the inactivation (killing) efficiency of different agents was clearly unrelated to their point-mutagenic efficiency should have indicated a difference in the produced types of DNA alterations. But this difference was virtually disregarded, presumably because so many other cellular reactions, apart from DNA alterations, can possibly lead to cell death. It would therefore be ideal to measure both effects, the one leading to point mutations and the other causing inactivation, on isolated DNA. This was possible by the use of transforming DNA (see Herriott, this volume) and in particular the system of linked mutation induction in the tryptophan region (Freese and Strack, 1962; Freese and Freese, 1966). In this system it was clearly shown that the same agent can induce two different types of DNA changes: mutagenic DNA alterations, which induce point mutations, and inactivating DNA alterations, which only inactivate DNA but induce hardly any point mutations. These two types of alterations will be analyzed in the following more thoroughly. Their relative frequency for different agents can best be compared by a plot of the frequency of point mutations versus the number of lethal hits (which is the sum of inactivating DNA alterations and of mutagenic DNA alterations in vital genes (see Fig. 5). Whereas high concentrations of hydroxylamine or nitrous acid are highly point-mutagenic, H_2O_2 or other radical-producing agents induce predominantly inactivating alterations (are lethal). Alkylating agents are intermediate in effect.

C. Test Systems for Mammalian Investigations

The duplication time of mammals is so long and the number of organisms one can handle so small that it is usually not possible to detect small mutagenic effects. One therefore uses high doses of the agents under investigation and employs detection techniques that have been developed specifically for mammalian systems. Such tests are necessary for environmental agents because some compounds are enzymically activated to mutagens inside the cell and other compounds are inactivated.

Theoretically, mammalian tissue culture systems could be developed in which mutation frequencies should be measurable, similar to microbial cultures, in reasonably short times and with statistically significant numbers. Such developments look particularly promising for the tissue culture systems of Chinese hamsters (see Chu, in Volume 2). However, many chromosome anomalies have been observed to arise in tissue cultures spontaneously, and some mutagenic agents have been found effective in tissue cultures but at least so far not in intact animals (e.g., caffeine in the dominant lethal tests; see Bateman and Epstein, in Volume 2). Until media can be found in which tissue cultures can develop more stably, other tests employing live animals appear more reliable. These tests, described in detail elsewhere in these volumes, include the cytological investigation of chromosomes and dominant lethal tests in mice, both of which discover mainly the effect of agents inducing chromosome aberrations, and the test for host-induced modifications, which is preferably useful for the detection of agents inducing point mutations. It is important to realize that no single test can detect all mutagenic effects. A negative finding is significant only when several tests are used which together can discover all types of large alterations and point mutations in the mammalian system.

IV. PRIMARY DNA ALTERATIONS AND THEIR GENOTYPIC AND PHENOTYPIC CONSEQUENCES

Chemical or enzymic DNA alterations (produced in resting or duplicating DNA) can be divided into three types, both by their presumptive effect on DNA synthesis and by their genotypic and phenotypic consequences: nonhereditary, mutagenic, and inactivating DNA alterations (see Fig. 6).

A. Nonhereditary DNA Alterations

Nonhereditary DNA alterations neither prevent the duplication of DNA nor induce changes in the DNA information (mutations). They usually involve the chemical or enzymic modification of the DNA bases at sites that do not interfere with base pairing, such as the methylation or hydroxy-methylation of pyrimidines of the 5 position. Although important as control mechanisms (DNA can be specifically cut at methylated bases), they do not concern us here.

B. Mutagenic DNA Alterations

Mutagenic DNA alterations also do not prevent the duplication of DNA but usually give rise to the change of one or a few nucleotide pairs (point mutations) in some of the progeny DNA. They do not induce large

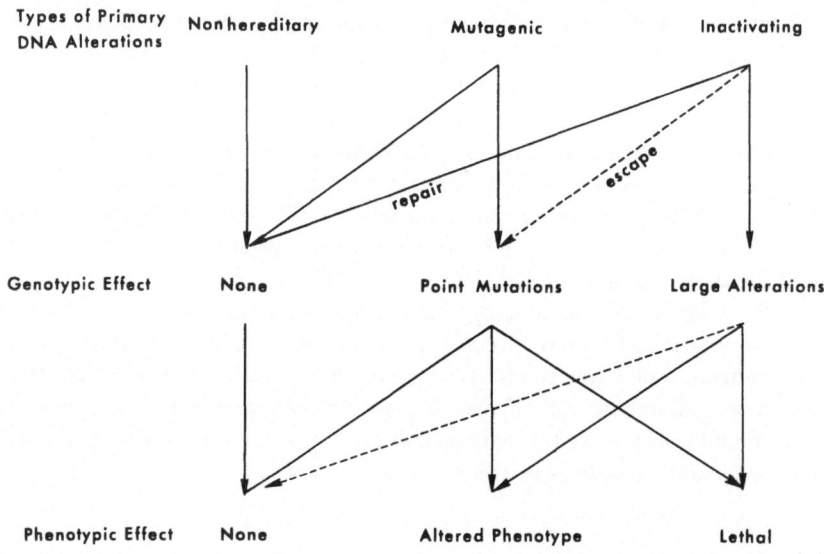

FIGURE 6. Types of primary DNA alterations and their possible genotypic and phenotypic consequences.

chromosome alterations. Mutagenic alterations either consist of minor base
modifications, which alter the specificity of hydrogen bonding to complementary bases (see nitrous acid), or they are caused by base-pairing mistakes induced by some agents during DNA duplication (see bromouracil).
The resulting point mutation in DNA (= genotypic effect) may produce
no phenotypic effect (unobservable mutation), or a phenotypic change (observable mutation), or it may be lethal to the cell if it alters a vital gene.

C. Inactivating DNA Alterations

Inactivating DNA alterations prevent DNA duplication across the
altered site, unless they have been eliminated by repair. The block in DNA
replication has been extensively studied in bacteria after UV irradiation,
which produces thymine- (and other pyrimidine-) dimers (review: Smith,
1966). Inhibition of DNA synthesis has been demonstrated for many chemicals, e.g., alkylating agents (Yamamoto *et al.*, 1966; Iyer and Szybalski,
1963; Tanaka, 1965). That this replication block is caused by a reaction
with DNA itself is inferred from the fact that transforming or bacteriophage
DNA loses its biological activity without induction of point mutations when
it is treated outside the cell and the agent is removed before cells are added.
Various types of inactivating DNA alterations, which will be discussed in

more detail in section V, are summarized in Fig. 7. Agents inducing in-activating DNA alterations also induce chromosome breaks and chromosome mutations, as would be expected as a consequence of the block in DNA synthesis. Most types of inactivating DNA alterations probably can be re-paired by specific enzyme systems, as is shown by results for UV- and alkylation-induced breaks in bacteria and by the cytologically observed re-stitution of chromosome breaks. When two inactivating DNA alterations occur on chromosome strands that are (or move) close to each other, intra- and interstrand exchange reactions occur which lead to chromosome aber-rations, as described earlier.

It seems necessary to repeat that both mutagenic and inactivating DNA alterations induce mutations, though of different types at the molecular level.

Most viruses also produce chromosome breaks, probably by producing or liberating some DNase; they can be considered in this connection as inducing inactivating DNA alterations. Following infection by some viruses

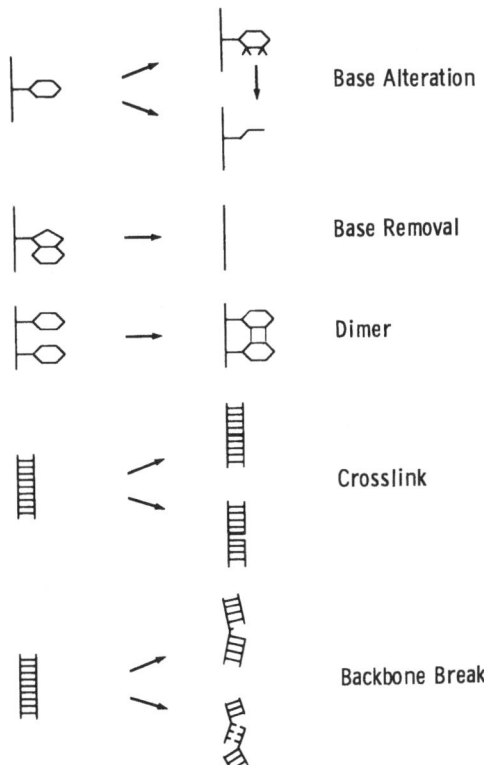

FIGURE 7.   Inactivating DNA alterations.

(e.g., SV40 and measles or rubella), many chromosome breaks in each cell are only very slowly restituted; the virus apparently inhibits also the repair (restitution) mechanism (Moorhead and Saksela, 1965). Consequently, exchange figures of chromosomes can be found only after long times. Other viruses produce both chromosome breaks and exchange figures (for details see Nichols, 1969; Cohen and Hirschhorn, in Volume 2). For reasons outlined above, viruses with drastic chromosome-breaking and therefore lethal effects on cells are teratogenic (Sever, 1967) if they affect early development, but at other stages of development they are less harmful to the organism than viruses that do not break so many chromosomes and occasionally produce mutations. This consideration points out a *potential danger of live virus vaccines:* whereas such virus mutants produce fewer immediate disease symptoms and are less lethal for cells, some of them might induce more chromosome exchange figures and hence mutations or neoplastic changes. This side-effect would be discovered, without a conscious statistical effort, only if the frequency of neoplasms or of hereditary defects increased drastically as a consequence of vaccines used in man. The danger might not be detected in animal screening tests, because viruses are very host-specific and probably would have quite different effects in animals than in man.

D. Most Agents Induce Several Primary DNA Alterations

All agents which induce mutagenic DNA alterations can also induce vinactiating DNA alterations by other reactions with DNA. For example, 5-bromouracil is extremely sensitive to light, heat, or reducing chemicals which inactivate DNA containing this base analog; alkylating agents induce mutagenic DNA alterations by the alkylation of the 7 position of G and inactivating alterations either by the removal of the alkylated purines or by backbone breakage; hydroxylamine induces point mutations by reaction with C, but, in the presence of trace metals and oxygen, it also produces radicals which rapidly inactivate DNA (especially at low hydroxylamine concentrations, at which the compound cannot reduce the radicals again). To measure the relative frequency of mutagenic or inactivating DNA alterations from the biological effects caused by a given agent, an elaborate analysis involving either several genetic tests or conditions which emphasize one or the other of the two reactions is therefore required.

In contrast, some agents (e.g., radical-producing) which cause inactivating DNA alterations produce very few (or no) mutagenic DNA alterations, but they apparently induce point mutations at a very small frequency (e.g., one point mutation per 10^4 inactivating DNA alterations), either by mistakes during repair or recombination processes or by occasionally overcoming the replication block through the incorporation of any base without base pairing. Nevertheless, the frequency of mutations observed in a cellular organism may sometimes appear high, if the particular type of inactivating DNA alterations (e.g., induced by UV) is very efficiently repaired.

E. Correlation of Agents Inducing Inactivating DNA Alterations with Other Effects

As has been pointed out previously (Freese and Freese, 1966), agents which induce inactivating DNA alterations have also been found to induce the liberation of latent viruses, e.g., phage λ in *Escherichia coli* (see Heinemann, this volume), and to produce neoplasms (cancer) or leukemia. It is not known whether the primary neoplastic effect is due to the production of mutations, the liberation of a latent virus, or some other effect of the agents on cellular control mechanisms. But the correlation between the inactivating effects on DNA, chromosome-breaking effects, and carcinogenic effect is so strong and continuously increasing that neoplasms are very likely due to some alteration of the genetic material. A meaningful correlation obviously can be obtained only if it refers to the ultimately effective mutagen or carcinogen; some agents (e.g., H_2O_2) never reach the nucleus of normal cells, and other agents (e.g., urethan and aromatic amines) have to be enzymically converted into the active agent (reviews: Uehlecke, 1965; Miller and Miller, 1969; and Miller and Miller, this volume).

Several chromosome-breaking agents have been shown to induce teratogenic effects if they are added during a critical period of early development (see Kalter, 1968, and this volume).

Since most inactivating DNA alterations are lethal to the cell, agents inducing them can usually be detected also by their toxicity. But many other agents are highly toxic without much genetic effect. Using toxicological tests, an environmental agent causing a relatively small toxic effect might be regarded as harmless, whereas the same agent might produce a high frequency of induced mutations. Even worse, the toxicology test fails completely for agents inducing predominantly mutagenic DNA alterations.

Many mutagenic (e.g., alkylating) agents react not only with DNA but also with cellular proteins. Forming haptens, they can initiate antibody synthesis to the altered (and thus foreign) protein; thereby the allergic responses (rashes, coughing, etc.) are produced which are often found among industrial or laboratory personnel.

V. THE EFFECT OF DIFFERENT AGENTS ON DNA AND CHROMOSOMES

Many chemicals as well as ionizing and UV radiations produce mutagenic effects. It would be impossible to mention all these agents, but it is possible to point out the chemical structures that render a compound potentially mutagenic.

One can distinguish agents that mutate only replicating DNA (base analogs, intercalating agents) and others that alter resting DNA and exert their effect even when they are removed before DNA duplication is allowed to proceed (alkylating and radical-producing agents).

A. Incorporation of Base Analogs into DNA

Apart from naturally occurring bases (5-methyl cytosine, 5-hydroxy-methyl uracil, uracil, etc.), only few base analogs are incorporated into DNA. Others are rejected by the various enzymes that have to attach to the base the deoxyribose moiety and the three phosphate groups in order to form a deoxynucleoside triphosphate. But even in the triphosphate form, nucleotides are not necessarily incorporated into DNA, e.g., nucleotides having ribose instead of deoxyribose are rejected, as can be seen from the absence of uracil in normal DNA. The DNA replicating system apparently can recognize both the deoxyribose and the triphosphate moieties and presumably check the position and angle of the incoming deoxyribose–triphosphate groups with respect to the already present DNA strands. Thus even deoxynucleoside triphosphates containing base analogs which do not undergo the normal base pairing by hydrogen bonding to the complementary base could be rejected without enzymic recognition of the base (Freese and Freese, 1967). This agrees with *in vitro* experiments which have indicated that bacterial DNA polymerase has only one binding site for which all four deoxynucleoside triphosphates compete (Englund *et al.*, 1969) (for complications see subsection L, *Mutator Genes*).

Uracils, halogenated in the 5 position (5-bromouracil, 5-chlorouracil, and 5-iodouracil), can be incorporated into DNA in the place of thymine. 5-Bromouracil (BU) is most effectively incorporated, presumably because bromine has about the same van der Waals radius as the methyl group in thymine. The amount of BU incorporated into DNA is often higher when the deoxynucleoside of BU is used and when thymine starvation is produced by aminopterin, 5-fluorodeoxyuridine, etc. BU and its deoxynucleoside are highly mutagenic, probably because BU undergoes, at a much higher frequency than T, a tautomeric shift which causes it to hydrogen bond to G instead of the normal complementary base A (see Fig. 8). Figure 9 illustrates the way in which BU can induce a base–pair change during DNA duplication. The change

$$\begin{array}{ccc} G & & A \\ C & \longrightarrow & T \end{array}$$

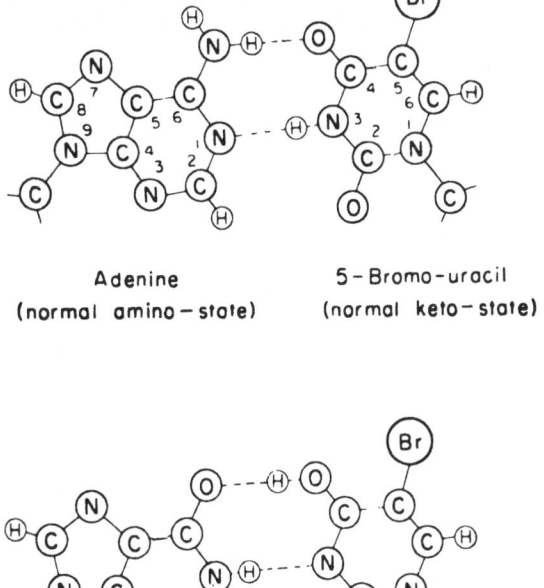

FIGURE 8. Normal and rare base pairs for 5-bromouracil.

occurs when BU is occasionally incorporated into DNA in place of C (mistake in incorporation, Fig. 9, upper), whereas the opposite change occurs
when BU is first incorporated in place of T and then occasionally undergoes
a mistake by pairing in one of the following DNA duplications with G instead of A (mistake in replication, Fig. 9, lower). Both pair changes are
transitions and BU should be able to induce the back-mutation (reversion
to original genotype and phenotype) of all mutations induced by it. These
expectations have all been experimentally verified in several microbial
systems. In addition to the mutagenic effect, it has been observed that cells
whose DNA contains BU or any halogenated base are rapidly inactivated
by X-rays, UV, fluorescent light, or by hydroxylamine, because the halogenated base is labile. Thus BU can induce inactivating DNA alterations
in addition to mutagenic DNA alterations. In fact, BU can give rise to

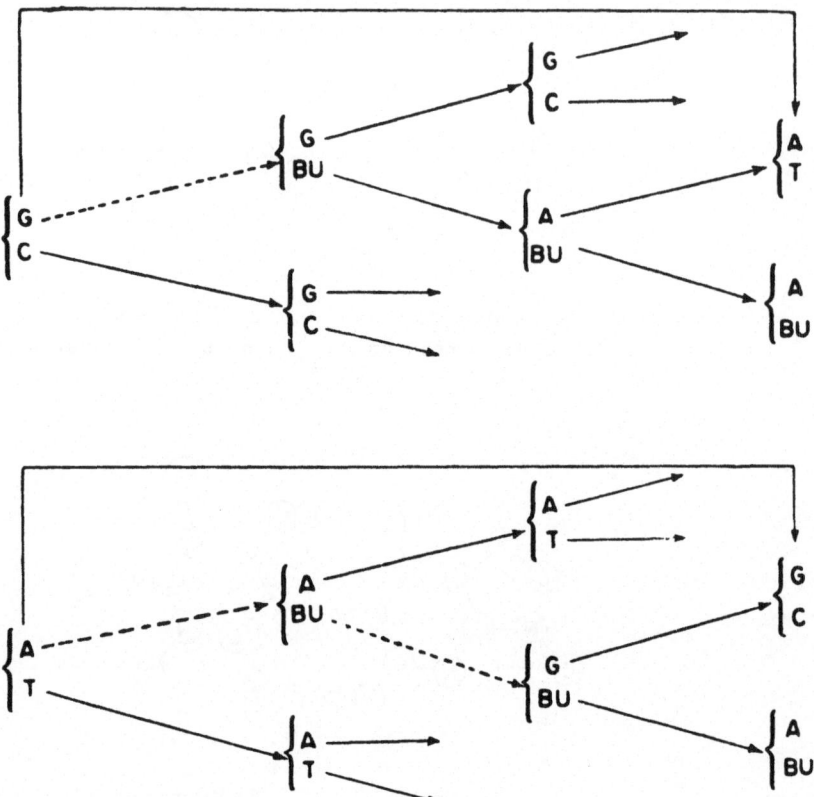

FIGURE 9.   Base-pairing mistakes of 5-bromouracil leading to transitions. Upper scheme:
Mistakes in incorporation, Lower scheme: Mistakes in replication.

chromosome breaks and exchange figures in tissue cultures (Hsu and Somers, 1961). Probably for this reason, iododeoxyuridine has been successfully employed against *Herpes simplex* in the eye.

2-Aminopurine is also incorporated into DNA in certain organisms
(phage T4, *E. coli*), although much less than BU. In these organisms it is
highly mutagenic. Presumably, the cellular enzymes of some organisms let
a small amount of 2-aminopurine slip through to the deoxynucleoside
triphosphate state and then cannot stop its incorporation into DNA; other
cells may have more exacting enzymes. 2-Aminopurine can form two hydrogen bonds to T and, after a tautomeric shift, two hydrogen bonds to C.
It thus induces base-pair transitions, also in both directions, similar to
BU. Apart from the mutagenic effect, high concentrations of 2-aminopurine
inhibit growth, presumably by interfering with some step in the purine

metabolism. 2,6-Diaminopurine is also slightly mutagenic for phage T4 (for more information see Freese, 1963).

B. Effect of Nitrous Acid on Resting DNA

When sodium nitrite ($NaNO_2$) is exposed to low pH (pK = 3.3), free nitrous acid is produced which deaminates the bases G, C, and A at rates decreasing in this order. The deamination of C to U and of A to hypoxanthin is mutagenic, producing base-pair transitions (in both directions), whereas the base xanthine, resulting from deamination of G, can no longer pair with either C or T (Michelson and Monny, 1966) and represents an

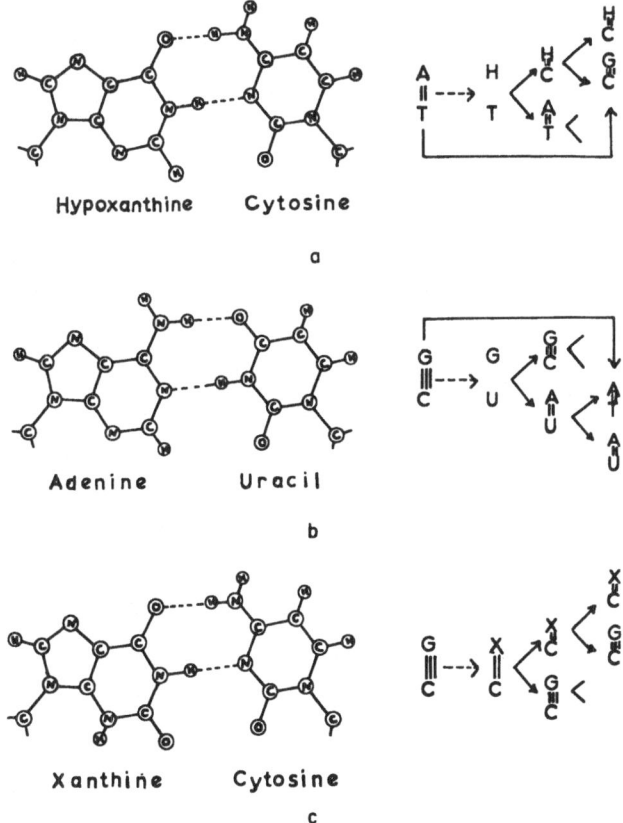

FIGURE 10. Effect of nitrous acid on base pairing of deaminated bases, A ⟶ hypoxanthine, C ⟶ U, G ⟶ xanthine, and the resulting base-pair transisions.

inactivating DNA alteration (see Fig. 10). Nitrous acid induces inactivating DNA alterations (Fig. 5) also by crosslinking DNA strands. The relative frequency of the mutagenic versus the lethal effect in phage T2 decreases with increasing pH of treatment (Vielmetter and Schuster, 1960), which indicates the existence of two different reactions with DNA. The inactivating effect is probably responsible for the induction of extended deletions by nitrous acid in phage T4 (Tessman, 1962).

Nitrates are used in large quantities as fertilizer and they are found

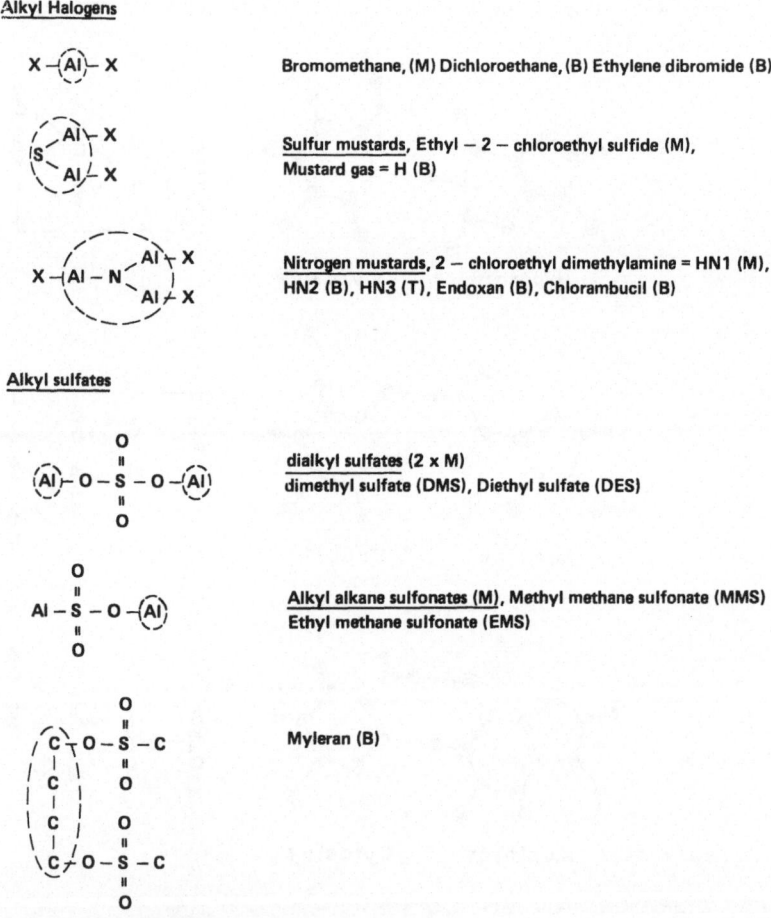

FIGURE 11. Types of alkylating agents. To simplify the above formulas, all hydrogens attached to carbon atoms were left out. Al = alkyl group; R = any organic group;

to some extent in food. Plants and some bacteria reduce nitrates to nitrites, which are known to have a toxic effect in man. Small quantities of these compounds are found in vegetables harvested from areas rich in nitrate fertilizer (see Commoner, 1968). In addition, both nitrates and nitrites are added at low concentrations to stored meat and fish in order to avoid spoilage, i.e., to prevent growth of microorganisms (perhaps by reacting with their genetic material). It is not known whether the nitrites consumed with food are partially responsible for ulcers and intestinal or other cancers.

Unstable 3 – membered rings

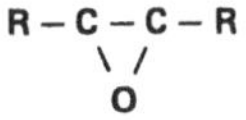

$$R - C - C - R$$
$$\backslash\ /$$
$$O$$

Epoxides (M), Diepoxybutane, Epichlorhydrin, Ethylene oxide, Propylene oxide

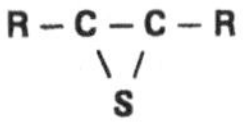

$$R - C - C - R$$
$$\backslash\ /$$
$$S$$

Ethylene sulfide   (M)

$$C - C$$
$$\backslash\ /$$
$$N$$
$$|$$
$$R$$

Ethylene imines, Apholate (T), Ethylene imine (M), TEPA (T), Triethylene melamine (T)

Unstable lactones

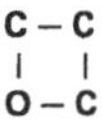

$$C - C$$
$$|\ \ |$$
$$O - C$$

4 membered lactone ring (M), β – propiolactone

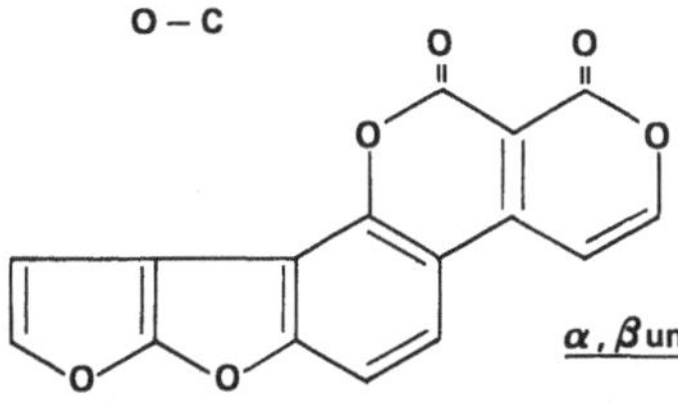

α, β unsaturated lactone, aflatoxin G_1

Diazo Compounds

$$N \equiv N = R$$

Azaserine, Diazomethane,
Diazouracil, Ethyldiazoacetate

X = Cl or Br. M = monofunctional, B = bifunctional, T = trifunctional agent.

C. Alkylating Agents

Alkylating agents (review: Stacey, 1958; Freese, 1963; Kihlman, 1966; Loveless, 1966; Lawley, 1966) include a large number of molecules that carry one, two, or more of the alkylating groups shown in Fig. 11, and are consequently called mono-, di-, or polyfunctional alkylating agents. If the alkylating groups in a polyfunctional alkylating agent are covalently linked (as in the sulfur or nitrogen mustards but not in dialkyl sulfates), they can crosslink DNA strands (either the two strands of one DNA molecule = intrastrand crosslink, or two strands of different DNA molecules = interstrand crosslink). Even monofunctional agents can occasionally cause crosslinking, owing to reactive DNA ends created by the alkylation-induced backbone breakage. Negatively charged alkylating agents, like chlorambucil, react less with the negatively charged DNA than positively charged agents, but when DNA is neutralized by histones or protamine (in nucleoprotein), it is readily attacked by either alkylating agent (Alexander and Stacey, 1958).

Most, and probably all, alkylating agents have some mutagenic effect, whether they are mono- or polyfunctional; this has been shown in many genetic systems, including maize, *Vicia*, *Drosophila*, *Neurospora*, bacteria, phages, and transforming DNA. The agents induce point mutations (mainly transitions), chromosome breaks, and chromosome mutations. One functional group is sufficient to induce mutations (at least point mutations).

In polar solvents, alkylating agents give rise to positive carbonium ions, e.g.,

$$CH_3 \overset{\overset{\displaystyle O}{\|}}{\underset{\underset{\displaystyle O}{\|}}{S}} O\text{-}CH_3 \longrightarrow CH_3 \overset{\overset{\displaystyle O}{\|}}{\underset{\underset{\displaystyle O}{\|}}{S}} O^- + CH_3{}^+$$

which react readily with nucleophilic groups such as sulfhydryl, thioester, ionized acid, or nonionized amino groups. In addition, alkylating agents hydrolyze, forming an alcohol and an acid. If the pH of the medium is not properly checked, some of the observed reactions may in fact be caused or at least influenced by low pH. In DNA the agents alkylate the phosphate groups, forming semistable triesters, and the ring nitrogens of the bases. The effects of the alkyl sulfates and sulfonates have been examined most thoroughly. The point mutagenic effect, in which mainly transitions from

$$\begin{array}{c} G \\ C \end{array} \longrightarrow \begin{array}{c} A \\ T \end{array}$$

are observed, apparently is caused by the alkylation of G at the 7-position

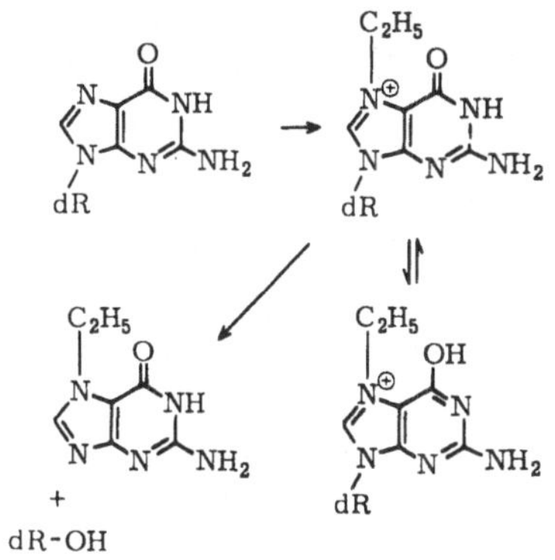

FIGURE 12. Alkylation (ethylation) of guanine (at the 7N position) and resulting tautomeric shift (lower right) and depurination.

(see Fig. 12). Chemically, G is the preferably alkylated base in double-stranded DNA, whereas the alkylation of A or C is minor. [In single-stranded or denatured DNA, in which the bases are not hydrogen bonded, alkylation of A at the N-1 position is the major reaction (Lawley and Brookes, 1963).] Eventually, the alkylated G (or A) hydrolyzes from DNA and the backbone of the unstable depurinated DNA breaks (by β-elimination, see Fig. 12). Following the discovery of the alkylation of G and its liberation from DNA, this effect has been tacitly considered the major (if not only) mechanism by which alkylating agents cause a DNA backbone breakage. However, more recent experiments with oligodeoxynucleotides of T (a base which does not react with alkyl sulfonates) have clearly shown that the phosphate triesters can also cause backbone breakage (and do not always merely liberate the alkyl group) (Rhaese and Freese, 1969). About one half of the methylation reactions leading to backbone breakage (perhaps the methylation of G but not the triester formation) can be repaired in *Bacillus subtilis* (Strauss and Robbins, 1968).

The removal of G from DNA is clearly lethal, because depurination of transforming DNA by low *p*H inactivates DNA. But it is not known whether the chromosome-breaking effect of alkylating agents is mainly due to the removal of G or whether it occurs only when the DNA backbone has been ruptured.

Alkylating agents are not only known to induce mutations and chromosome aberrations, but they are also toxic, teratogenic, and they cause cancer. Human should therefore not be exposed even to small quantities of these agents in (or on) food, in the form of pesticides in the house or in nature, or in industry. Most of the highly reactive alkylating agents are carefully handled because of their toxicity, but some alkylating agents are sufficiently slowly reactive that they are still used for the conservation of fruits (ethylene dibromide), as pesticides (ethylene oxide, endrin, dieldrin), as antibiotics (griseofulvin, streptonigrin), or they are considered as pesticide sprays (apholate, TEPA, thio-TEPA, METEPA, Aramite). Human exposure to such a chemical should be permitted only if an overwhelming evidence of animal tests indicates that the compound is harmless (which is very unlikely because it is used to kill insects or bacteria by a genetic reaction).

Recently a carcinogenic α,β-unsaturated lactone, aflatoxin, was discovered on stored seeds, where it is produced by the mold *Aspergillus*. Other potential alkylating agents include the epoxide scopolamine and some phosphate triesters which might transalkylate with nucleic acids.

D. N-Nitroso Compounds

N-nitroso compounds are considered separately, because they are all highly carcinogenic, induce chromosome breaks and aberrations, and some of them (MNNG) are extremely potent point mutagens in cellular systems (Wheeler, 1962; Marquardt *et al.*, 1964; Kihlman, 1966). The structures of some typical N-nitroso compounds are shown in Fig. 13. The dialkyl nitrosamines (uppermost compound in Fig. 13) are stable compounds which can act on DNA only after enzymic activation, (e.g., removal of one alkyl group). They induce mutations in *Drosophila*, whereas they are not effective in bacteria, yeast, or *Neurospora*, unless they are activated either chemically (Malling, 1966) or enzymically in the host-mediated assay (see Legator, this book). The activated compound is not the diazoalkane, because 7-methyl-G, produced by deuterated dimethylnitrosamine in rats, contains a $-CD_3$ group (Lijinsky *et al.*, 1968).

The acyl nitrosamides and other compounds (the lower four in Fig. 13), having electrophilic groups attached to the nitrogen, are more or less labile; they decompose at alkaline pH, giving rise to alkylating diazocompounds. For example, MNNG (see Fig. 13) produces the alkylating agent diazomethane at a rate which increases with the pH when that is above 7. However, the strong point-mutagenic effect observed with MNNG has a pH optimum of about 6, which would seem to rule out mutation induction via free diazomethane (Süssmuth and Lingens, 1969). Treatment at pH 5–7 of transforming DNA by MNNG produced mainly a lethal effect and very few point mutations (Freese and Freese, 1966). Bacterial mutations apparently are produced by MNNG only while DNA duplicates and then predominantly in the replicating fork (Cerdá-Olmedo *et al.*, 1968). Both transitions and transversions, but no frameshift mutations, are induced (Whitfield *et al.*, 1966). Different mutagenic mechanisms are conceivable: point mutations may arise only by reaction with single-stranded DNA, by

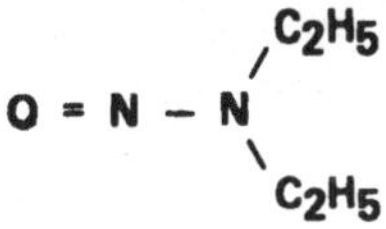

FIGURE 13. Typical N-nitroso compounds.

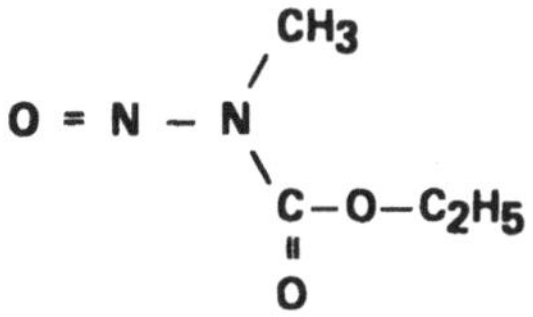

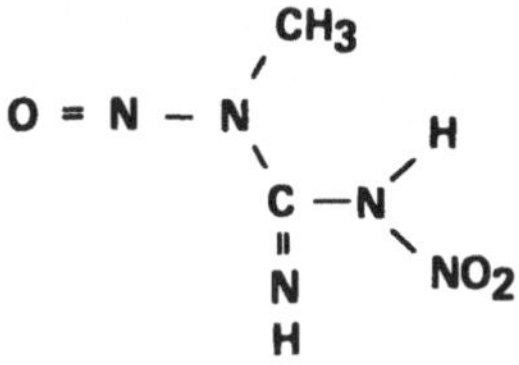

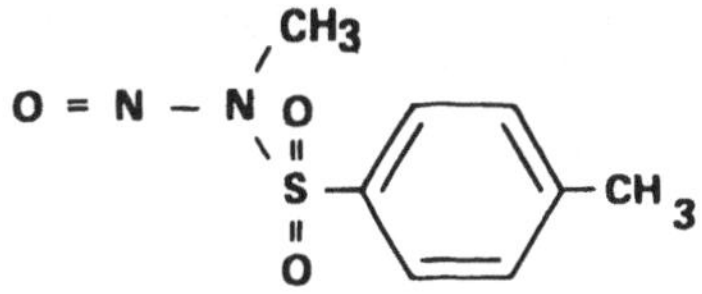

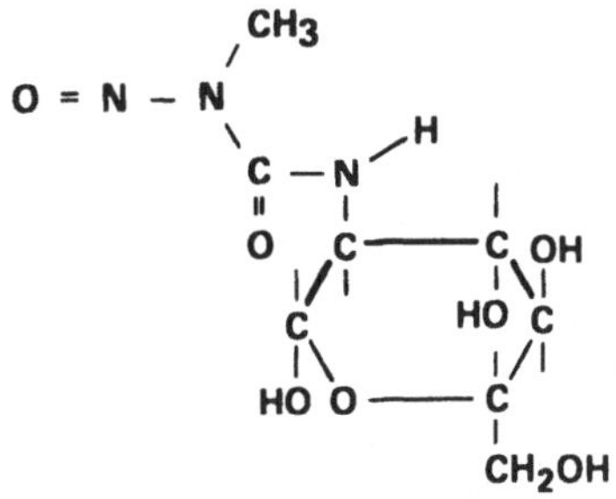

interference with the DNA replicating system, or only after cellular conversion of MNNG into an active compound. In fact, sulfhydryl groups have been found to increase DNA methylation by MNNG with a pH maximum of 6, and at that pH a higher rate of methylation has been observed *in vivo* than *in vitro* (Süssmuth and Lingens, 1969). But it is still unclear whether direct alkylation of DNA or some indirect effect is responsible for the induction of point mutations. The same reaction could cause chromosome breakage, but that effect could also result from one of the other reactions of N-nitroso compounds, including DNA alkylation by diazo compounds or radical reactions.

N-nitroso compounds have to be used with care, even in the laboratory, because allergic rashes result from repeated exposure to the liberated diazoalkane.

The broad-spectrum antibiotic streptozotocin (see Fig. 13) is an N-nitroso compound produced by *Streptomyces achromogenes*. It induces tumors, cancer, latent phages, mutations, and produces diabetes (see Legator, this volume). A compound with a structure similar to the N-nitroso compounds is the glycoside cycosin

$$(H_3C—N{=}N—CH_2O{\cdot}C_6H_{11}O_5)$$
$$\downarrow$$
$$O$$

which is produced by the cycad nut; it was consumed by inhabitants of Guam until its carcinogenic effect was discovered. This compound is activated by the intestinal flora (bacteria) to the alkylating aglycone methylazoxymethanol

$$(H_3C—N{=}N—CH_2OHC_6H_{11}O_5)$$
$$\downarrow$$

E. Hydroxylamines

Hydroxylamine $(HA{=}H_2NOH)$, N-methyl-HA(CH_3HNOH), phenyl-HA$(\phi\text{-}NHOH)$, N-hydroxycarbamates

$$\begin{array}{c} O \\ \parallel \\ (—C—C—NOH) \\ | \end{array}$$

and N-hydroxyureas

$$\begin{array}{c} O \\ \parallel \\ (—N—C—N—OH) \\ |\qquad\; | \end{array}$$

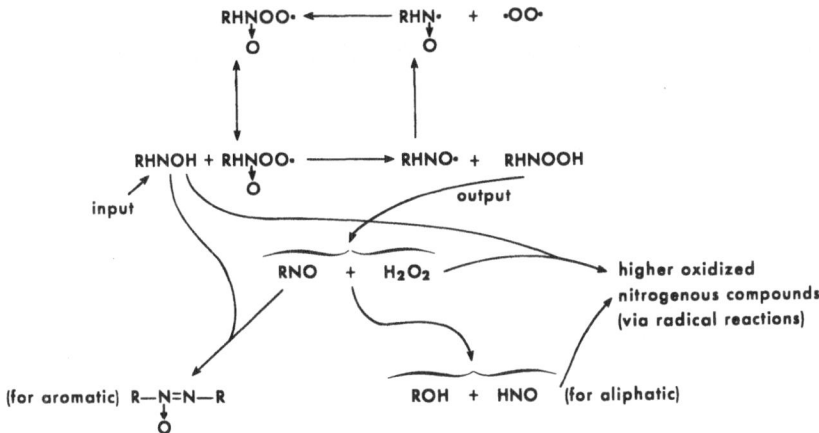

FIGURE 14. Probable radical mechanisms leading to oxidation of hydroxylamines and production of H_2O_2.

are mild reducing agents that form free radicals in the presence of oxygen and trace metals. (reviews: Freese and Freese, 1966; Phillips and Brown, 1967) They produce hydrogen peroxide, presumably by the radical reaction circle, involving peroxy intermediates, shown in Fig. 14. Hydrogen peroxide reacts further with the HAs (via radical intermediates) to higher oxidized forms of nitrogen compounds. The H_2O_2 production or consumption can be prevented by agents which quench the radical reactions, such as pyrophosphate, cyanide, excess of EDTA, etc. The free radicals or H_2O_2 (via radicals), produced by the HAs, inactivate transforming DNA and phages (Freese *et al.*, 1967). HA, *N*-methyl-HA, *N*-hydroxyurea, and *N*-hydroxyurethan induce chromosome aberrations in hamster tissue cultures and in root tips of *Vicia* and *Allium* (Kihlman, 1966). Other *N*-hydroxycarbamates inactivate bacteria, producing long threads of non-dividing cells, similar to the effects of UV irradiation (De Giovanni-Donnelly *et al.*, 1967).

At least HA and *N*-methyl HA (CH_3HNOH) also react specifically with cytosine (C), producing derivatives which are *N*-hydroxylated in the 4-, 6-, or both positions. These derivatives are in a different tautomeric state than C so that they can form hydrogen bonds to A in place of the original G. Thus treatment of transforming DNA or DNA viruses with high concentrations of HA induces mutagenic DNA alterations which give rise to base-pair transitions

$$\begin{array}{ccc} G \\ C \end{array} \longrightarrow \begin{array}{ccc} A \\ T \end{array}$$

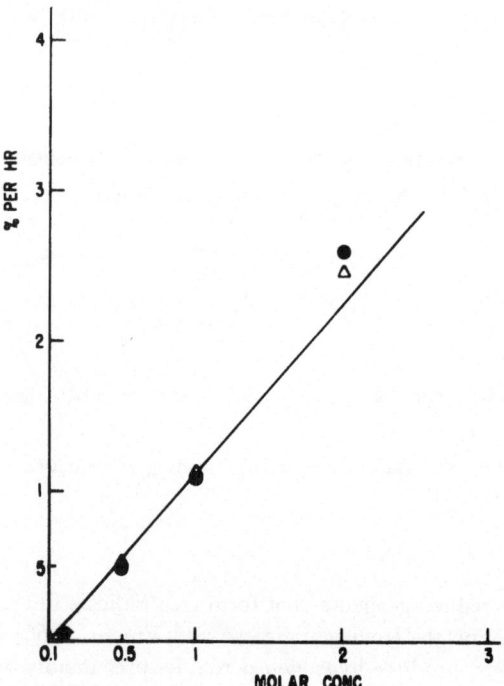

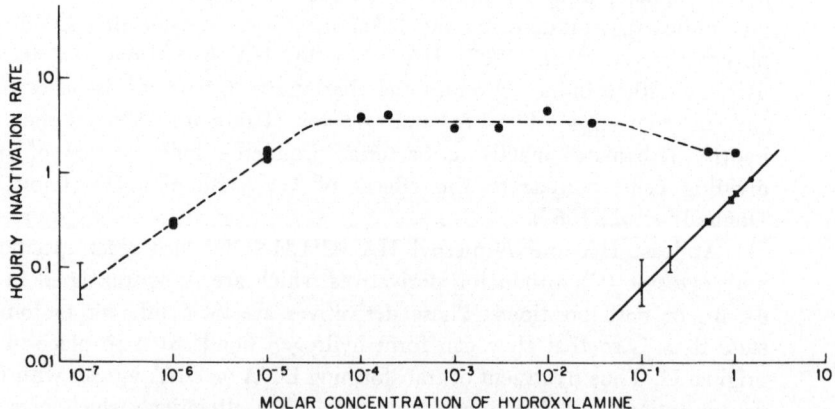

FIGURE 15. Concentration dependence of the effect of
hydroxylamine on (transforming) DNA. Upper: Point
mutagenic effect. Lower: Inactivation in the presence
(upper curve) and absence (straight line) of oxygen.

and result in point mutations. It is apparent that HA cannot induce the back-mutation to the original phenotype. Actually, HA is the most specific point-mutagenic agent known.

HA is ideal to demonstrate separately the induction of both mutagenic and inactivating DNA alterations in transforming DNA. Figure 15 shows the concentration dependence of inactivation and mutation rates in air and in nitrogen or in 0.05 M sodium pyrophosphate. The rate at which point mutations are induced increases linearly with the HA concentration and is not affected by pyrophosphate, in agreement with a direct reaction of HA with C. However, the rate of inactivation in air (and the concentration of H_2O_2 produced) shows a zero-order concentration dependence when the HA concentration ranges over several powers of ten. This inactivating effect is caused by the continuous production and destruction of H_2O_2, both reactions being directly proportional to the HA concentration. The reaction rate decreases only when HA becomes rate limiting at low HA concentrations or oxygen becomes rate limiting at high HA concentrations. In the absence of oxygen or the presence of pyrophosphate as radical quencher, the inactivation rate increases linearly with the HA concentration, in parallel to the mutation curve; the residual inactivation is then probably caused by point mutations induced in a vital gene by the direct reaction of HA with C. The permeability barrier of most cells (and the ability to metabolize HA) will usually permit HAs to be present inside the cell only in low concentrations; they will therefore mostly produce inactivating DNA alterations and chromosome breaks (unless the radical formation is very efficiently quenched by cellular components).

F. Agents Acting after Conversion to Hydroxylamines or Hydroxamates

Urethan and certain other carbamates cause mutations, chromosome breaks and aberrations, teratogenic effects, and cancer in certain test organisms (reviews: Wheeler, 1962; Freese, 1967).

These compounds do not directly inactivate DNA but only after they are oxidized, by microsomal enzymes present in many organisms including man, into the HA (and perhaps nitroso) derivatives (Freese et al., 1967). The HA derivatives of carbamates (N-hydroxyurethan) or aromatic amines (phenylhydroxylamine, hydroxylamine, etc.) are usually more potent chromosome-breaking or carcinogenic agents than the original amines (Uehlecke, 1965; Miller and Miller, 1969). Depending on the chemical groups attached to the NOH moiety, some HAs react rapidly, others slowly with free oxygen (and perhaps other oxidizing compounds in the cell) to form free radicals. The free radicals induce inactivating DNA alterations if they happen to encounter DNA during their short life span. Alternatively, the radicals can react with each other or with other cell components to produce labile intermediates (biradicals, peroxides, esters, etc.) which have more time to move to the DNA site and exert their effect.

Some aromatic HAs can be enzymically further converted to labile acetic acid esters which apparently act as alkylating agents because they induce both point mutations and inactivating DNA alterations. This has been demonstrated for the carcinogen 2-acetylaminofluorene, which is first N-hydroxylated to the more potent carcinogen N-hydroxy-acetylaminofluorene. This compound is further acetylated to N-acetoxy-acetylaminofluorene, which, in contrast to the HA compound, induces point mutations in transforming DNA (Maher *et al.*, 1968). The last compound may be the ultimate carcinogen (see Miller and Miller, 1969, and this volume).

Instead of oxidation of amines, reduction of nitro ($-NO_2$) or nitroso ($-NO$) compounds can also produce hydroxylamines. Among aromatic compounds the hydroxylamines usually are more carcinogenic than the nitro compounds (4-nitrodiphenyl, 4-nitrostilbene, 2-nitrofluorene).

A number of 4-nitro- or 4-hydroxylamino-quinolines (which form radicals with unpaired electrons localized on the ring or the NO_2 nitrogen) are carcinogenic, photodynamically active, induce prophages, and inactivate mitochondria in yeast (Epstein and St. Pierre, 1969). It is not known whether these compounds, or free radicals produced by them, can directly act on DNA or whether they have to be enzymically converted into active compounds.

Apart from the carcinogenic effect, aromatic amines or nitro compounds also produce allergies; this indicates that the compounds can be enzymically activated to react with tissue proteins and form haptens.

Many pesticides are carbamates, ureas, or other amino, nitro, or nitroso compounds (for a list see Neumayer *et al.*, 1969). Some of them are known to break chromosomes in plants (Brian, 1964; Kihlman, 1966). Varnishes are often composed of poycarbamates (produced by polymerization of tolyl diisocyanate and bifunctional alcohols and diols). Some antianxiety compounds are amides (oxanamide) or carbamates (tybamate). No information about the mutagenic effects of any of these compounds in mammals is available.

The sweetener cyclamate is converted, by the intestinal flora of some people, to cyclohexylamine, which has been shown to cause cancer or teratogenic effects in animals (see Legator, Volume 2).

G. Other Free-Radical-Producing Agents

Radicals are produced by ionizing radiations, by short-wave ultraviolet light, and by certain chemicals. The rate of free radical formation by chemicals depends on the presence of transition metals (Fe, Cu, etc.) and can often be greatly increased by heat or light. Radicals are also naturally produced in oxidation–reduction reactions, but they remain then bound to enzymes. DNA can be attacked only by free radicals that are produced immediately adjacent to it. But radicals can open many chemical bonds, and, by reacting with another radical, produce new covalent bonds. Although these new bonds are more stable than the radicals, some of them are quite reactive (certain esters or amines) or they are unstable and can decompose again into radicals (peroxides) or alkylating agents. Such compounds can diffuse and react with DNA (or other compounds) when they encounter reactive (nucleophilic) groups. Since many physical and chemical reactions produce free radicals, cells must have evolved a number of protective mechanisms. Known are catalase or peroxidase which eliminate hydrogen peroxide and certain organic peroxides, a nuclear membrane which protects chromosomal DNA during the synthetic (S) period, and cellular-reducing compounds; also repair mechanisms probably exist which eliminate radical-induced DNA damage.

Small radical concentrations can be measured by the initiation rate of polymerization from monomers, by the destruction of light-absorbing stable radicals (e.g., diphenylpicryhydrazyl), or by the production of fluorescence; higher radical concentration can be measured by paramagnetic resonance. For biological examinations it is more important to have some indications for *in vivo* experiments whether an agent acts by a radical mechanism or not. Such a mechanism is indicated but not proven by one of the following findings: (1) The agent is a radical or can produce radicals. (2) The effect is oxygen dependent. (3) A dependence on oxygen is suggested when mild reducing agents such as cysteine and cysteamine reduce the effect; but strong reducing agents, such as ascorbic acid, produce radicals, in the presence of oxygen, and inactivate DNA. (4) Radical scavengers, (Fe^{3+}, isodine, hydroquinone) prevent radical formation even if no oxygen is involved. (5) Radical production often is inhibited by excess of chelating agents. But note that low concentrations of EDTA can even enhance the effect of Fe^{2+}, since the combination acts like catalase (Thullier, 1959; Mader, 1960; von Selke and Krause, 1963).

The results of these tests are sometimes difficult to interpret when cells or cellular organisms are studied, because such biological systems require oxygen (and intermediate radical formation) also for oxidative phosphorylation, which is inhibited by the same agents. In addition, repair mechanisms occur during and immediately after the treatment, so that agents which inhibit repair would act opposite to expectation. The above tests for radical involvement do become significant, however, when the

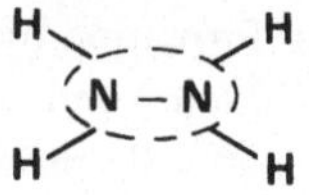

FIGURE 16. Some hydrazines and hydrazides.

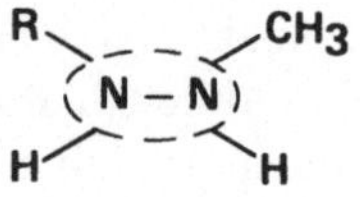

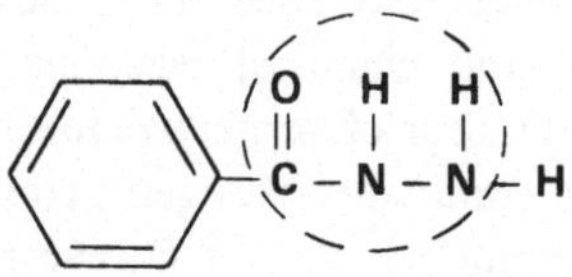

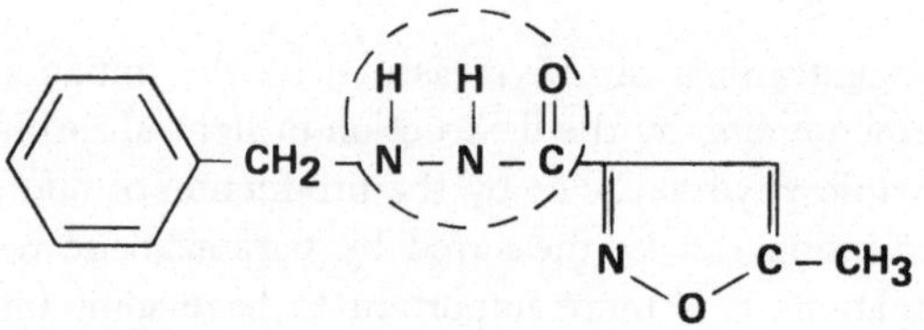

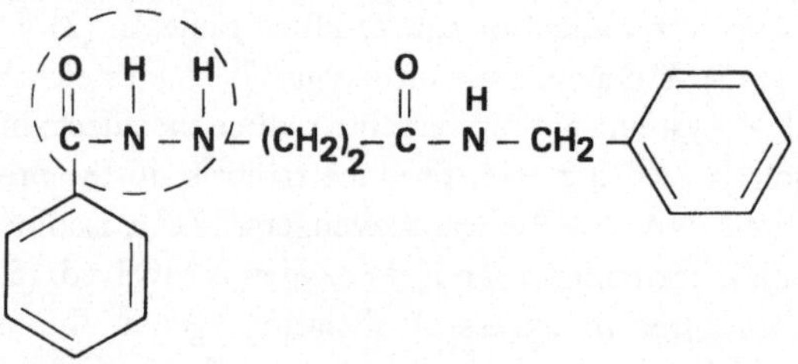

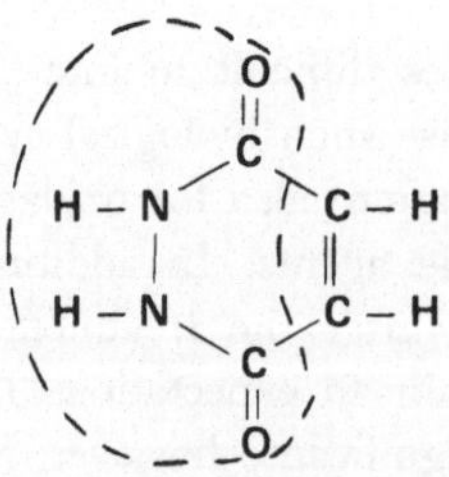

effect of an agent is examined on isolated viruses or transforming DNA, because all reagents can be removed before the nucleic acid is added to cells.

In the following, a few agents that produce radicals will be discussed in more detail.

1. Hydrazines and Hydrazides

Hydrazines or hydrazides (see Fig. 16) react, similar to hydroxylamines, with oxygen in the presence of trace metals and produce radicals and hydrogen peroxide (review: Freese and Freese, 1966). At least hydrazines and methylhydrazines also react directly with the pyrimidine bases, especially at high pH, breaking the pyrimidine ring and causing the removal of the base from DNA.

In bacteria, hydrazines and methylhydrazines induce mutations at a low frequency (compared to the lethal effect). In isolated transforming DNA or phages, the hydrazines or hydrazides shown in Fig. 16 induce predominantly inactivating DNA alterations, with a concentration dependence similar to that observed with hydroxylamine (not yet checked for maleic hydrazide) (Freese et al., 1967, 1968). Some hydrazines (1-methyl-2-benzyl-hydrazine and other methyl hydrazines: Rutishauser and Bollag, 1963) are known to induce chromosome breaks in mouse ascites tumor, and isoniazid (Fig. 16) rather specifically inactivates tubercle bacilli. Other hydrazines (see Fig. 16 and phenelzine) are used as antidepressants and for that reason alone must be lipophilic enough to enter cells, since they can pass the blood–brain barrier; these hydrazines do produce H_2O_2 and inactivate DNA, but their effects on chromosomes have not been examined (Freese et al., 1968). Certain hydrazines apparently are employed as components in rocket fuels.

Azo dyes (ar—N=N—ar) with nonpolar substituents can enter cells and some of them are well-known carcinogens (e.g., butter yellow, 2,2'-azonaphthalene) (Weisburger and Weisburger, 1966) and teratogens (Beck and Lloyd, 1966). These activities are found if the aromatic moieties are substituted at certain alternative positions; the side groups either influence the probability of free radical formation or they are needed for enzymic activation of the compounds.

Maleic hydrazide (Fig. 16) breaks chromosomes in many different plants and does that more frequently in the prsence of oxygen (Kihlman, 1966). Its effect on DNA or the production of H_2O_2 has not been measured nor does its effect on chromosomes in animal cells seem to be known.

2. Ionizing Radiations

Although α-, β-, γ-, or x-rays usually are treated as if they produced DNA or chromosome alterations by directly hitting the hereditary material, this is almost certainly not true for most alterations obtained in hydrated cells. Alterations in dehydrated cells, such as seeds or spores, may be more dependent on direct DNA hits but these events occur at a much lower frequency (see Sussman and Halvorson, 1966). Ionizing radiations knock single electrons out of molecules or they excite molecules by radiation transfer to decompose into two radicals. The liberated electron is trapped by water to form the hydrated electron e^-_{aq}, which acts like other radicals. Any of these radicals can react with water or other molecules. If they encounter the biradical oxygen ($\cdot O\text{—}O\cdot$), they tend to react with it preferentially to form peroxy radicals ($R\text{—}O\text{—}O\cdot$), which in turn can react with another radical to produce the chemically more stable peroxides.

It has often been observed that the lethal or chromosome-breaking effect of X-rays is greatly increased in the presence of oxygen. X-radiation of DNA in the presence of oxygen produces breakdown products of DNA bases and causes backbone breakage. Some of the reaction products of pyrimidines are hydroxy-hydroperoxy bases (see Fig. 16); the T derivative is stable, whereas the C derivative decomposes into 5-hydroxycytosine or isobarbituric acid. Both the base alterations and the backbone breakage constitute inactivating DNA alterations, which apparently are caused to a large extent by radical reactions.

3. Hydrogen Peroxide (H_2O_2) and Organic Peroxides ($R_1\text{—}O\text{—}O\text{—}R_2$)

Many of the previously mentioned radical reactions produced hydrogen peroxide. This compound decomposes into two $\cdot OH$ radicals spontaneously, but the decomposition is greatly accelerated by heat, UV, or trace amounts of transition metals (Fe, Cu, etc.). In the presence of reduced transition metals, H_2O_2 probably decomposes also via $HOO\cdot$, which by reaction with OH radicals produces O_2. The reaction of H_2O_2 with DNA or its components occurs via radical intermediates, because it depends on the presence of trace metals (increased by Fe^{2+}, cyanide, pyrophosphate) or excess of chelating compounds (EDTA). H_2O_2 causes the alteration and liberation of DNA bases and backbone breakage. It reacts with pyrimidines, causing the opening of the 5,6 double bond and may produce hydroxy-hydroperoxy pyrimidines either directly or by the mechanism described in Fig. 17 which involves O_2. The altered pyrimidines apparently are not able to undergo normal hydrogen bonding to any complementary bases, for one

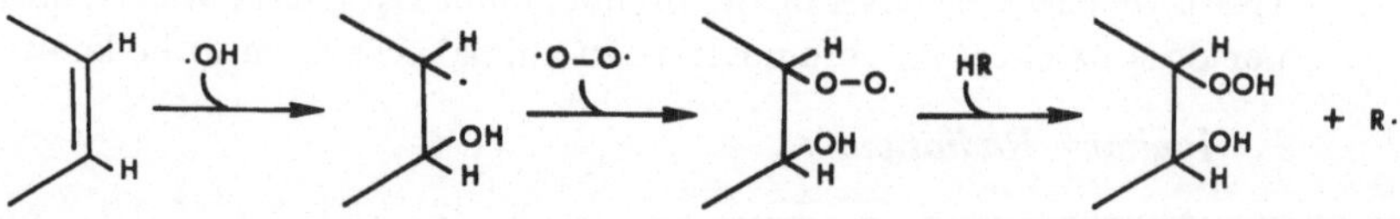

FIGURE 17. Production of hydroxyhydroperoxypyrimidines by radiation in air or by OH radicals.

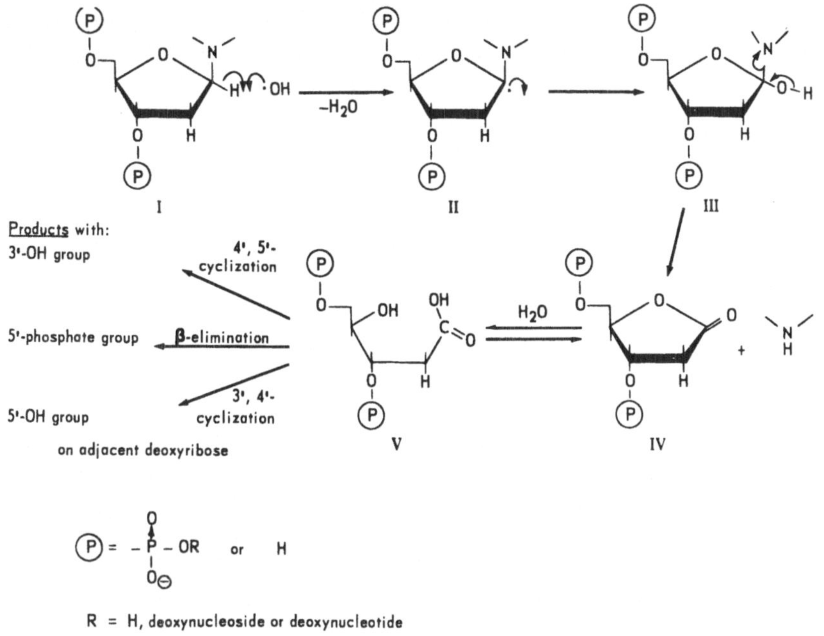

FIGURE 18. Mechanism of DNA backbone breakage by H_2O_2.

reason because the opening of the double bond seems to destroy the planarity of the ring and makes it impossible to have two hydrogen bonds in one plane, and because at least the altered C (and probably also some altered T) decomposes further. These base alterations inactivate DNA but do not induce point mutations. H_2O_2 also reacts with A, producing 7-N-hydroxy A from which the OH group migrates to the 8-position at elevated pH. Both 7- or 8-N-hydroxy A should still be able to make hydrogen bonds with T (it is not known whether this altered base can increase the frequency of point mutations). Finally, Fig. 18 shows how H_2O_2 breaks the N—C bond between the base and the deoxyribose, giving rise to deoxyribonic acid which is unstable and results in DNA backbone breakage by the β-elimination reaction with the 5'-phosphate group (Rhaese and Freese, 1968).

Hydrogen peroxide rapidly inactivates transforming DNA or phages and produces practically no point mutations. Most cells are efficiently protected against H_2O_2 by a catalase. Presumably for that reason no chromosome breaks induced by H_2O_2 have been observed. But bacteria lacking catalase are rapidly inactivated by H_2O_2 and chromosome breaks have been observed in tissue cultures having greatly reduced amounts of catalase (Schöneich, 1967).

Several organic peroxides (di-t-butyl peroxide, cumene peroxide, disuccinyl peroxide) induce mutations in and are lethal for *Drosophila* and microorganisms. Only disuccinyl peroxide, however, inactivates transforming DNA by a reaction depending on trace metals (Freese *et al.*, 1967). This indicates that the other two compounds have to be activated inside the cell before they can display their radical-producing effect. Peroxides and epoxides are produced by X- or UV radiation of organic matter (Adler, 1963; Philpot, 1963), and some peroxides produce cancer (Kotin and Falk, 1963). The chemical reactions of organic peroxide radicals with DNA have not been investigated in much detail but are probably similar to those of H_2O_2.

4. Aldehydes and Phenols

Formaldehyde induces mutations at a low rate in microorganisms and in early larval spermatocytes of *Drosophila*. Aldehydes slowly autoxidize by a radical mechanism (Walling, 1957). The mutagenic effect of formaldehyde is enhanced in the presence of H_2O_2 or UV, which by themselves are little effective. Catalase inhibitors also enhance the effect. The oxidation products of formaldehyde (formic acid, performic acid, dihydroxymethyl peroxide) do not inactivate or mutate transforming DNA, whereas dihydroxydimethyl peroxide is a fairly potent mutagen in *Drosophila* (but not in *Aspergillus*). This indicates that the genetic effects of formaldehyde result from an intermediate oxidation product of formaldehyde, probably a free radical peroxide (Sobels, 1963; Freese *et al.*, 1967).

Several phenols (e.g., cresol, hydroquinone, phenol, pyrocatechol, pyrogallol) are mild chromosome-breaking agents in plants (Levan and Tjio, 1948; Biesele, 1958). The most reactive phenols are readily oxidizable. The chromosome reaction probably involves either a radical reaction or some enzymic activation, because the phenols themselves do not react with DNA, but they are known to produce radicals and peroxides by autoxidation (Walling, 1957).

Several habit-forming psychotropic drugs have phenolic groups (morphine, codeine, hydromorphone, hydrocodone) or aldehyde groups (benzaldehyde), and at least morphine has been shown to break chromosomes in plants (Oehlkers, 1956).

H. Ultraviolet Light

UV affects DNA both directly and indirectly. The indirect effect is not observed when buffer or minimal media are irradiated, but only when the media contain UV-absorbing compounds that can be converted into radicals and organic peroxides (Stone *et al.*, 1947; Wyss *et al.*, 1948). The ultimate indirect effect of UV on DNA is therefore similar to that of the indirect effect of ionizing radiations.

The direct effect of UV gives rise to photoproducts such as the dimers between pyrimidine bases (for TT a cyclobutane formation has been established, for the other dimers it is likely), dihydrothymine, cytosine hydrate, and relatively few intra- as well as interstrand crosslinks between DNA strands (see Smith, 1966). The pyrimidine dimers, which seem to be the major stable reaction products, represent inactivating DNA alterations, as is shown by the absence of point mutations after treatment of transforming DNA or bacteriophages (Drake, 1969). The dimers can be reverted to the original bases by visible light (photoreactivation) or they can be enzymically cut out of DNA and the gaps repaired as discussed earlier (Howard-Flanders, 1968). Mutations induced by UV apparently result from aberrant dimer repair or recombination processes, because their appearance can be avoided by photoreactivation (see Drake, 1969).

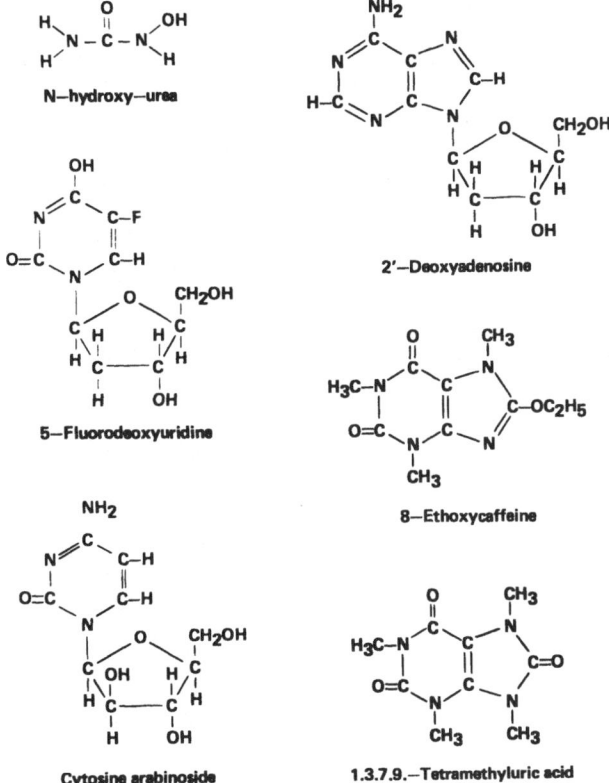

FIGURE 19. Some inhibitors of DNA synthesis.

I. Inhibitors of DNA Synthesis

A number of agents inhibit DNA synthesis. Some of them are shown in Fig. 19. They induce chromosome breaks in plant or animal cells and in human leukocytes (Kihlman, 1966). For the latter reason they are used as antineoplastic agents.

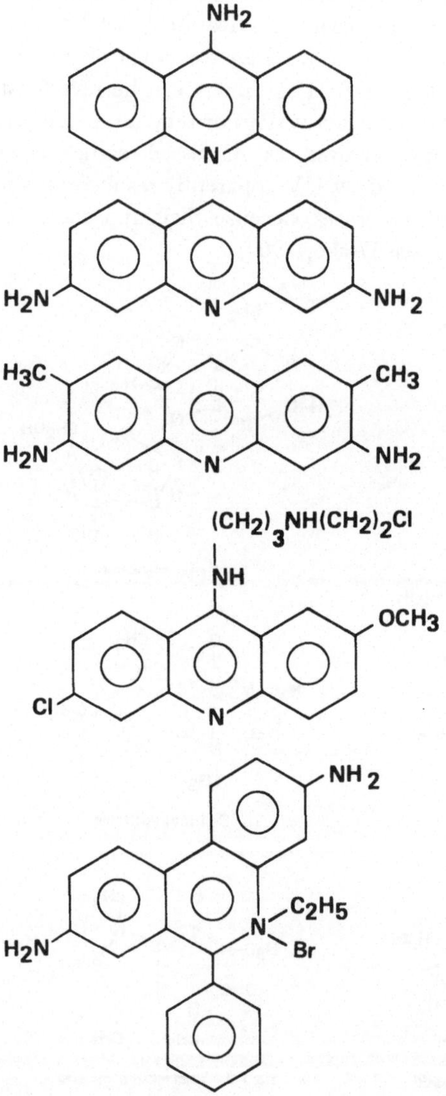

5—Aminoacridine

Proflavine
(2, 8 — Diaminoacridine)

Acridine Yellow
(2, 8 Diamino — 3, 7 Dimethyl— acridine)

IRC 191

Ethidium Bromide
(Homidium Bromide)

FIGURE 20.   Intercalating agents. (From Lerman, 1966.)

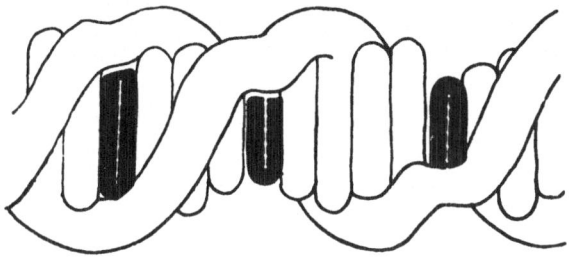

FIGURE 21. Acridine intercalation into DNA.

N-hydroxyurea reversibly inhibits DNA synthesis by a mechanism that is not completely understood; *in vitro* it inhibits irreversibly the reduction of CDP to dCDP (Rosenkranz and Jacobs, 1968). This compound may induce chromosome breaks also by the free radical mechanism of hydroxylamines described earlier.

5-Fluorodeoxyuridine is phosphorylated and the nucleotide inhibits the conversion of dUMP to dTMP (Cohen *et al.*, 1958). This methylation step can also be inhibited by aminopterin and amethopterin, which inhibit the formation of tetrahydrofolic acid.

Cytosine arabinoside becomes phosphorylated inside the cell and the diphosphate poorly inhibits the reduction of CDP to dCDP. More effective is the inhibition of DNA polymerase in animal (but not bacterial) cells (Furth and Cohen, 1968). The CDP reductase is also inhibited by dATP, which is formed from AdR (Reichard *et al.*, 1961). A itself, as well as azaserine, inhibits an early step of the purine metabolism and thereby prevents both RNA and DNA synthesis.

The lack of one of the deoxynucleotide triphosphate arrests DNA synthesis. When an incorrect base is incorporated, a point mutation may result; alternatively, degradative enzymes, involved in DNA synthesis or repair, may cut the exposed DNA strands (presumably at the replicating point) and thus insure chromosome breaks and large alterations.

A new potent inhibitor of DNA synthesis is 2′,3′-deoxyadenosine. Its triphosphate not only inhibits DNA polymerase, but it is incorporated into the terminus of DNA and thereby apparently blocks DNA synthesis irreversibly (Toji and Cohen, 1969).

Ethoxycaffeine, tetramethyluric acid (Fig. 19), and other methylated oxypurines which cause chromosome breaks, as well as ethidium bromide (Fig. 20), can intercalate between DNA bases, reduce the DNA melting temperature, and apparently thereby interfere with normal DNA and RNA synthesis (Tso *et al.*, 1962; Radloff *et al.*, 1967).

J. Intercalating Agents

Certain acridines (Fig. 20) intercalate between two purines in DNA (see Fig. 21) (reviews: Lerman, 1966; Orgel, 1965; Drake, 1969). Apparently due to the intercalation, frameshift point mutations (deletions or insertions of one or more bases) are produced during the duplication of bacteriophages or during meiosis of yeast. Acridines that are positively charged at neutral pH are particularly effective. The mutations apparently arise as mistakes in repair processes that occur during phage duplication when the ends of different DNA chains join, or during yeast meiosis in the region of recombination. In bacteria, certain acridines with alkylating side-chains are strong frameshift mutagens. The repair process might be initiated either by DNA alkylation or by recognition of the acridine moiety as a foreign element in DNA; in the latter case, any group (even nonalkylating) which protrudes from the acridine moiety would stimulate repair and enhance the mutation frequency.

Several antimalaria compounds (chloroquine, quinacrin) apparently intercalate into DNA and inhibit both DNA and RNA synthesis (O'Brien *et al.*, 1966).

K. Metals

Certain heavy metals (Co, Ni, Cr, Zn) were long known to induce chromosome breaks and aberrations and mutations in plants (Gläss, 1956). Mercury compounds, including those used as pesticides, also induce chromosome breaks and mitotic aberrations (Ramel, 1969). The reason for these effects is not known, except that mercury tends to bind to the purines in DNA.

Manganese is mutagenic for bacteria and phages and produces mainly transitions (Orgel and Orgel, 1965). Since Mn also interferes with DNA polymerase, causing the incorporation of RNA nucleotides into DNA (Berg *et al.*, 1963), it may cause mutations by altering the accuracy of the base-copying mechanism during DNA duplication.

L. Mutator Genes

The frequency of spontaneous (i.e., not knowingly induced) mutations is heritable and must have decreased during evolution from a high frequency (needed for rapid evolution of lower organisms) to a low frequency (required for relative stability of a species and sufficient longevity of the individual organism to reproduce). Mutations which alter this frequency can be isolated in microorganisms, and for some of the "mutator genes" it is known which types of mutations they control (review: Drake, 1969). In yeast, a mutation in a mutator gene specifically increases both the frequency of frameshift mutations and the sensitivity to X-rays (von Borstel *et al.*, 1968). In bacteria, mutations are known which specifically increase the frequency of either transitions or transversions (Yanofsky *et al.*, 1966). In bacteriophage T4, both increased and decreased mutation frequencies have been observed for mutants altered in the gene for DNA polymerase; in some of these mutants, the rate of mutation induction by external agents is almost the same as for the standard phage; in others, it apparently is altered (Drake, 1969). To explain these phenomena, it is sufficient to to assume that the mutant polymerase is less or more exacting than the normal polymerase in recognizing the correct structure and steric location of the incoming deoxynucleoside triphosphate. But it is not excluded that the polymerase can recognize also the structure of the incoming and the preexisting bases and accept (or reject) incorrect ones. In any case, the recognition mechanism apparently is not sufficient to eliminate the need for hydrogen bonding of complementary bases during DNA duplication, because otherwise some base analogs that cannot base pair should have been found to be incorporated into DNA and to induce mutations.

The frequency of mutations or the susceptibility to produce cancer apparently is genetically determined also in humans, as can be demonstrated most clearly in persons with some hereditary diseases. The case of UV sensitivity in persons with xeroderma pigmentosum (Cleaver, 1969) has been already mentioned. Mongoloids have a tenfold higher incidence of leukemia than the general population, and people with Bloom's syndrome or Fanconi's anemia (having a mutation in an autosome) show a high frequency of neoplasia that is preceded by chromosome breaks and aberrations observable in many cells (Hirschhorn and Bloch-Shtacher, 1969). A not quite so drastic variability can be expected for "normal" people with respect to both the spontaneous incidence of mutations, teratogenic effects, or cancer, and the sensitivity toward mutagenic or carcinogenic agents (e.g., the inducible hydrocarbon hydroxylase which inactivates benzpyrene may be produced by some people and not by others).

VI. REFERENCES

Adler, H. J. (1963), *Radiation Res. Suppl. 3*, 110.

Ahmed, A., Case, M. E., and Giles, N. H. (1964), *Brookhaven Symp. Biol. 17*, 53.

Alexander, P., and Stacey, K. A. (1958), *Ann. N.Y. Acad. Sci. 68*, 1225.

Altman, P. L., and Dittmer, D. S., eds. (1964), "Biology Data Book," Federation of American Societies for Experimental Biology, Washington, D.C.

Bautz, L., and Bujard, H. (1969), Genetics (in press).

Beadle, G. W., and Tatum, E. L. (1941), *Proc. Natl. Acad. Sci. U.S. 27*, 499.

54 **Ernst Freese**

Beck, F., and Lloyd, J. B. (1966), in "Advances in Teratology" (D. H. M. Woollham, ed.), p. 131, Academic Press, Inc., New York.

Benzer, S. (1961), *Proc. Natl. Acad. Sci. U.S. 47*, 403.

Berg, P., Fancher, H., and Chamberlin, M. (1963), in "Informational Macromolecules (H. J. Vogel, V. Bryson, and J. O. Lampen, eds.), p. 467. Academic Press, Inc., New York.

Biesele, J. J. (1958), "Mitotic Poisons and the Cancer Problem," Elsevier, New York.

Boyland, E., and Nery, R. (1965), *Biochem. J. 94*, 198.

Brian, R. C. (1964), *in* "The Physiology and Biochemistry of Herbicides" (U. L. Audus, ed.), p. 1, Academic Press, Inc., New York.

Carr, D. H. (1967), *J. Obstet. Gynec. 97*, 283.

Cerdá-Olmedo, E., Hanawalt, P. C., and Guerola, N. (1968), *J. Mol. Biol. 33*, 705.

Cleaver, J. E. (1969), *Proc. Natl. Acad. Sci. U.S. 63*, 428.

Cohen, S. S., Flaks, J. G., Barner, H. D., Loeb, M. R., and Lichtenstein, J. (1958), *Proc. Natl. Acad. Sci. U.S. 44*, 1004.

Commoner, B. (1968), *Scientist and Citizen 10*, 9.

Crick F. H. C., Barnett, L., Brenner, S. and Watts-Tobin, R. J. (1961), *Nature 192*, 1227.

Davis, R. W., and Davidson, N. (1968), *Proc. Natl. Acad. Sci. U.S. 60*, 243.

De Giovanni-Donnelly, R., Kolbye, S. M., and Dipaolo, J. A. (1967), *Mutation Res. 4*, 543.

Drake, J. W. (1969), *Ann. Rev. Genet. 3*, 247.

Englund, P. T., Huberman, J. A., Javin, T. M., and Kornberg, A. (1969), *J. Biol. Chem. 244*, 3038.

Epstein, S. S., and St. Pierre, J. A. (1969), *Toxicol. Appl. Pharmacol. 15*, 451.

Evans, H. J. (1962), *Internal. Rev. Cytol. 13*, 221.

Freese, E. (1959), *J. Mol. Biol. 1*, 87.

Freese, E. (1963), *in* "Molecular Genetics" (H. Taylor, ed.), p. 207, Academic Press, Inc., New York.

Freese, E., and Freese, E. B. (1966), *Radiation Res. Suppl. 6*, 97.

Freese, E., and Strack, H. B. (1962), *Proc. Natl. Acad. Sci. U.S. 48*, 1796.

Freese, E., Sklarow, S., and Freese, E. B. (1968). *Mutation Res. 5*, 343.

Freese, E. B. (1967), *Mol. Gen. Genet. 100*, 150.

Freese, E. B., and Freese, E. (1967), *Proc. Natl. Acad. Sci. U.S. 57*, 650.

Freese, E. B., Gerson, J., Taber, H., Rhaese, H. J., and Freese, E. (1967), *Mutation Res. 4*, 517.

Furth, J. J., and Cohen, S. S. (1968), *Cancer Res. 28*, 2061.

Gläss, E. (1956), *Z. Botanik 44*, 1.

Hayes, W. (1964), "The Genetics of Bacteria and Their Viruses," John Wiley & Sons, New York.

Hirschhorn, K., and Bloch-Shtacher, N. (1969), Genetics concepts and neoplasia, *in* "Proceedings of the 23rd Annual Symposium on Fundamental Concepts of Cancer," Houston, Texas (unpublished).

Howard-Flanders, P. (1968), *Ann. Rev. Biochem. 37*, 175.

Hsu, T. C., and Somers, C. E. (1961), *Proc. Natl. Acad. Sci. U.S. 47*, 396.

Hsu, T. C., Dewey, W. C., and Humphrey, R. M. (1962), *Exptl. Cell Res. 27*, 441.

Iyer, V. N., and Szybalski, W. (1963), *Proc. Natl. Acad. Sci. U.S. 50*, 355.

Kalter, H. (1968), "Teratology of the Central Nervous System," University of Chicago Press, Chicago.

Kihlman, B. A. (1966), "Actions of Chemicals on Dividing Cells," Prentice-Hall, Inc., Englewood Cliffs, N.J.

Kotin, P., and Falk, H. (1963), *Radiation Res. Suppl. 3*, 193.

Lawley, P. D. (1966), *in* "Progress in Nucleic Acid Research and Molecular Biology," (J. H. Davidson and W. E. Cohn, eds.), Vol. 5, p. 89, Academic Press, Inc., New York.

Lawley, P. D., and Brookes, P. (1963), *Biochem. J. 89*, 127.

Lerman, L. S. (1966), *J. Cell. Comp. Physiol. 64* (Suppl. 1), 1.

Levan, A., and Tjio, J. H. (1948), *Hereditas 34*, 453.

Lijinsky, W., Loo, J., and Ross, A. E. (1968), *Nature 218*, 1174.

Loveless, A. (1966), "Genetic and Allied Effects of Alkylating Agents," Pennsylvania State University Press, University Park, Pa.

Mader, P. M. (1960), *J. Am. Chem. Soc. 82*, 2956.

Maher, V. M., Miller, E. C., Miller, J. A., and Szybalski, W. (1968), *Mol. Pharmacol. 4*, 411.

Malling, H. V. (1966), *Mutation Res. 3*, 357.

Margoliash, E., and Smith, E. L. (1965), *in* "Evolving Genes and Proteins" (V. Bryson and H. J. Vogel, eds.), p. 221, Academic Press, Inc., New York.

Marquardt, H., Zimmermann, F. K., and Schwaier, R. (1964), *Z. Vererb. 95*, 82.

Mellin, G. W. (1963), *in* "Birth Defects" (M. Fishbein, ed.), p. 1, Lippincott, Philadelphia.

Michelson, A. M., and Monny, C. (1966), *Biochim. Acta 129*, 460.

Miller, J. A., and Miller, E. C. (1969), *in* "Physico-Chemical Mechanisms of Carcinogenesis" (E. D. Bergmann and B. Pullman, eds.), "The Jerusalem Symposia in Quantum Chemistry and Biochemistry," Vol. 1, p. 237.

Moorhead, P. S., and Saksela, E. (1965), *Hereditas 52*, 271.

Neumayer, J., Gibbons, D., and Trask, H. (1969), *Chem. Week 104*, 37 (Apr. 12), and *104*, 37 (Apr. 26).

Nichols, W. W. (1969), *in* "Handbook of Molecular Cytology" (A. Lima-de-aria, ed.), p. 733, North-Holland Publishing Co., Amsterdam.

O'Brien, R. L., Olenick, J. G., and Hahn, F. (1966), *Proc. Natl. Acad. Sci. U.S. 55*, 1511.

Oehlkers, F. (1956), *Z. Vereb. 87*, 584.

Orgel, A., and Orgel, L. E. (1965), *J. Mol. Biol. 14*, 453.

Orgel, L. E. (1965), *Advan. Enzymol. 27*, 289.

Phillips, J. H., and Brown, D. M. (1967), *in* "Progress in Nucleic Acid Research and Molecular Biology" (J. N. Davidson and W. E. Cohn, eds.) Vol. 7, p. 399, Academic Press, Inc., New York.

Philpot, J. St. L. (1963), *Radiatlon Res. Suppl. 3*, 55.

Radloff, R., Bauer, W., and Vinograd, J. (1967), *Proc. Natl. Acad. Sci. U.S. 57*, 1514.

Ramel, C. (1969). *Hereditas 61*, 208.

Reichard, P., Canellakis, Z. N., and Canellakis, E. S. (1961), *J. Biol. Chem. 236*, 2514.

Revell, S. H. (1959), *Proc. Roy. Soc. London, Series B 150*, 563.

Rhaese, H. J., and Freese, E. (1968), *Biochim. Biophys. Acta 155*, 476.

Rhaese, H. J., and Freese, E. (1969), *Biochim. Biophys. Acta 190*, 418.

Rosenkranz, H. S., and Jacobs, S. J. (1968), *Monograph 6*, 15.

Rutishauser, A., and Bollag, W. (1963), *Experientia 19*, 131.

Schöneich, J. (1967). *Mutation Res. 4*, 385.

Sever, J. L. (1967), *in* "Advances in Teratology" (D. H. M. Woollam, ed.), p. 127, Academic Press, Inc., New York.

Smith, K. C. (1966), *Radiation Res. Suppl. 6,* 54.

Sobels, F. H. (1963), *Radiation Res. Suppl. 3,* 171.

Sober, H. A., ed.(1968), "Handbook of Biochemistry," Chemical Rubber Co., Cleveland, Ohio.

Srb, A. M., Owen, R. D., and Edgar, R. S. (1965), "General Genetics," 2nd ed., Freeman and Co., San Francisco.

Stacey, K. A., Cobb, M., Consens, S. F., and Alexander, P. (1958), *Ann. N. Y. Acad. Sci. 68,* 682.

Stone, W. S., Wyss, O., and Haas, F. (1947), *Proc. Natl. Acad. Sci. U.S. 33,* 59.

Strauss, B. (1968), *Current Topics Microbiol. Immunol. 44,* 1.

Strauss, B., and Robbins, M. (1968), *Biochim. Biophys. Acta 161,* 68.

Sussman, A. S., and Halvorson, H. O. (1966), "Spores, Their Dormancy and Germination," Harper & Row, New York.

Süssmuth, R., and Lingens, F. (1969), *A. Naturforsch. 246,* 903.

Swanson, C. P. (1957), "Cytology and Cytogenetics," Prentice-Hall, Englewood Cliffs, N.J.

Tanaka, N. (1965), *J. Antibiotics, Series A 18,* 111.

Taylor, J. H. (1953), *Exptl. Cell Res. 9,* 164.

Terzaghi, E., Okada, Y., Streisinger, G., Emrich, J., Inouye, M., and Tsugita, A. (1966), *Proc. Natl. Acad. Sci. U.S. 56,* 500.

Tessman, I. (1962), *J. Mol. Biol. 5,* 442.

Thullier, G. (1959), *Bull. Soc. Chim. France 10,* 1431.

Toji, L., and Cohen, S. S. (1969), *Proc. Natl. Acad. Sci. U.S. 63,* 871.

Tso, P. O. P., Helmkamp, G. K., and Sander, C. (1962), *Proc. Natl. Acad. Sci. U.S. 48,* 686.

Uehlecke, H. (1965), *Arzneimittel Forsch. 8,* 1.

Vielmetter, W., and Schuster, H. (1960), *Biochem. Biophys. Res. Commun. 2,* 324.

Von Borstel, R. C., Graham, D. E., La Brot, K. J., and Resnick, M. S. (1968), *Genetics 60,* 233.

Von Selke, R., and Krause, H. W. (1963), *J. Prakt. Chem. 22,* 319.

Walling, C. (1957), "Free Radicals in Solution," John Wiley & Sons, New York.

Warburton, D., and Fraser, F. C. (1964), *Am. J. Human Genet. 16,* 1.

Weisburger, J. H., and Weisburger, E. K. (1966), *Chem. Eng. News 44,* 124.

Wheeler, G. P. (1962), *Cancer Res. 22,* 651.

Whitfield, J., Martin, R. G., and Ames, B. N. (1966), *J. Mol. Biol. 21,* 335.

Witkin, E. M. (1966), *Radiation Res. Suppl. 6,* 30.

Witkin, E. M. (1969), *Ann. Rev. Genet. 3,* 525.

Wolff, S., ed. (1963), "Radiation Induced Chromosome Aberrations," Columbia University Press, New York.

World Health Organization Bulletin (166), *Memorandum 34,* 765.

Wyss, O., Clark, J. B., Haas, F., and Stone, W. S. (1948), *J. Bacteriol. 56,* 51.

Yamamoto, N., Naito, T., and Shimkin, M. B. (1966) *Cancer Res. 26,* 2301.

Yanofsky, C. (1965), *Biochem. Biophys. Res. Commun. 18,* 898.

Yanofsky, C., Cox, E. C., and Horn, V. (1966), *Proc. Natl. Acad. Sci. U.S. 55,* 274.

CHAPTER 2

Correlation Between Teratogenic and Mutagenic Effects of Chemicals in Mammals

Harold Kalter*

Children's Hospital Research Foundation and Department of Pediatrics
University of Cincinnati College of Medicine
Cincinnati, Ohio

I. INTRODUCTION

Teratology is the discipline dealing with the causes and the development of congenital malformations. As used here, *congenital malformations* are gross structural anomalies of prenatal origin. In the widest sense, *teratogens* include endogenous factors (i.e., point mutations or chromosomal aberrations) as well as exogenous agents. There has been a tendency to forget these facts, as teratology has come to be associated with "environment," whereas properly it should carry no such connotation.

Congenital malformations, caused by gene mutations and chromosome aberrations or in whose origin these factors are major determinants, are hereditary. In contrast, congenital malformations produced by environmental interference with normal embryonic processes are *modifications*, and not hereditary.

External agents are teratogenic if they can be shown to cause malformations in offspring of treated pregnant females. Prenatal death, as indicated

---

* Supported in part by NIH grant HD03502.

TABLE 1. Chemicals or Classes of Chemicals Tested for Mutagenicity in Mammals, and Their Teratogenicity Status (Numbers Are References)

Chemicals	Mutagenic	Nonmutagenic	Teratogenic	Nonteratogenic
A. Alkylating agents				
1. Nitrogen mustards				
a. Mechlorethamine (nitrogen mustard; HN2)	29, 68–70		28, 58, 71–83, 86	84, 85
b. Mechlorethamine oxide HCl (Nitromin)	22			
c. Acetyl derivative of HN2			78, 87	
d. Fluoroacetyl derivative of HN2			78, 87	
e. Isopropyl-bis(β-chloroethyl)amine		22		
f. n-Butyl-bis(β-chloroethyl)amine		22		
g. Melphalan (L-3-[p-(bis[2-chloroethyl]-amino)phenyl]alanine)		88		
h. Merphalan (DL-3-[p-(bis[2-chloroethyl]-amino)phenyl]alanine; Sarcolysin)			89	
i. Chlorambucil (4-[p-(bis[2-chloroethyl]amino)-phenyl]butyric acid)		90	28, 58, 91–94	
j. Cyclophosphamide (2-[bis(2-chloroethyl)amino]-tetrahydro-2H-1,3,2-oxazaphosphorine 2-oxide)	95, 96		1, 80, 81, 93, 97–110	
k. Mannomustine (Degranol)				
l. p-(N,N-[bis(2-chloroethyl)amino]phenyl-N-[p-carboxyphenyl]carbamate)			87	111
m. ICR-170 (2-methoxy-6-chloro-9-[3-(ethyl-2-chloroethyl)aminopropylamino]acridine HCl)		112, 113		
2. Ethyleneimines				
a. TEM (triethylenemelamine; 2,4,6-tris(1-aziridinyl)-s-triazine)	30, 32, 33, 47, 48, 70, 88, 114–122		28, 72, 80, 123–127	90, 111, 128
b. ThioTEPA (triethylenethiophosphoramide; tris(1-aziridinyl)phosphine sulfide)	88, 122, 129		28, 80, 130–136	111
c. TEPA (triethylenephosphoramide; tris(1-aziridinyl)phosphine oxide)	122		123	

d.	METEPA (tris[2-methyl-1-aziridinyl]phosphine oxide)	122, 137		137
e.	Triaziquone (2,3,5-tris[1-aziridinyl]-p-benzoquinone; Trenimon)	49, 50, 95, 138	31	
f.	Inproquone (2,5-bis[1-aziridinyl]-3,-6-dipropoxy-p-benzoquinone; E39)			139, 140
g.	Apholate (2,2,4,4,6,6-hexakis[1-aziridinyl]-2,2,4,4,6,6-hexahydro-1,3,5,2,3,6-triazatriphosphorine)			141
h.	Ethyleneimine	129, 142, 143	22	261
i.	N-methyleneimine		22	
j.	Monoethyleneurea	88		
k.	Diethyleneurea	88		
l.	N-methyleneurea	22		
m.	N,N-dimethyleneurea	22		
n.	N-ethoxycarbonylethyleneimine	88		
o.	N-toluene-p-sulfonylethyleneimine		88	
p.	6-Chloro-2,4-diethyleneiminopyrimidine	88		
q.	Diethyleneiminosulfoxide	88		
r.	1,3-Di(ethylenesulfamoyl)propane	88		
s.	Tetraethylenepyrophosphoramide	22		
t.	PN6	22		
3.	Methanesulfonic esters			
a.	Busulfan (1,4-butanediol dimethanesulfonate; Myleran)	88, 144–149	28, 80, 145, 146, 148, 150–158	90
b.	1,4-dimethyl busulfan	159		
c.	MMS (methylmethanesulfonate)	112, 113, 122, 159–162	158	
d.	MES (methylethanesulfonate)	147, 159		
e.	EMS (ethylmethanesulfonate)	51, 112, 147, 159	158	

TABLE 1. (Continued)

Chemicals	Mutagenic	Nonmutagenic	Teratogenic	Nonteratogenic
f. IMS (isopropylmethanesulfonate)	147, 159, 162, 164		158	
g. n-Propylmethanesulfonate	159, 164			
h. β-Chloroethylmethanesulfonate	159	70		
i. Methylenedimethanesulfonate			158	
j. Dimethylenedimethanesulfonate	162	162	158	
k. Trimethylenedimethanesulfonate	162			158
4. Other alkylating agents				
a. β-Propiolactone		88		
b. MNNG (N-methyl-N'-nitro-N-nitrosoguanidine)	113	112		
c. Diepoxybutane	70	122		
d. Diethyl sulfate	143			
B. Miscellaneous				
1. Acriflavine HCl (Trypaflavine)	165	69 122		
2. Methenamine (hexamethylenetetramine)	165			
3. Aflatoxin	122	166	167	166, 167, 262
4. Benzo(a)pyrene	122		168	169
5. Trimethylphosphate	170			
6. p-Benzoquinone		171		
7. Hydroxyurea		122	58, 172–179	
8. Dimethylsulfoxide		162	180–183	158, 184
9. Urethan (ethyl carbamate)		88, 122	124, 175, 185–198	172, 176, 199–201
10. Hydroxyurethan			172, 176	
11. Methylnitrosourethan				
12. N-propyl carbamate			195	200
13. β-Hydroxyethyl carbamate			195	
14. Ethyl N-methyl carbamate			195	

No.	Compound				
15.	Ethyl N-hydroxy carbamate		195		
16.	Diethyl carbonate		195		
17.	Allyl carbamate	195			
18.	n-Butyl carbamate	195			
19.	Ethyl N,N-dimethyl carbamate	195			
20.	Hydroxylamine		140, 176		
21.	Methylhydroxylamine			122	
22.	5-Bromodeoxyuridine		202–204	122	
23.	5-Iododeoxyuridine		80	122	
24.	Aminopterin	213–215	72, 75, 79, 205–212	122	
25.	Griseofulvin		216	122	
26.	Nicotine	221–225	218–220	217	
27.	Caffeine	232	230–233	122, 226–229	
28.	Theophyllin		234		
29.	Dibenzanthracene			88	
30.	Nitrofurantoin			88	
31.	Nitrofurazone			88	
32.	Benzanthrone			122	
33.	Dimethylnitrosamine			122	
34.	3-Methylcholanthrene				235
35.	Captan	236, 237		122	
36.	Chlorpromazine	238–242	81	122	
37.	DDT			122	
38.	Formaldehyde	243, 244		122	
39.	Butylated hydroxytoluene	245		122	
40.	Maleic hydrazide			122	
41.	Colchicine	91, 124, 257–259	248–256	247	246
42.	Fusel oil				260

by reduced fertility, decreased litter size, and increased resorption, does not of itself prove teratogenicity, since deformity and lethality may be uncorrelated[1-3] and due to different mechanisms.

Mutations are changes in the genetic material. When they occur in the germ-cell line they are hereditary, i.e., transmissible from one generation to another, and they may be revealed by the appearance of new hereditary traits not due to segregation of preexisting alleles or attributable to spontaneous mutations. Mutations involve changes in one or very few nucleotides, i.e., single genes, or they consist of gross chromosomal alterations of structure or number.

Parenthetically, it seems to me that the convention of calling single-gene changes and chromosomal aberrations by the same term—mutation—is overdue for reconsideration. The possible theoretical reasons for this mixing are far outweighed by the confusion and error the practice may give rise to. First, there is as yet little knowledge of the causes of these phenomena in nature, and the possibility cannot be discounted that the causes of gene changes are usually different from the causes of chromosomal aberrations; second, conditions due to gene changes and those due to chromosomal aberrations have their own patterns of hereditary transmission and thus demand separate clinical and other consideration; last, the genetic significance of induced chromosomal anomalies is not yet clear. Thus, it may be wise for the present to admit the existence of these problems by restricting the term "mutation" to single-gene changes, and to use a different word to refer to chromosomal aberrations, e.g., allochrome (Gr. *allos*, other, different).

Excluded from discussion here are chromosomal aberrations produced *in vitro*, those studied only in somatic cells, and those for which there is at present no evidence of hereditary transmission. This article is further limited to dealing with experimental mammals and with the effects of chemical substances only; it does not concern itself with ionizing radiation, the mutagenic and teratogenic effects of which have been amply reviewed,[4-8] or with chromosomal aberrations produced by postfertilization application of physical or chemical agents[9] to females.

Only two *in vivo* methods have thus far been used in studies of the potential mutagenicity of chemicals in mammals: the so-called dominant lethal[10,11] and the specific locus[12] procedures, of which the former has been far more often employed; and only two mammals—mice and rats—have as yet been the subjects of such studies, with the former more often and lately almost exclusively used.

Snell[13,14] was the first to demonstrate conclusively a heritable variation in a mammal as a result of exposure of a parent to an environmental influence, in this case X-irradiation. When male mice were X-rayed and mated to untreated females before the onset of the sterile period some of

TABLE 2. Correspondence Between Mutagenicity and Teratogenicity. (For References see Table 1.)
(M = Mutagenic; T = Teratogenic; nM = Nonmutagenic; nT = Nonteratogenic)

M and T	nM and T	M and sometimes[d] T	nM and sometimes T
1. Nitromin	14. Chlorambucil	19. Mechlorethamine	24. Dimethylsulfoxide
2. Cyclophosphamide	15. Hydroxyurea	20. TEM	25. Urethan
3. TEPA	16. 5-Bromodeoxyuridine	21. ThioTEPA	26. Aminopterin
4. METEPA	17. 5-Iododeoxyuridine	22. Busulfan	27. Nicotine
5. MMS	18. Griseofulvin	23. Benzo(a)pyrene	28. Theophyllin
6. EMS			29. Chlorpromazine
7. IMS	M and nT		
8. MDS[a]	None		Sometimes M and sometimes T
9. TDS[b]			
			31. Aflatoxin
nM and nT			
10 DDS[c]			
11. Captan			
12. Formaldehyde			
13. Butylated hydroxytoluene			

[a] Methylenedimethanesulfonate.
[b] Trimethylenedimethanesulfonate.
[c] Dimethylenedimethanesulfonate.
[d] That is, reports are conflicting.

them and some of their viable offspring tended to produce litters of decreased size. The term *dominant lethal mutation* to describe the reduced litter size in this type of study was apparently first used by Strandskov.[15] Such individuals were called *semisterile*, and the condition was inherited as a quasi-dominant with some of the viable offspring of semisterile animals also showing the trait. The decreased litter size resulted mostly from early post-implantation death of a proportion of embryos (see references given by Russell[9,16]). It was pointed out[13] that the results fitted the interpretation that semisterility was due heterozygosity for a reciprocal chromosomal translocation, which was confirmed genetically and cytologically.[17-19]

In cytogenetics a translocation is the exchange of parts between two nonhomologous chromosomes. Individuals heterozygous for a translocation produce several kinds of gametes, so far as chromosomal structure is concerned.[20] Some gametes contain all normal chromosomes, and these of course give rise to normal young. Other gametes contain the normal complement of genes but with exchanged chromosome parts. These, combined with normal gametes, produce individuals heterozygous for the translocation, which are semisterile like one of the parents. Finally, some gametes contain unbalanced sets of chromosomes, i.e., with deficiencies and duplications of chromosomal segments, and these may give rise to inviable and malformed offspring. In the radiation studies mentioned above, the malformations, which occurred only in the F_2 and later generations, and almost always involved the central nervous system, affected about 5% of living offspring.[8]

Many chemical substances have been tested for dominant lethality in mammals, but they have comprised only few chemical classes,[21] mostly alkylating agents.[22] Chemicals tested for dominant lethality and in a few cases point mutagenicity are listed in Table 1. Table 1 also indicates which of these chemicals have been found to be teratogenic and which not.

The table, however, does not list numerous substances or classes of substances that have not been tested for mutagenicity in mammals but which are teratogenic. These constitute a large number of many varieties of drugs and other chemicals, dyes, hormones, vitamins, antibiotics and bacteriostatics, minerals and elements, different types of antagonists and analogs, and alkaloids of plant origin; also absent from the table are many teratogenic procedures, e.g., deficiency of nutrients, fasting, gases, abnormal temperature, abnormal atmospheric pressure, infectious agents, noise, surgical techniques, immunological agents, and various physiological modifications.[8,23-27]

Table 2 shows that among the agents that have been explored for both mutagenicity and teratogenicity there is some overlap between these effects, but it is quite apparent that the correspondence between them is not impressive; i.e., numerous mutagens are not teratogens or not consistently so and vice versa. This leads us to ask whether we should expect closer correspondence and prompts a brief discussion of the similarities and differences between the characteristic properties and actions of mutagens and teratogens.

II. FREQUENCY OF EFFECTS OF CHEMICALS

The frequency of congenital malformations induced by environmental teratogens sometimes varies with the lethality for pregnant animals and conceptuses of the agent employed.[28] In absolute terms the frequency of induced malformations ranges from just above the spontaneous level to 100% of surviving offspring. (Incidentally, the number of the latter need be little affected or may be variably reduced.) Indeed, it is not uncommon for a large proportion of offspring to be malformed; actually, it may be necessary to produce high frequencies of abnormalities for certain establishment of the teratogenicity of an agent, since spontaneous anomalies sometimes occur at an appreciable frequency.

The different varieties of mutation need to be considered separately since for different agents the frequency of induced point mutations is not correlated with that of dominant lethal mutations. Recessive or dominant visible effects, i.e., apparent point mutations, induced by chemicals have been quite rare,[29,30] or not found at all,[31] while dominant lethal mutations have been frequently observed, as gauged by either reduced mean litter size or increased frequency of cytologically abnormal or semisterile F_1 offspring.[32]

In such work, as in teratogenicity studies, the toxicity of the tested substance cannot be neglected in assessing its action. For example, the yield of structural chromosomal changes and semisterility was fairly small in the progeny of nitrogen mustard-treated males,[29] probably because the toxicity of this agent prevented using more effective doses. A higher yield of abnormalities was obtained with TEM,[33] perhaps because it is less toxic than nitrogen mustard and larger doses could be used.

A word of admonition should be expressed here. Demonstration of reduced fecundity of treated animals, as determined by decreased litter size or increased fetal death rate, is insufficient for establishing that dominant lethal mutations are induced, since altered physiology may conceivably be its cause. The presence of chromosomal aberrations must also be demonstrated. In addition, further generations should be examined for fertility, chromosomal aberrations, and other effects. On the one hand, without such testing, a partial or perhaps distorted picture of the nature and load of the effects produced may be gotten, since, for example, the destroyed F_1 conceptuses may screen out severe chromosomal abnormalities, leaving only minimal effects in survivors. But, on the other hand, some phenomena may occur only in later generations, as Snell[34] noted for congenital malformations.

III. SPECIFICITY OF ACTION OF CHEMICALS

Although a certain specificity of action of chemical mutagens can be discerned, the phenotypic outcome of these alterations usually appears to be unspecific. Thus the chemical structure of mutagens may sometimes determine the type of molecular change produced in DNA, and certain properties of chemicals may influence the cell stage most affected during gametogenesis. But there is little evidence—aside from the possible differences in the distribution of mutations among seven mouse loci produced by TEM and ionizing radiation[30,35] and the possible characteristic *in vitro* effects of FUDR[36]—that genic or chromosomal sites are hit other than randomly. For example, most prenatal deaths due to chromosomal aberrations occur at about the same time, *viz.*, soon after implantation, regardless of the type of aberration; in every instance of congenital malformation resulting from radiation-induced aneuploidy, the only apparent defect was of the anterior central nervous system, again regardless of the chromosomes or types of defect involved. [37,38]

However, specific effects seem to be produced by at least some teratogens. These appear to cause distinctive pleiotropic states—variable perhaps in expressivity and penetrance[2,39-42]—but their effects are diagnostic and permit identification of the teratogen with a high degree of certainty. Other teratogens produce single, apparently identical malformations, e.g., cleft palate, [43-46] and thus, in the absence of other evidence (e.g., morphogenetic), seem to lack specificity of action.

Thus there is the paradoxical situation of chemical mutagens causing molecular changes in the DNA fine structure and sometimes possibly inducing characteristic gross chromosomal effects both of which have apparently random phenotypic consequences; while, in contrast, some teratogens seem to cause distinctive types or combinations of congenital malformations, but their specific proximate site or manner of action is presently unknown.

IV. DOSE

The amounts of a chemical needed to produce dominant lethal mutations and teratogenic effects are not always the same. A few examples are listed in Table 3. For some chemicals there seems to be a clear-cut difference in the quantities producing these two effects. For example, a larger dose of

TABLE 3. Comparison of Mutagenic and Teratogenic Doses (mg/kg)
(References in Parentheses)

Chemical	Mutagen		Teratogen	
	Mice	Rats	Mice	Rats
HN2	2.4, 3.2 (29)		1–2 (73)	0.5–1 (71)
Cyclophosphamide	60, 210 (95)		20 (109)	7–10 (27)
TEM	0.2 (33)	0.2, 0.4 (47)	1.5–1.65 (125)	0.5–0.75 (80)
ThioTEPA	5 (122)		3–5 (135)	3–5 (80)
Busulfan	10–40 (149)	4–10 (147)	25 (158)	18–34 (80)
MMS	50 (161)			100 (158)
EMS	240 (163)			200 (158)
IMS		50 (147)		50 (158)

busulfan and TEM had to be administered to cause malformations than to induce dominant lethals; while for nitrogen mustard dominant lethality was more readily induced; and the same quantity of thioTEPA appeared to induce both effects.

V. SPECIES, SEX, AND STRAIN DIFFERENCES

Species differences in degree and type of sensitivity have been noted frequently for teratogenic chemicals; probably the most publicized in this regard is thalidomide.[8] A few examples are also available for chemicals inducing dominant lethal mutations. Bateman[47] noted, first, that male mice are more resistant to the dominant lethal effect of TEM than male rats and, second, that midstage spermatids are the most sensitive germ-cell stage in rats but are the most resistant in mice. Another example—a difference between mice and rats in sensitivity to the dominant lethal effects of busulfan—is noted in Table 2.

Few studies have been made of chemical mutagenesis in adult females,[48-51] but in almost all it was found that the rate of induced dominant lethals was lower in females than males. Large strain differences have also been noted.[51]

No sex differences have been noted in experimental teratology, i.e., induced malformations occur in both sexes with approximately equal frequencies, with the exception, of course, of those of genital and associated structures. Strain and stock differences in teratology, however, are very frequent.[8,52,53]

VI. TIMING

If the outcome of a uniform stimulus varies with some temporal transformation, then the time of application of the stimulus must be carefully ascertained. Timing is important for mutagenesis, since some stages of the cell-division cycle and of the gametogenetic cycle can be more liable to mutation or chromosome breakage than others, with respect to particular agents. Timing is also important for teratogenesis; here the transitory processes involved are embryonic differentiation, morphogenesis, and growth, which play large roles in determining the type and degree of vulnerability of parts to maldevelopment.[54]

VII. ARE INDUCED MALFORMATIONS DUE TO MUTATIONS?

It has been suggested[55] that ionizing irradiation, a long-known and potent teratogen,[4,8] produces congenital malformations through the induction of somatic mutations. But it has been pointed out[56] that to account for the consistency of teratogenic response it is necessary to assume either directed mutations or a high rate of gene changes with dominant cell-lethal effects. The former assumption is contrary to the findings of radiation genetics, and the latter is untenable, because the radiation-induced somatic mutation rate is far below that needed to account for observed frequencies of malformations.[57] The same arguments hold against chemically induced malformations being of point-mutational origin.

The possibility that induced congenital malformations may be due to or associated with chromosomal damage in deformed embryonic tissues is unlikely for some potent teratogens, since they produce relatively low frequencies of or no fetal chromosomal aberrations.[58] In other cases, such an association has been claimed[59,60] but is so far not entirely convincing.

VIII. MUTAGENICITY AND TERATOGENICITY TESTING

The problems urgently faced by scientists during the past several years in devising predictive drug teratogenicity tests[61] are now confronting those attempting to do the same for mutagenicity. It is a commonplace that all drugs are toxic for some persons, at some dosages, under some circumstances. What is needed are estimates of the magnitude of risk so that they may be weighed against the benefits of the substance in question. This weighing of relative risk and benefit is done in the evaluation of every pharmaceutical product. The means, imperfect though they may sometimes be, consist of preliminary conventional animal toxicity studies, followed by investigative clinical trials, with subsequent collection of reports of adverse reaction among the general population. For teratogenicity—and mutagenicity—testing, however, the second step noted above—clinical trial—cannot be taken. Hence what is demanded are animal procedures of an order of predictiveness far beyond that ordinarily necessary.

The problem thus consists of finding the species and methods most appropriate for uncovering the deleterious potential of the test substance. It is felt by some investigators[62,63] that because of phylogenetic considerations and similarities to human beings in reproductive features the higher nonhuman primates are the best teratogenicity test subjects. The fact,[64] however, is that no species of animal can possibly serve as a pharmacological alter ego of another, since each is probably unique so far as its pattern of drug responses is concerned. Probably no *a priori* considerations are of much help in finding the best animal model for study of a given chemical in relation to man. That can only be decided on the basis of pharmacological knowledge of ourselves, which may permit a rational choice of models most like us for each agent. Obviously, this is a gross oversimplification since it ignores certain imponderables that must be reckoned with, e.g., intraspecific genetic variation in drug response and embryonic metabolism; but anticipated difficulties must not discourage first steps in a promising direction.

Mutagenicity testing has its special problems, the foremost being that, aside from chromosomal aberrations of unknown hereditary significance, induced mutations have not been detected, and obviously are difficult to detect, in human populations, as has been noted in extensive radiation studies.[65-67]

IX. CONCLUSION

It appears from the spotty overlap in mutagenicity and teratogenicity of the relatively few chemicals that have been tested for both effects in mammals (remembering that in all but one or two such studies the only "mutational" effect investigated was dominant lethality) and from the other dissimilarities discussed above, that a relation between these effects of chemicals has so far not been established.

X. REFERENCES

1. J. G. Wilson, Teratogenic interaction of chemical agents in the rat, *J. Pharmacol. Exp. Therap. 144*, 429–436 (1964).
2. H. Kalter, *in* "Teratology: Principles and Techniques" (J. G. Wilson and J. Warkany, eds.) pp. 57–80, University of Chicago Press, Chicago (1965).

3. M. J. Edwards, Congenital defects in guinea pigs: Fetal resorptions, abortions, and malformations following induced hyperthermia during early gestation, *Teratology 2*, 313–328 (1969).

4. L. B. Russell, *in* "Radiation Biology" (A. Hollaender, ed.) Vol. 1, Part 2, pp. 861–918, McGraw-Hill, New York (1954).

5. W. L. Russell, Studies in mammalian radiation genetics, *Nucleonics 23*, 53–56, 62 (Jan., 1965).

6. S. P. Hicks and C. J. D'Amato, Effects of ionizing radiations on mammalian development, *Adv. Teratol. 1*, 196–250 (1966).

7. E. L. Green, Genetic effects of radiation on mammalian populations, *Ann. Rev. Genet. 2*, 87–120 (1968).

8. H. Kalter, "Teratology of the Central Nervous System," University of Chicago Press, Chicago (1968).

9. L. B. Russell, Chromosome aberrations in experimental animals, *Prog. Med. Genet. 2*, 230–294 (1962).

10. A. J. Bateman, Testing chemicals for mutagenicity in a mammal, *Nature 210*, 205–206 (1966).

11. G. Röhrborn, Mutagenicity tests in mice. I. The dominant lethal method and the control problem, *Humangenetik 6*, 345–361 (1968).

12. W. L. Russell, X-ray-induced mutations in mice, *Cold Spring Harbor Symp. Quant. Biol. 16*, 327–336 (1951).

13. G. D. Snell, X-ray sterility in the male house mouse, *J. Exp. Zool. 65*, 421–441 (1933).

14. G. D. Snell, The induction by X-rays of hereditary changes in mice, *Genetics 20*, 545–567 (1935).

15. H. H. Strandskov, Effects of X-rays in an inbred strain of guinea pigs, *J. Exp. Zool. 63*, 175–202 (1932).

16. W. L. Russell, *in* "Radiation Biology" (A. Hollaender, ed.) Vol. 1, Part 2, pp. 825–859, McGraw-Hill, New York (1954).

17. P. C. Koller and C. A. Auerbach, Chromosome breakage and sterility in the mouse, *Nature 148*, 501–502 (1941).

18. P. C. Koller, Segmental interchange in mice, *Genetics 29*, 247–263 (1944).

19. G. D. Snell, An analysis of translocations in the mouse, *Genetics 31*, 157–180 (1946).

20. A. B. Griffen, *in* "Biology of the Laboratory Mouse" (E. L. Green, ed.) 2nd ed., pp. 51–85, McGraw-Hill, New York (1966).

21. C. Auerbach, The chemical production of mutations, *Science 158*, 1141–1147 (1967).

22. H. Jackson, The effects of alkylating agents on fertility, *Brit. Med. Bull. 20*, 107–114 (1964).

23. A. Fave, Les embryopathies provoquées chez les mammifères, *Thérapie 19*, 43–164 (1964).

24. H. Nishimura, "Chemistry and Prevention of Congenital Anomalies," Charles C Thomas, Springfield, Ill. (1964).

25. D. A. Karnofsky, Drugs as teratogens in animals and man, *Ann. Rev. Pharmacol. 5*, 447–472 (1965).

26. F. Vichi, P. L. Masi, P. Pierleoni, S. Orlando, L. Pagni, and I. Tollaro, "I Fattori Metagenetici Esogeni nelle Malformazioni Congenita (Oro-maxillo-faciali): Teratogenesi Sperimentale e Riferimenti Clinici," Minerva Medica, Saluzzo (1966).

27. S. Chaube and M. L. Murphy, Teratogenic effects of the recent drugs active in cancer chemotherapy, *Adv. Teratol. 3*, 181–237 (1968).

28. M. L. Murphy, A. Del Moro, and C. Lacon, Comparative effects of five polyfunctional alkylating agents on the rat fetus, with additional notes on the chick embryo, *Ann. N.Y. Acad. Sci. 68*, 762–781 (1958).

29. D. S. Falconer, B. M. Slizynski, and C. Auerbach, Genetical effects of nitrogen mustard in the house mouse, *J. Genet. 51*, 81–88 (1952).

30. B. M. Cattanach, Chemically induced mutations in mice, *Mutation Res. 3*, 346–353 (1966).

31. G. Röhrborn and F. Vogel, A search for dominant mutations in F_1 progeny of male mice treated with trenimone (triethyleneiminobenzoquinone-1,4), *Humangenetik 7*, 43–50 (1969).

32. B. M. Cattanach, The sensitivity of the mouse testis to the mutagenic action of triethylenemelamine, *Z. Vererb. 90*, 1–6 (1959).

33. B. M. Cattanach, Induction of translocations in mice by triethylenemelamine, *Nature 180*, 1364–1365 (1957).

34. G. D. Snell, Induction by roentgen rays of hereditary changes in mice, *Radiology, 36*, 189–194 (1941).

35. W. L. Russell and L. B. Russell, The genetic and phenotypic characteristics of radiation-induced mutations in mice, *Radiat. Res. Suppl. 1*, 296–305 (1959).

36. C. Ockey, T. Hsu, and L. Richardson, Chromosome damage induced by 5-fluoro-2′-deoxyuridine (FUDR) in relation to the cell cycle of the Chinese hamster, *J. Nat. Cancer Inst. 40*, 465–475 (1968).

37. G. D. Snell, E. Bodemann, and W. Hollander, A translocation in the house mouse and its effect on development, *J. Exp. Zool. 67*, 93–104 (1934).

38. E. M. Otis, Prenatal Mortality Rates of Seventeen Radiation Induced Translocations in Mice, University of Rochester Atomic Energy Report UR-291 (1953).

39. J. Warkany and R. C. Nelson, Skeletal abnormalities in offspring of rats reared on deficient diets, *Anat. Rec. 79*, 83–100 (1941).

40. J. Gillman, C. Gilbert, T. Gillman, and I. Spence, A preliminary report on hydrocephalus, spina bifida, and other congenital anomalies in the rat produced by trypan blue, *S. Afr. J. Med. Sci. 13*, 47–90 (1948).

41. H. Kalter and J. Warkany, Congenital malformations in inbred strains of mice induced by riboflavin-deficient, galactoflavin-containing diets, *J. Exp. Zool. 136*, 531–566 (1957).

42. H. Kalter and J. Warkany, Experimental production of congenital malformations in strains of inbred mice by maternal treatment with hypervitaminosis A, *Am. J. Pathol. 38*, 1–21 (1961).

43. H. Baxter and F. C. Fraser, The production of congenital defects in the offspring of female mice treated with cortisone, *McGill Med. J. 19*, 245–249 (1950).

44. J. J. Buresh and T. J. Urban, The teratogenic effect of the steroid nucleus in the rat, *J. Dent. Res. 43*, 548–554 (1964).

45. K. M. Massey, Teratogenic effects of diphenylhydantoin sodium, *J. Oral Therap. Pharmacol. 2*, 380–385 (1966).

46. K. T. Szabo, Teratogenicity of lithium in mice, *Lancet 2*, 849 (1969).

47. A. J. Bateman, The induction of dominant lethal mutations in rats and mice with triethylenemelamine (TEM), *Genet. Res. 1*, 381–392 (1960).

48. B. M. Cattanach, The effect of triethylene-melamine on the fertility of female mice, *Internat. J. Radiat. Biol. 3*, 288–292 (1959).

49. G. Röhrborn, Über einen Geschlechtsunterschied in der mutagenen Wirkung von Trenimon bei der Maus, *Humangenetik 2*, 81–82 (1966).

50. G. Röhrborn and H. Berrang, Dominant lethals in young female mice, *Mutation Res. 4*, 231–233 (1967).

51. W. M. Generoso and W. L. Russell, Strain and sex variations in the sensitivity of mice to dominant-lethal induction with ethyl methanesulfonate, *Mut. Res. 8*, 589–598 (1969).

52. C. P. Dagg, *in* "Biology of the Laboratory Mouse" (E. L. Green, ed.) 2nd ed., pp. 309–328, McGraw-Hill, New York (1966).

53. M. Smithberg, Teratogenesis in inbred strains of mice, *Adv. Teratol. 2*, 257–288 (1967).

54. J. Wilson, Embryological considerations in teratology, *Ann. N.Y. Acad. Sci. 123*, 219–227 (1965).

55. J. G. Wilson, Differentiation and reaction of rat embryos to radiation, *J. Cell. Comp. Physiol. 43*, Suppl., 11–37 (1954).

56. L. B. Russell and W. L. Russell, An analysis of the changing radiation response of the developing mouse embryo, *J. Cell. Comp. Physiol. 43*, Suppl., 103–147 (1954).

57. L. B. Russell and M. H. Major, Radiation-induced presumed somatic mutations in the house mouse, *Genetics 42*, 161–175 (1957).

58. S. Soukup, E. Takacs, and J. Warkany, Chromosome changes in embryos treated with various teratogens, *J. Embryol. Exp. Morphol. 18*, 215–228 (1967).

59. T. H. Ingalls, E. F. Ingenito, and F. J. Curley, Acquired chromosomal anomalies induced in mice by injection of a teratogen in pregnancy, *Science 141*, 810–812 (1963).

60. A. Endo and T. H. Ingalls, Chromosomal anomalies in embryos of diabetic mice, *Arch. Env. Hlth.*, *16*, 316–325 (1968).

61. Commission on Drug Safety, "Report. Conference on Prenatal Effects of Drugs," Chicago (1963).

62. H. Vagtborg, ed., "Use of Nonhuman Primates in Drug Evaluation," University of Texas Press, Austin (1968).

63. C. O. Miller, ed., "Proceedings. Conference on Nonhuman Primate Toxicology," U.S. Department of Health, Education, and Welfare, Washington, D.C. (1968).

64. B. B. Brodie and W. D. Reid, Is man a unique animal in response to drugs? *Am. J. Pharm. 141*, 21–27 (1969).

65. J. V. Neel, "Changing Perspectives on the Genetic Effects of Radiation," Charles C Thomas, Springfield, Ill. (1963).

66. J. V. Neel, Atomic bombs, inbreeding, and Japanese genes: The Russel Lecture for 1966, *Univ. Mich. Med. Cent. J. 32*, 107–116 (1966).

67. W. J. Schull, J. V. Neel, and A. Hashizume, Some further observations on the sex ratio among infants born to survivors of the atomic bombings of Hiroshima and Nagasaki, *Am. J. Hum. Genet. 18*, 328–338 (1966).

68. C. Auerbach and D. S. Falconer, A new mutant in the progeny of mice treated with nitrogen mustard, *Nature 163*, 678–679 (1949).

69. R. G. Edwards, The experimental induction of gynogenesis in the mouse. III. Treatment of sperm with trypaflavine, toludine blue, or nitrogen mustard, *Proc. Roy. Soc. London, Series B 149*, 117–129 (1958).

70. B. M. Cattanach, *in* "Effects of Ionizing Radiation on the Reproductive System" (W. D. Carlson and F. X. Gassner, eds.) pp. 415–431, Macmillan, New York (1964).

71. D. Haskin, Some effects of nitrogen mustard on the development of external body form in the fetal rat, *Anat. Rec. 102*, 493–511 (1948).

72. S. P. Hicks, Some effects of ionizing radiation and metabolic inhibition on developing mammalian nervous system, *J. Pediat. 40*, 489–513 (1952).

73. C. H. Danforth and E. Center, Nitrogen mustard as a teratogenic agent in the mouse, *Proc. Soc. Exp. Biol. Med. 86*, 705–707 (1954).

74. O. Thalhammer and E. Heller-Szöllösy, Exogene Bildungsfehler ("Missbildungen") durch Lostinjection bei der graviden Maus, *Z. Kinderheilk. 76*, 351–365 (1955).

75. M. L. Murphy and D. A. Karnofsky, Effect of azaserine and other growth-inhibiting agents on fetal development of the rat, *Cancer 9*, 955–962 (1956).

76. M. L. Murphy, C. P. Dagg, and D. A. Karnofsky, Comparison of teratogenic chemicals in the rat and chick embryos, *Pediatrics 19*, 701–714 (1957).

77. H. Nishimura and S. Takagaki, Congenital malformations in mice induced by nitrogen mustard, *Acta Schol. Med. Univ. Kioto Jap. 36*, 20–26 (1959).

78. A. Jurand, Further investigations on the cytotoxic and morphogenetic effects of some nitrogen mustard derivatives, *J. Embryol. Exp. Morphol. 9*, 492–506 (1961).

79. R. L. Brent, B. T. Bolden, and J. B. Franklin, The evaluation of teratogenic agents by means of the uterine vascular clamping technique, *Am. J. Dis. Child. 104*, 464–654, abst. (1962).

80. M. L. Murphy, Teratogenic effects in rats of growth inhibiting chemicals, including studies on thalidomide, *Clin. Proc. Child. Hosp. (Washington, D.C.) 18*, 307–322 (1962).

81. N. Brock and T. von Kreybig, Experimenteller Beitrag zur Prüfung teratogener Wirkungen von Arzneimitteln an der Laboratoriumsratte, *Naunyn-Schmiedebergs Arch. Exp. Pathol. 249*, 117–145 (1964).

82. M. Müller and N. Škreb, Does nitrogen mustard mimic the X-ray effects in any case? *Experientia 20*, 70–71 (1964).

83. M. Müller, Does nitrogen mustard affect the foetus directly or secondarily by its effects on the mother? *Experientia 22*, 247 (1966).

84. R. A. Hettig, G. G. Robertson, and D. T. Cline, Effect of nitrogen mustard injections on the embryos of pregnant rats, *J. Lab. Clin. Med. 36*, 833–834, abst. (1950).

85. I. Hayashi, T. Yama, and H. Fujii, Experimental studies on the relationship between embryonic environment and developmental anomalies, *Acta Pathol. Jap. 7*, 444–445, abst. (1957).

86. K. Okano, J. Esumi, S. Ito, S. Kashiyama, H. Fujita, T. Toba, and H. Ito, Influences of nitromin on rat embryo, *Acta Pathol. Jap. 8*, 561, abst. (1958).

87. A. Jurand, Anti-mesodermal activity of a nitrogen mustard derivative. *J. Embryol. Exp. Morphol. 11*, 689–696 (1963).

88. J. Jackson, B. W. Fox, and A. W. Craig, The effect of alkylating agents on male rat fertility, *Brit. J. Pharmacol. 14*, 149–157 (1959).

89. V. A. Alexandrov, Characteristics of the pathogenic action of sarcolysin on the embryogenesis of rats, *Dokl. Akad. Nauk. SSSR 171*, 746–749 (1966).

90. H. Jackson, Antifertility substances, *Pharmacol. Rev. 11*, 135–172 (1959).

91. K. Didcock, D. Jackson, and J. M. Robson, The action of some nucleotoxic substances on pregnancy, *Brit. J. Pharmacol. 11*, 437–441 (1956).

92. I. W. Monie, Chlorambucil-induced abnormalities of the urogenital system of rat fetuses, *Anat. Rec. 139*, 145–153 (1961).

93. H. Tuchmann-Duplessis and L. Mercier-Parot, Répercussions des neuroleptiques et des antitumoraux sur le développement prénatal, *Bull. Acad. Suisse Sci. Med. 20*, 490–526 (1964).

94. S. Chaube, G. Kury, and M. L. Murphy, Teratogenic effects of cyclophosphamide (NSC-26271) in the rat, *Cancer Chemotherap. Rep. 51*, 363–376 (1967).

95. D. Brittinger, Die mutagene Wirkung von Endoxan bei der Maus, *Humangenetik 3*, 156–165 (1966).

96. E. Schleiermacher, Über den Einfluss von Trenimon und Endoxan auf die Meiose der männlichen Maus. II. Cytogenetische Befunde nach Behandlung mit Trenimon und Endoxan, *Humangenetik 3*, 134–155 (1966).

97. P. Gerlinger, Action du cyclophosphamide injecté à la mère sur la réalisation de la forme du corps des embryons de lapin, *Compt. Rend. Soc. Biol. 158*, 2154–2157 (1964).

98. P. Gerlinger and J. Clavert, Action du cyclophosphamide injecté à des lapines gestantes sur les gonades embryonnaires, *Compt. Rend. Acad. Sci. 258*, 2899–2901 (1964).

99. P. Gerlinger and J. Clavert, Action du cyclophosphamide injecté à différentes pèriodes de la gestation sur les cellules sexuelles embryonnaires de lapin, *Compt. Rend. Soc. Biol. 158*, 2464–2466 (1964).

100. P. Gerlinger and J. Clavert, Etude de l'évolution des cellules sexuelles d'embryons issus de lapines traitées au cyclophosphamide, *Compt. Rend. Soc. Biol. 159*, 1386–1389 (1965).

101. P. Gerlinger and J. Clavert, Anomalies observées chez des lapins issues de mères traitées au cyclophosphamide, *Compt. Rend. Soc. Biol. 159*, 1462–1466 (1965).

102. I, Hackenberger and T. von Kreybig, Vergleichende teratologische Untersuchungen bei der Maus und der Ratte, *Arzneimittelforschung 15*, 1456–1460 (1965).

103. T. von Kreybig, Die teratogene Wirkung von Cyclophosphamid während der embryonalen Entwicklungsphase bei der Ratte, *Naunyn-Schmiedebergs Arch. Exp. Pathol. 252*, 173–195 (1965).

104. T. von Kreybig, Zür Wirkung von Teratogenen auf frühe Stadien der vorgeburtlichen Entwicklung der Ratte,, *Naunyn-Schmiedebergs Arch. Exp. Pathol. 252*, 196–204 (1965).

105. T. von Kreybig, Extremitätsmissbildungen im Tierexperiment, *Arzneimlttelforschung 15*, 1213–1217 (1965).

106. R. Shoji and E. Ohzu, Effect of endoxan on developing mouse embryos, *J. Fac. Sci. 15*, 662–665 (1965).

107. T. von Kreybig and W. Schmidt, Zur chemischen Teratogenese bei der Ratte, *Arzneimittelforschung 16*, 989–1000 (1966).

108. T. von Kreybig and W. Schmidt, Chemisch induzierte Fetopathien bei der Ratte. Experimentelle Untersuchungen über die Wirkung von Cyclophosphamid und N-Methyl-N-nitroso-Harnstoff nach der Gabe am 15. oder 16. Tag des Gestation, *Arzneimittelforschung 17*, 1093–1099 (1967).

109. J. E. Gibson and B. A. Becker, The teratogenicity of cyclophosphamide in mice, *Cancer Res. 28*, 475–480 (1968).

110. J. E. Gibson and B. A. Becker, Effect of phenobarbital and SKF 525-A on the teratogenicity of cyclophosphamide in mice, *Teratology 1*, 393–398 (1968).

111. C. E. Adams, M. F. Hay, and C. Lutwak-Mann, The action of various agents upon the rabbit embryo, *J. Embryol. Exp. Morphol. 9*, 468–491 (1961).

112. U. H. Ehling, R. B. Cumming, and H. V. Malling, Induction of dominant lethal mutations by alkylating agents in male mice, *Mutation Res. 5*, 417–428 (1968).

113. W. M. Generoso, Chemical induction of dominant lethals in female mice, *Genetics 61*, 461–470 (1969).

114. H Jackson and M. Bock, The effect of triethylamine on the fertility of rats, *Nature 175*, 1037–1038 (1955).

115. M. Bock and H. Jackson, The action of triethylenemelamine on the fertility of male rats, *Brit. J. Pharmacol. 12*, 1–7 (1957).

116. B. M. Cattanach and R. G. Edwards, The effects of triethylenemelamine on the fertility of male mice, *Proc. Roy. Soc. Edinburgh, Series B, 67*, 54–64 (1958).

117. E. Steinberger, W. O. Nelson, A. Boccabella, and W. J. Dixon, A radiomimetic effect of triethylenemelamine on reproduction in the male rat, *Endocrinology 65*, 40–50 (1959).

118. J. K. Sherman and E. Steinberger, Effect of triethylenemelamine on reproductive capacity of mouse spermatozoa, *Proc. Soc. Exp. Biol. Med. 103*, 348–350 (1960).

119. B. M. Cattanach, A chemically-induced variegated-type position effect in the mouse, *Z. Vererb. 92*, 165–182 (1961).

120. B. M. Cattanach, Induction of paternal sex-chromosome losses and deletions and of autosomal gene mutations by the treatment of mouse post-meiotic germ cells with triethylenemelamine, *Mutation Res. 4*, 73–82 (1967).

121. B. N. Hemsworth and H. Jackson, *in* "Embryopathic Activity of Drugs" (J. M. Robson, F. M. Sullivan, and R. L. Smith, eds.) pp. 116–137, Little, Brown, Boston (1965).

122. S. S. Epstein and H. Shafner, Chemical mutagens in the human environment, *Nature 219*, 385–386 (1968).

123. J. B. Thiersch, Effect of 2,4,6 triamino-"s"-triazene (TR) 2,4,6 "tris" (ethyleneimino)-"s"-triazene (TEM) and N, N′, N″-triethylenephosphoramide (TEPA) on rat litter in utero, *Proc. Soc. Exp. Biol. Med. 94*, 36–40 (1957).

124. H. Tuchmann-Duplessis and L. Mercier-Parot, Sur l'action tératogène de quelques substances antimitotiques chez le rat, *Compt. Rend. Acad. Sci. 247*, 152–154 (1958).

125. A. Jurand, Action of triethanomelamine (TEM) on early and late stages of mouse embryos, *J. Embryol. Exp. Morphol. 7*, 526–539 (1959).

126. M. Kageyama, Multiple development anomalies in offspring of albino mice mice injected with triethylenemelamine (TEM) during pregnancy [in Japanese with English summary], *Acta Anat. Nippon. 36*, 10–23 (1961).

127. M. Kageyama and H. Nishimura, Developmental anomalies in mouse embryos induced by triethylenemelamine (TEM), *Acta Schol. Med. Univ. Kioto Jap. 37*, 318–327 (1961).

128. S. Sobin, Experimental creation of cardiac defects, *Proc. M. R. Pediat. Res. Conf. 14*, 13–16 (1955).

129. N. I. Nuzhdin and G. V. Nizhnik, Fertilization and embryonic development of rabbits following in vitro treatment of spermatozoa with chemical mutagens [in Russian], *Dokl. Akad. Nauk. SSSR 181*, 224–227 (1968).

130. K. Okano, H. Fujita, I. Ito, S. Kashiyama, K. Esumi, H. Ito, and T. Toba, Effects of thio-TEPA on fetus of albino rats, *Acta Pathol. Jap. 9*, 644–645 abst. (1959).

131. T. Tanimura and H. Nishimura, Teratogenic effect of thio-TEPA, a potent antineoplastic compound, upon the offspring of pregnant mice [in Japanese], *Acta Anat. Nippon. 37*, 66–67 (1962).

132. H. Nanjo, Maldevelopment of the fetuses caused by maternal administration of thio-TEPA in relation to maternal age [in Japanese with English summary], *Acta Anat. Nippon. 39*, 258–262 (1964).

133. M. Nishikawa, Teratogenic effect of combined administration of fasting and thio-TEPA upon mouse embryos [in Japanese with English summary], *Acta Anat. Nippon. 39*, 252–257 (1964).

134. K. Takano, T. Tanimura, and H. Nishimura, The susceptibility of the offspring of alloxan-diabetic mice to a teratogen, *J. Embryol. Exp. Morphol. 14*, 63–73 (1965).

135. H. Nishimura, M. Terada, and M. Yasuda, Mitigated teratogenicity of thio-TEPA in the goldthioglucose obese mice, *Proc. Soc. Exp. Biol. Med. 124*, 1190–1193 (1967).

136. T. Tanimura, Relationship of dosage and time of administration to teratogenic effects of thio-TEPA in mice, *Okaj. Folia Anat. Jap. 44*, 337–355 (1968).

137. T. B. Gaines and R. D. Kimbrough, The sterilizing, carcinogenic and teratogenic effects of metepa in rats, *Bull. World Health Org. 34*, 317–320 (1966).

138. G. Röhrborn, Die mutagene Wirkung von Trenimon bei der männlichen Maus, *Humangenetik 1*, 576–578 (1965).

139. S. Sokół, The influence of the drug Bayer E-39 [Improquone] solubile on the embryonic development of the rat, *Folia Biol. 14*, 317–330 (1966).

140. W. Zimmermann and G. H. M. Gottschewski, Die Wirkung bestimmter Substanzen auf die DNS- und Proteinsynthese in der frühen Embryonalentwicklung des Kaninchens, *Bull. Schweiz. Akad. Med. Wiss. 22*, 166–183 (1966).

141. R. L. Younger, Probable induction of congenital anomalies in a lamb by apholate, *Am. J. Vet. Res. 26*, 991–995 (1965).

142. A. M. Malashenko and I. K. Yegorov, The induction of dominant lethals in inbred mice by ethylene imine [in Russian with summary in English], *Genetika 3*, 59–67 (1967).

143. A. M. Malashenko and I. K. Yegorov, The induction of dominant lethals in male mice with ethylene and diethylsulphate [in Russian with summary in English], *Genetika 4*, 21–27 (1968).

144. W. Bollag, Cytostatica in der Schwangerschaft, *Schweiz. Med. Woch. 84*, 393–395 (1954).

145. B. N. Hemsworth and H. Jackson, Effect of busulphan on the developing gonad of the male rat, *J. Reprod. Fert. 5*, 187–194 (1963).

146. B. N. Hemsworth and H. Jackson, Effect of busulphan on the developing ovary in the rat, *J. Reprod. Fert. 6*, 229–233 (1963).

147. M. Partington and H. Jackson, The induction of dominant lethal mutations in rats by alkane sulphonic esters, *Genet. Res. 4*, 333–345 (1963).

148. A. E. Light, Additional observations on the effects of busulfan on cataract formation, duration of anesthesia, and reproduction in rats, *Toxicol. Appl. Pharmacol. 10*, 459–466 (1967).

149. U. H. Ehling and H. V. Malling, 1,4 di(methane-sulfonoxy) butane (Myleran) as a mutagenic agent in mice, *Genetics 60*, 174–175, abst. (1968).

150. B. H. Hemsworth and H. Jackson, in "Embryopathic Activity of Drugs" (J. M. Robson, F. M. Sullivan, and R. L. Smith, eds.) pp. 116–137, Little, Brown, Boston (1965).

151. V. A. Alexandrov, Pathological effect of myelosan on embryogenesis [in Russian], *Dokl. Akad. Nauk. SSSR 159*, 918–920 (1964).

152. V. A. Alexandrov, The role of allantois injury in the pathogeny of antenatal death and certain embryopathies in rats treated with myelosane (mylerane). The critical period of allantois development in rats [in Russian], *Dokl. Akad. Nauk. SSSR 162*, 232–235 (1965).

153. V. A. Alexandrov, Teratogenous effect of antileukemic drug—myelosan (mileran) on rat embryo [in Russian with English summary], *Arkh. Anat. 44*, 87–94 (1965).

154. V. A. Alexandrov, Analysis of the lethal effect of 'myleran' on rat embryos, *Nature 209*, 1215–1216 (1966).

155. V. A. Alexandrov, Characteristics of disturbance of embryogenesis under the action of myelosan in the early stages of the development of the rat [in Russian], *Biull. Eksp. Biol. Med. 61*, 81–84 (1966).

156. J.-G. Forsberg and H. Olivecrona, The effect of prenatally administered busulphan on rat gonads, *Biol. Neonat. 10*, 180–192 (1966).

157. Y. Kameyama, T. Sugawara, T. Ogawa, K. Hoshino, and U. Murakami, Effects of myleran on the development of the skeletal system in the rat and mouse embryos, *Ann. Rep. Res. Inst. Environ. Med. Nagoya Univ. 14*, 61–78 (1966).

158. B. N. Hemsworth, Embryopathies in the rat due to alkane sulphonates, *J. Reprod. Fert. 17*, 325–334 (1968).

159. H. Jackson, B. W. Fox, and A. W. Craig, Antifertility substances and their assessment in the male rodent, *J. Reprod. Fert. 2*, 447–465 (1961).

160. H. Jackson, M. Partington, and A. L. Walpole, Production of heritable partial sterility in the mouse by methyl methanesulphonate, *Brit. J. Pharmacol. Chemotherap. 23*, 521–528 (1964).

161. M. Partington and A. J. Bateman, Dominant lethal mutations induced in male mice by methylmethanesulphonate, *Heredity 19*, 191–200 (1964).

162. B. N. Hemsworth, Effect of alkane sulphonic esters on ovarian development and function in the rat, *J. Reprod. Fert. 18*, 15–20 (1969).

163. B. M. Cattanach, C. E. Pollard, and J. H. Isaacson, Ethyl methanesulfonate-induced chromosome breakage in the mouse, *Mutation Res. 6*, 297–307 (1968).

164. U. H. Ehling, D. G. Doherty, and H. V. Malling, Differential mutagenic action of n- and iso-propyl methanesulfonate in mice, *Proc. XII Int. Cong. Genet. 1*, 103, abst. (1968).

165. K. H. Baldermann, G. Röhrborn, and T. M. Schroeder, Mutagenitätsuntersuchungen mit Trypaflavin und Hexamethylentetramin am Säuger in vivo und in vitro, *Humangenetik 4*, 112–126 (1967).

166. H. F. Hintz, H. Heitman, Jr., A. N. Booth, and W. E. Gagne, Effects of aflatoxin on reproduction in swine, *Proc. Soc. Exp. Biol. Med. 126*, 146–148 (1967).

167. J. A. DiPaolo, J. Elis, and H. Erwin, Teratogenic response by hamsters, rats and mice to aflatoxin B1, *Nature 215*, 638–639 (1967).

168. R. H. Rigdon and E. G. Rennels, Effect of feeding benzpyrene on reproduction in the rat, *Experientia 20*, 224–226 (1964).

169. R. H. Rigdon and J. Neal, Effects of feeding benzo(a)pyrene on fertility, embryos, and young mice, *J. Nat. Cancer Inst. 34*, 297–306 (1965).

170. H. Jackson and A. R. Jones, Antifertility action and metabolism of trimethylphosphate in rodents, *Nature 220*, 591–592 (1968).

171. G. Röhrborn and F. Vogel, Mutationen durch chemische Einwirkung bei Säuger und Mensch. 2. Genetische Untersuchungen an der Maus, *Deut. Med. Woch. 92*, 2315–2321 (1967).

172. M. L. Murphy and S. Chaube, Preliminary survey of hydroxyurea (NSC-32065) as a teratogen, *Cancer Chemotherap. Rep. 40*, 1–7 (1964).

173. M. L. Murphy, *in* "Teratology: Principles and Techniques" (J. G. Wilson and J. Warkany, eds.) pp. 161–184, University of Chicago Press, Chicago (1965).

174. V. H. Ferm, Teratogenic activity of hydroxy urea, *Lancet 1*, 1338–1339 (1965).
175. V. H. Ferm, Severe developmental malformations. Malformations induced by urethane and hydroxyurea in the hamster, *Arch. Pathol. 81*, 174–177 (1966).
176. S. Chaube and M. L. Murphy, The effects of hydroxyurea and related compounds on the rat fetus, *Cancer·Res. 26*, 1448–1457 (1966).
177. T. von Kreybig, R. Preussmann, and W. Schmidt, Chemische Konstitution und teratogene Wirkung bei der Ratte. I. Carbonsäureamide, Carbonsäurehydrazide und Hydroxamsäuren, *Arzneimittelforschung 18*, 645–657 (1968).
178. V. Lauro, F. E. S. Giornelli, A. Fanelli, and C. Giornelli, Inibitori metabolici e fertilità: Indagine sugli effecti dell'idrossiurea nella ratta, *Arch. Ostet. Ginec. 73*, 451–462 (1968).
179. R. Roll and F. Bar, Untersuchingen über die teratogene Wirkung von Hydroxyharnstoff während der frühen und embryonalen Entwicklung der Maus, *Arch. Toxik. 25*, 150–168 (1969).
180. V. H. Ferm, Teratogenic effect of dimethyl sulphoxide, *Lancet 1*, 208–209 (1966).
181. V. H. Ferm, Congenital malformations induced by dimethyl sulphoxide in the golden hamster, *J. Embryol. Exp. Morphol. 16*, 49–54 (1966).
182. M. Marin-Padilla, Mesodermal alterations induced by dimethyl sulfoxide, *Proc. Soc. Exp. Biol. Med. 122*, 717–720 (1966).
183. F. M. E. Caujolle, D. H. Caujolle, S. B. Cros, and M.-M. J. Calvet, Limits of toxic and teratogenic tolerance of dimethyl sulfoxide, *Ann. N.Y. Acad. Sci. 141*, 110–125 (1967).
184. M. B. Juma and R. E. Staples, Effect of maternal administration of dimethyl sulfoxide on the development of rat fetuses, *Proc. Soc. Exp. Biol. Med. 125*, 566–569 (1967).
185. J. G. Sinclair, A specific transplacental effect of urethane in mice, *Texas Rep. Biol. Med. 8*, 623–632 (1950).
186. E. K. Hall, Developmental anomalies in the eye of the rat after various experimental procedures, *Anat. Rec. 116*, 383–394 (1953).
187. H. Nishimura and M. Kuginuki, Congenital malformations induced by ethylurethan in mouse embryos, *Folia Anat. Jap. 31*, 1–10 (1958).
188. S. L. Kauffman, Early morphologic changes in mouse embryo neural tube following transplacental exposure to urethane, *Ped. Proc. 23*, 128, abst. (1964).
189. K. Tsuchikawa and A. Akabori, Differences of the response to teratogens between inbred strains of mice. (1) Differences of the response to ethylurethane between strain CBA and C3HeB/Fe, *Proc. Congen. Anom. Res. Ass. Jap. 4*, 48, abst. (1964).
190. K. Tsuchikawa and A. Akabori, Strain differences of susceptibility to induced congenital anomalies in mice, *Proc. Congen. Anom. Res. Assoc. Jap. 5*, 2, abst. (1964).
191. S. Takekoshi, Effects of urethan on the teratogenic action of hypervitaminosis A, *Gumma J. Med. Sci. 14*, 210–212 (1965).
192. J. G. Sinclair and B. E. Abreu, Transplacental effects of drugs in mice, *Texas Rep. Biol. Med. 23*, 849–853 (1965).
193. S. Takaori, K. Tanabe, and K. Shimamoto, Developmental abnormalities of skeletal system induced by ethylurethan in the rat, *Jap. J. Pharmacol. 16*, 63–73 (1966).
194. T. Tanaka, Postnatal fate of polydactyly induced by administration of ethylurethane to pregnant mice, *Acta Anat. Nippon. 41*, 299–305 (1966).
195. J. A. DiPaolo and J. Elis, Comparison of teratogenic and carcinogenic effects of some carbamate compounds, *Cancer Res. 27*, 1969–1702 (1967).

196. S. L. Kauffman and L. Herman, Ultrastructural changes in embryonic mouse neural tube cells after urethane exposure, *Dev. Biol. 17*, 55–74 (1968).

197. S. L. Kauffman, Cell proliferation in embryonic mouse neural tube following urethane exposure, *Dev. Biol. 20*, 146–157 (1969).

198. S. Peters and M. Strassburg, Stress als teratogener Faktor. Tierexperimentelle Untersuchingen zur Erzeugung von Gaumenspalten, *Arzneimittel Forschung 19*, 1106–1111 (1969).

199. N.-J. Höglund, Effects of ethyl urethane on reproduction in mice, *Acta Pharmacol. 8*, 82–84 (1952).

200. T. von Kreybig, Verschiedene Wirkmechanismen in teratologischen Experiment, *Naunyn-Schmiedebergs Arch. Exp. Pathol. 251*, 197–198 (1965).

201. S. D. Vesselinovitch, N. Mihailovich, and G. Pietra, The prenatal exposure of mice to urethan and the consequent development of tumors in various tissues, *Cancer Res. 27*, 2333–2337 (1967).

202. J. A. DiPaolo, Polydactylism in the offspring of mice injected with 5-bromodeoxyuridine, *Science 145*, 501–503 (1964).

203. P. R. Ruffolo and V. H. Ferm, The teratogenicity of 5-bromodeoxyuridine in the pregnant Syrian hamster, *Life Sci. 4*, 633–637 (1965).

204. P. R. Ruffolo and V. H. Ferm, The embryocidal and teratogenic effects of 5-bromodeoxyuridine in the pregnant hamster, *Lab. Invest. 14*, 1547–1553 (1965).

205. G. Sansone and C. Zunin, Embriopatie sperimentali da deficienza di acido folico nel ratto, prodotte mediante somministrazione di aminopterina, *Bull. Soc. Ital. Biol. Sper. 29*, 1697–1699 (1953).

206. G. Sansone and C. Zunin, Embriopatie sperimentali da somministrazione di antifolico, *Acta Vitamin. 8*, 73–79 (1954).

207. H. J. K. Obbink, Bijdrage tot de kennis omtrent de oorzaken van aangeboren misvormingen van niet-erfelijke oorsprong. Een experimenteel onderzoek bij ratten over het verband tussen foliumzuurdeficientie tijdens de zwangerschap en het onstaan van aangeboren hydrocephalus [in Dutch with English summary], *Mededel. Lab. Physiol. Chem. Univ. Amsterdam Ned. Inst. Volksvoed 16*, 1–132 (1955–57).

208. S. Kotani, E. Araki, K. Yukioka, T. Inaba, S. Nakamura, and T. Miyoshi, Experimentally produced congenital anomalies through folic acid deficiency [in Japanese with English summary], *J. Osaka City Med. Cent. 7*, 353–359 (1958).

209. C. S. Kinney and L. M. Morse, Effect of a folic acid antagonist, aminopterin, on fetal development and nucleic acid metabolism in the rat, *J. Nutr. 84*, 288–294 (1964).

210. V. S. Baranov, The peculiar features of the injuring effect of aminopterin at different stages of rat embryogenesis [in Russian], *Dokl. Akad. Nauk. SSSR 163*, 1032–1035 (1965).

211. V. S. Baranov, The specificity of the teratogenic effect of aminopterin as compared to other teratogenic agents [in Russian with English summary], *Biull. Eksp. Biol. Med. 61*, 77–82 (1966).

212. L. F. James and R. F. Keeler, Teratogenic effects of aminopterin in sheep, *Teratology 1*, 407–412 (1968).

213. J. B. Thiersch and F. S. Philips, Effect of 4-aminopteroylglutamic acid (aminopterin) on early pregnancy, *Proc. Soc. Exp. Biol. Med. 74*, 204–208 (1950).

214. M. L. Murphy, Effects of antimetabolites on development of the fetal rat, *Am. J. Dis. Child. 96*, 533–534, abst. (1958).

215. M. L. Murphy, *in* "Teratology: Principles and Techniques" (J. G. Wilson

and J. Warkany, eds.) pp. 145–161, University of Chicago Press, Chicago (1965).

216. N. N. Slonitskaya, Teratogenic effect of griseofulvin-forte on rat fetus [in Russian with English summary], *Antibiotiki 14*, 44–47 (1969).

217. B. M. Cattanach, Lack of effect of nicotine on the fertility of male and female mice, *Z. Vererb. 93*, 351–355 (1962).

218. H. Nishimura and K. Nakai, Developmental anomalies in offspring of pregnant mice treated with nicotine, *Science 127*, 877–878 (1958).

219. H. Nanjo, Maldevelopment of the fetuses caused by maternal administration of nicotine during pregnancy in relation to maternal age [in Japanese with English summary], *Acta. Anat. Nippon. 39*, 212–216 (1964).

220. J. M. Mallette and B. C. Y. Man, Some effects of maternal intake of nicotine on fetal mice, *Am. Zool. 6*, 611, abst. (1966).

221. D. Vara and O. Kinnunen, Effect of nicotine on the female rabbit and developing fetus, *Ann. Med. Exp. Biol. Fenniae 29*, 202–213 (1951).

222. L. M. Geller, Failure of nicotine to affect development of offspring when administered to pregnant rats, *Science 129*, 212–214 (1959).

223. H. D. Mosier and M. K. Armstrong, Effects of maternal intake of nicotine on fetal and newborn rats, *Proc. Soc. Exp. Blol. Med. 116*, 956–958 (1964).

224. R. Becker, J. E. King, and C. R. D. Little, Delayed birth in experimental animals and its relation to the postmaturity syndrome, *Anat. Rec. 154*, 315, abst. (1966).

225. H. D. Mosier, Jr., and M. K. Armstrong, Effect of maternal nicotine intake on fetal weight and length in rats, *Proc. Soc. Exp. Biol. Med. 124*, 1135–1137 (1967).

226. M. F. Lyon, R. J. S. Phillips, and A. G. Searle, A test for mutagenicity of caffeine in mice, *Z. Vererb. 93*, 7–13 (1962).

227. B. M. Cattanach, Genetical effects of caffeine in mice, *Z. Vererb. 93*, 215–219 (1962).

228. W. Kuhlmann, H. Fromme, E. Heege, and W. Ostertag, The mutagenic action of caffeine in higher organisms, *Cancer Res. 28*, 2375–2389 (1968).

229. I. D. Adler, Does caffeine induce dominant lethal mutations in mice? *Humangenetik 7*, 137–148 (1969).

230. H. Nishimura and K. Nakai, Congenital malformations in offspring of mice treated with caffeine, *Proc. Soc. Exp. Biol. Med. 104*, 140–142 (1960).

231. C. Knoche and J. König, Zur pränatalen Toxizität von Diphenylpyralin-8-chlor-theophyllinat unter Berücksichtigung von Erfahrungen mit Thalidomid und Koffein, *Arzneimittel Forschung 14*, 415–424 (1964).

232. M. Bertrand, E. Schwam, A. Frandon, A. Vagne, and J. Alary, Sur un effet tératogène systématique et spécifique de la caféine chez les rongeurs, *Compt. Rend. Soc. Biol. 159*, 2199–2202 (1965).

233. T. Fujii, H. Sasaki, and H. Nishimura, Teratogenicity of caffeine in mice related to its mode of administration, *Jap. J. Pharmacol. 19*, 134–138 (1969).

234. A. Georges and J. Denef, Les anomalies digitales: Manifestations tératogéniques des dérivés xanthiques chez le rat, *Arch. Internat. Pharmacodyn. Thérap. 172*, 219–222 (1968).

235. B. K. Batra, The effect of methylcholanthrene painting of the ovaries on the progeny of mice, *Acta Union Internat. Contra. Cancer 15*, 128–133 (1959).

236. S. Fabro, R. L. Smith, and R. T. Williams, Embryotoxic activity of some pesticides and drugs related to phthalimide, *Food Cosmetol. Toxicol. 3*, 587–590 (1965).

237. G. Kennedy, O. E. Fancher, and J. C. Calandra, An investigation of the

teratogenic potential of captan, folpet and difolatan, *Toxicol. Appl. Pharmacol. 13*, 420–430 (1968).

238. Y. Chambon, Action de la chlorpromazine sur l'évolution et l'avenir de la gestation chez le rat, *Ann. Endocrinol. 16*, 912–922 (1955).

239. F. Bovet-Nitti and D. Bovet, Action of some sympatholytic agents on pregnancy in the rat, *Proc. Soc. Exp. Biol. Med. 100*, 555–557 (1959).

240. O. D. Murphree, B. L. Monroe, and L. D. Seager, Survival of offspring of of rats administered phenothiazines during pregnancy, *J. Neuropsychiat. 3*, 295–297 (1962).

241. J. M. Ordy, A. Latanick, R. Johnson, and L. Massopust, Chlorpromazine effects on pregnancy and offspring in mice, *Proc. Soc. Exp. Biol. Med. 113*, 833–866 (1963).

242. C. Rolsten, Effects of chlorpromazine and psilocin on pregnancy of C57BL/10 mice and their offspring at birth, *Anat. Rec. 157*, 311, abst. (1967).

243. S. Ranström, Stress and pregnancy, *Acta Pathol. Microbiol. Scand. Suppl. 111*, 113–114 (1956).

244. L. B. Schnürer, Maternal and foetal responses to chronic stress in pregnancy, *Acta Endocrinol. Suppl. 80*, 1–96 (1963).

245. D. J. Clegg, Absence of teratogenic effect of butylated hydroxyanisole (BHA) and butylated hydroxytoluene (BHT) in rats and mice, *Food Cosmetol. Toxicol. 3*, 387–403 (1965).

246. M. C. Chang, Artificial production of monstrosities in the rabbit, *Nature 154*, 150 (1944).

247. R. G. Edwards, Colchicine-induced heteroploidy in the mouse. I. The induction of triploidy by treatment of the gametes, *J. Exp. Zool. 137*, 317–348 (1958).

248. J. H. Van Dyke and M. G. Ritchey, Colchicine influence during embryonic development in rats, *Anat. Rec. 97*, 375, abst. (1947).

249. B. P. Wiesner, M. Wolfe, and J. Yudkin, The effects of some antimitotic compounds on pregnancy in the mouse, *Studies Fert. 9*, 129–136 (1958).

250. V. H. Ferm, Colchicine teratogenesis in hamster embryos, *Proc. Soc. Exp. Biol. Med. 112*, 775–778 (1963).

251. V. H. Ferm, Effect of transplacental mitotic inhibitors on the fetal hamster eye, *Anat. Rec. 148*, 129–138 (1964).

252. V. H. Ferm, The rapid detection of teratogenic activity, *Lab. Invest. 14*, 1500–1505 (1965).

253. R. Shoji and S. Makino, Preliminary notes on the teratogenic and embryocidal effects of colchicine on mouse embryos, *Proc. Jap. Acad. 42*, 822–827 (1966).

254. G. L. Vankin and H. J. Grass, Colcemid-induced teratogenesis in hybrid mouse embryos, *Am. Zool. 6*, 551, abst. (1966).

255. J. M. Morris, G. van Wagenen, G. D. Hurteau, D. W. Johnston, and R. A. Carlsen, Compounds interfering with ovum implantation and development. I. Alkaloids and antimetabolites, *Fert. Ster. 18*, 7–17 (1967).

256. T. H. Ingalls, F. J. Curley, and P. Zappasodi, Colchicine-induced craniofacial defects in the mouse embryo, *Arch. Environ. Health 16*, 326–332 (1967).

257. T. Kerr, On the effects of colchicine treatment of mouse embryos, *Proc. Zool. Soc. London 116*, 551–564 (1947).

258. C. H. Conaway, Embryo resorption and placental scar formation in the rat, *J. Mammalol. 36*, 516–532 (1955).

259. J. B. Thiersch, Effect of N-desacetyl thio colchicine (TC) and N-desacetyl-methyl-colchicine (MC) on rat fetus and litter in utero, *Proc. Soc. Exp. Biol. Med. 98*, 479–485 (1958).

260. W. Gibel, E Geibler, K. H. Lohs, H. Hilscher, G. P. Wildner, and S. Wittbrodt, Untersuchungen zur Frage einer möglichen mutagenen Wirkung von Fuselöl, *Arch. Geschwulstforsch. 33*, 49–54 (1969).
261. A. V. Bespamyatnova, S. D. Zaugalnikov. and Yu. Z. Sukhov, Embryotoxic and teratogenic action of ethylene-imine [in Russian with English summary]. *Farmakol. Toksik. 33*, 357–360 (1970).
262. E. Le Breton, C. Frayssment, C. Lafarge, and A. M. De Recondo. Aflatoxine—mécanisme de l'action, *Food Cosm. Toxic. 2*, 674–677 (1964).

CHAPTER 3

The Mutagenicity of Chemical Carcinogens: Correlations, Problems, and Interpretations

Elizabeth C. Miller and
James A. Miller

McArdle Laboratory for Cancer Research
University of Wisconsin Medical Center
Madison, Wisconsin

I. INTRODUCTION

Mutagenesis and carcinogenesis are complex cellular processes which are grossly similar in that each produces heritable changes in phenotype. Experimental mutagenesis by a pure chemical agent was first achieved by Auerbach and Robson (1944, 1946) and a variety of such agents are now known. Their mechanisms of action appear to be fairly well understood as consisting of expressible heritable changes in the molecular structure of the genetic material or DNA (Orgel, 1965; Freese and Freese, 1966; Auerbach, 1967; Freese, this book). Carcinogenesis by chemical agents was discovered first in humans almost 200 years ago, and today nearly a dozen agents active in the human and of widely varying structure are known (Clayson, 1962; Hueper and Conway, 1964; Miller, J. A., 1970). The experimental induction of cancer with a pure chemical compound was first achieved about 1930, and many chemical carcinogens of greatly differing structures are known today (see preceding references). However, it is not known at the molecular level how any chemical carcinogen induces the formation of cells which are not responsive to the growth controls of the host and which form neoplasms that grow more or less continuously and may invade normal tissues. This is also true of the carcinogenic action of viruses and radiation. Likewise, the molecular phenotype of no neoplasm is even partially understood.

The concept that neoplasms arise from "mutations" in somatic cells has long been a popular theory of cancer causation (Boveri, 1914; Bauer, 1928, 1963). The modern understanding of the chemical basis of mutation has enhanced its acceptance as an intellectually satisfying theory for the mechanism of action of chemical carcinogens (Clayson, 1962), particularly since chemical carcinogens or their metabolites have been frequently found to have mutagenic properties. However, other theories compete strongly today in directing experimental approaches on the genesis and nature of neoplasia. A prominent theory opposing the somatic mutation concept is the view that neoplasia results from quasi-permanent and heritable changes in genome expression without changes in the genome itself (Pitot and Heidelberger, 1963; Gelboin, 1967; Weinstein, 1970). This is the process by which cellular differentiation is generally considered to occur. Viral activity as the basis for the origin of chemically induced neoplasia is also a popular idea since many carcinogenic viruses are now known. According to this concept, the chemical carcinogen activates a latent carcinogenic virus or causes the expression of part or all of an integrated viral genome; this alteration in total genome content or expression can then induce neoplastic changes through one of the mechanisms discussed above. Another theory suggests that chemical carcinogens may alter selection pressures in the cellular environment (i.e., immunological factors) and thus permit latent tumor cells to grow and form gross neoplasms. There is no experimental basis at the present time for selecting any one of the above mechanisms as *the* mechanism by which chemicals induce neoplasia. In fact, each mechanism may be a valid explanation for the induction of tumors by chemicals in specific instances.

The present review is limited to an examination of the correlation of the mutagenic and carcinogenic properties of various chemicals and to a discussion of the extent to which such evidence supports a somatic mutation theory of chemical carcinogenesis. These problems were reviewed by Burdette in 1955, at which time he concluded that there was no correlation between the mutagenic and carcinogenic activities of those chemicals which had been assayed for both types of activity. In 1955 very little was known with regard to the structures of the ultimate reactive forms of chemical carcinogens *in vivo*, but Burdette recognized that the lack of correlation could have been due, among other factors, to differences in the metabolic fates of the chemicals in the different organisms used to assay carcinogenicity and mutagenicity. Our knowledge of the active forms of chemical carcinogens is still incomplete, but sufficient progress has been made since 1955 in this area, as well as in the conduct and assessment of carcinogenicity and mutagenicity assays, to make a reevaluation of the possible role of somatic mutations in carcinogenesis pertinent.

II. ULTIMATE CARCINOGENIC FORMS OF CHEMICAL CARCINOGENS AND THEIR REACTIVITIES

Considerable progress has been possible in the past decade as a consequence of the recognition that many chemical carcinogens are not active as such, but require conversion *in vivo* to metabolites which are the ultimate carcinogenic forms. Thus, much of the specificity of certain chemical carcinogens for particular species or tissues now appears to be a function of the amounts of the ultimate carcinogenic metabolites available to the tissues as a consequence of metabolic activation *in situ* or of transport from sites of activation. Because the initial molecular lesion(s) responsible for chemical carcinogenesis have yet to be defined, one can not determin easily the chemical natures of the ultimate carcinogens. Nevertheless, in a number of cases reasonable deductions as to the nature of the ultimate carcinogenic metabolite(s) have been possible from the *in vitro* and *in vivo* reactivities of known and probable metabolites of the carcinogens and from the structures of the protein- and nucleic acid-bound derivatives in tumor-susceptible tissues. The structures of the protein- and nucleic acid–bound derivatives have appeared to be useful leads, since it seems axiomatic that the heritable and at least quasi-permanent alterations in the cells undergoing neoplasia must arise as a consequence of interactions, directly or indirectly, of the chemical or its metabolite(s) with one or more of these macromolecules. Furthermore, in a number of cases correlations have been obtained between the levels of protein– or nucleic acid–carcinogen interactions and the likelihood of tumor development, and these correlations have suggested the possibility of a causal relationship (Miller, E. C., and Miller, J. A., 1966). In any event, these interactions serve as important indicators of the formation *in vivo* of reactive forms involved in the induction of neoplasia through the formation of specific macromolecule–carcinogen derivatives.

As indicated below, a survey of the literature suggests that most, if not all, chemical carcinogens are strong electrophilic (i.e., containing relatively electron-deficient atoms) reactants as administered or they are converted to potent electrophilic reactants *in vivo*. Concurrently, there is an increasing amount of evidence that these electrophilic reactants are responsible for the carcinogenic activity of the administered carcinogens. This generalization not only provides a unified view of structurally diverse chemical carcinogens, but also predicts some uniformity in the sites on the cellular macromolecules susceptible to their attack.

A. Alkylating Agents

Alkylating agents are recognized as electrophilic reactants *per se* (Fig. 1), and their reactivity nonenzymatically under physiological conditions with nucleophilic sites in proteins and nucleic acids has been well documented (Brookes and Lawley, 1960; Ross, 1962; Lawley and Brookes, 1963; Goldschmidt *et al.*, 1968; Price *et al.*, 1968; Shapiro, 1969). Where tested, these compounds undergo similar reactions on administration *in vivo* (Brookes and Lawley, 1960; Ross, 1962; Colburn and Boutwell, 1968*a,b*; Swann and Magee, 1968; Boutwell *et al.*, 1969). Since no metabolic activation is required, the carcinogenic activity of the alkylating agents is probably dependent in large part on their ability to penetrate to the intracellular site(s) critical for the induction of neoplasia before they are dissipated through

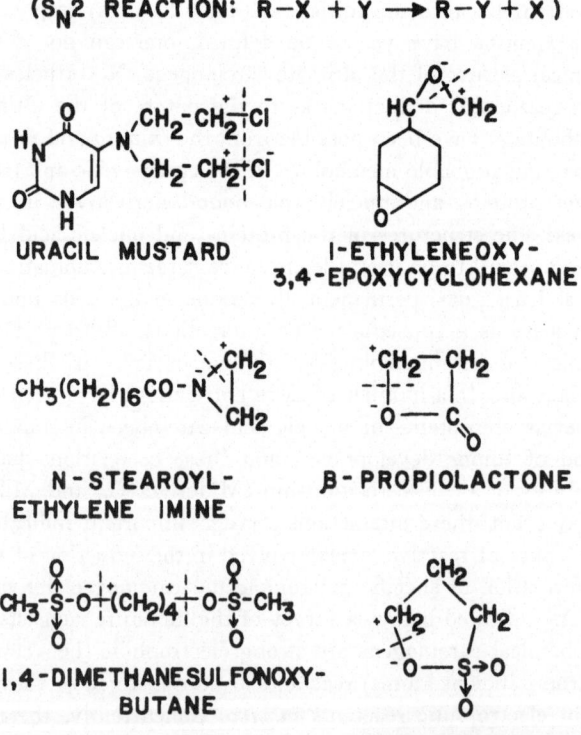

FIGURE 1. The electrophilic structures of some carcinogenic alkylating agents and their general mode of reaction with nucleophils (Y^-).

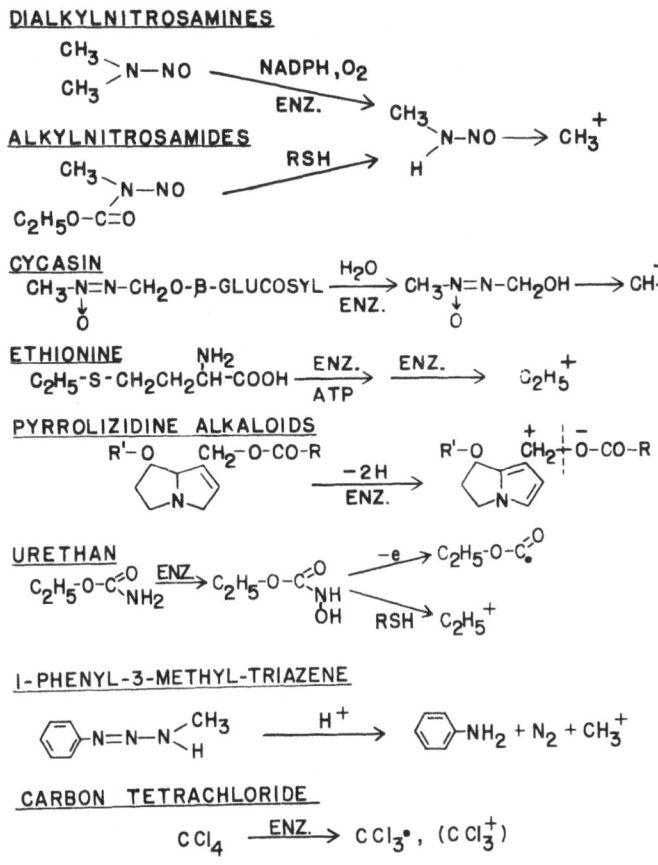

FIGURE 2. Electrophilic reactants derived from examples of various types of aliphatic chemical carcinogens. The carcinogen (precarcinogen) is shown at the left, and the possible carcinogenic electrophilic reactant (ultimate carcinogen) is shown at the right.

reaction with extracellular or noncritical nucleophilic cellular constituents.

B. Potential Alkylating Agents

A large group of chemical carcinogens are converted to alkylating agents *in vivo*, and their reactivity and carcinogenicity appear to be referable to the derived alkylating agents. These potential alkylating agents are exemplified by the large number of carcinogenic aliphatic nitrosamides and nitrosamines (Fig. 2) (Magee and Barnes, 1967; Druckrey *et al.*, 1967, 1969). Reaction with sulfhydryl groups at neutral *p*H converts the nitrosamides to monoalkyl-nitrosamines, which decompose spontaneously to yield carbonium ions or alkyl diazonium ions (Magee and Schoental, 1964; Lawley and Thatcher, 1970). Sulfhydryl groups occur ubiquitously in cells, and it is thus not surprising that tumors develop in a wide variety of tissues and species as a consequence of the administration of the nitrosamides. Similarly, the carcinogenicity of the dialkylnitrosamines for a variety of tissues is apparently mediated by their enzymatic oxidative dealkylation to unstable monoalkyl nitrosamines (Magee and Schoental, 1964; Magee and Barnes, 1967). The carcinogenic unsymmetrical dialkyl triazenes and symmetrical hydrazo-, azo-, and azoxydialkanes are similarly converted by microsomal oxidases to unstable intermediates which decompose to alkyl carbonium ions (Preussmann et al., 1969a,b). Alkylation in vivo has been observed with a number of these potential alkylating agents (Magee and Barnes, 1967; Lijinsky and Ross, 1969, Krüger et al., 1970). The levels of mixed-function oxidases in the tissues and the ease of oxidation of the alkyl groups appear to be critical factors in determining the carcinogenic potencies and tissue specificities of these compounds.

Closely akin to dimethylnitrosamine in its carcinogenic action is the naturally occurring compound cycasin (methylazoxymethanol-β-glucoside) (Fig. 2) (Laqueur and Spatz, 1968; Spatz, 1969). Orally administered cycasin is hydrolyzed by bacterial β-glucosidase in the intestinal tract to methylazoxymethanol, a carcinogen which decomposes spontaneously in vitro and in vivo to yield a methyl carbonium ion (Laqueur and Matsumoto, 1966; Matsumoto and Higa, 1966; Laqueur et al., 1967; Nagata and Matsumoto, 1969). In the absence of β-glucosidase (e.g., parenteral administration to conventional rats or oral administration to germ-free rats), cycasin is not carcinogenic and is excreted unchanged in the urine (Spatz, 1969).

Administration to rats of the hepatocarcinogen ethionine (Fig. 2) results in the ethylation of hepatic RNA and protein; ethylation of hepatic DNA, if it occurs, is much less extensive (Farber, 1963; Stekol, 1963, 1965; Farber et al., 1967; Orenstein and Marsh, 1968; Rosen, 1968; Ortwerth and Novelli, 1969). S-Adenosylethionine is also formed from ethionine in vivo, and, as an analog of the naturally occurring methylating intermediate S-adenosylmethionine, has been assumed to be the alkylating intermediate. However, recent studies by Ortwerth and Novelli (1969) suggest that other ethylating intermediates may also be formed from ethionine in vivo.

The hepatotoxic and hepatocarcinogenic pyrrolizidine alkaloids (Fig. 2) (Schoental, 1963, 1968a,b) form another class of alkylating agents. As first noted by Culvenor et al. (1962), these compounds have weak alkylating activity per se. However, recent studies by Mattocks (1968) and by Culvenor et al. (1969) have demonstrated the chemical formation from the pyrrolizidine alkaloids of pyrrole derivatives with much stronger alkylating activity and have also shown that similar pyrrole derivatives are formed from the alkaloids in vivo.

Several halogenated hydrocarbons [chloroform, carbon tetrachloride, and 2,2-bis-(p-chlorophenyl)-1,1,1-trichloroethane (DDT)] are carcinogenic for the mouse liver (Clayson, 1962; Innes et al., 1969), and carbon tetrachloride is also carcinogenic for the rat liver (Costa et al., 1963; Reuber and Glover, 1970) and hamster liver (Della Porta et al., 1961). These compounds are probably converted to halogenated alkylating agents in vivo. Administration of either $^{14}CCl_4$ or $C^{36}Cl_4$ yields protein-bound and possibly nucleic acid–bound radioactivity in the rat liver (Cessi et al., 1966; Reynolds, 1967; Rao and Recknagel, 1969). The identification of chloroform and hexachloroethane as metabolites of carbon tetrachloride (Butler, 1961; Fowler, 1969) suggests that a free radical is formed in vivo, although alkylation might also result from the formation of a carbonium ion.

C. Aromatic Amines and Amides

N-Hydroxylation is now recognized as the first step in the activation of aromatic amines and amides for carcinogenesis. The N-hydroxyamines and -amides have shown very little nonenzymatic reactivity with tissue nucleophils under physiological conditions, but esterification converts them to strong electrophils which arylate, arylamidate, or arylaminate nucleophils (Fig. 3) (Miller, J. A., and Miller, E. C., 1969a,b). The probable role of esters of N-hydroxy-2-acetylaminofluorene and of N-hydroxy-N-methyl-4-amino-azobenzene in the reactivity of 2-acetylaminofluorene and of N-methyl-4-aminoazobenzene in vivo was first deduced from the identity of some of the protein- and nucleic acid-bound derivatives from rats administered the parent compounds with derivatives formed by nonenzymatic reaction of the esters with tissue nucleophils (Scribner et al., 1965; Lotlikar et al., 1966; Miller, E. C., et al., 1966a, Kriek et al., 1967; Kriek, 1968, 1969; Lin et al., 1968, 1969; DeBaun et al., 1970a; Miller, J. A., et al., 1970). Furthermore, the acetic and benzoic acid esters of N-hydroxy-2-acetylaminofluorene each induced higher incidences of sarcomas at the site of subcutaneous injection in the rat than did N-hydroxy-2-acetylaminofluorene (Miller, J. A., and Miller, E. C., 1969a). Likewise, N-benzoyloxy-N-methyl-4-aminoazobenzene induced sarcomas at the injection site in rats, while N-methyl-4-amino-azobenzene was inactive at this site (Poirier et al., 1967).

The evidence strongly favors the sulfuric acid ester of N-hydroxy-2-acetylaminofluorene as the primary ultimate carcinogenic derivative of N-hydroxy-2-acetylaminofluorene in rat liver. Thus, rat liver contains sulfotransferase(s) which transfer the sulfonate group from 3'-phosphoadenosine-5'-phosphosulfate to N-hydroxy-2-acetylaminofluorene, and the level of this transferase activity correlates well with the susceptibility of the livers of rats under various hormonal conditions and of the livers of various species to hepatic carcinogenesis by N-hydroxy-2-acetylaminofluorene

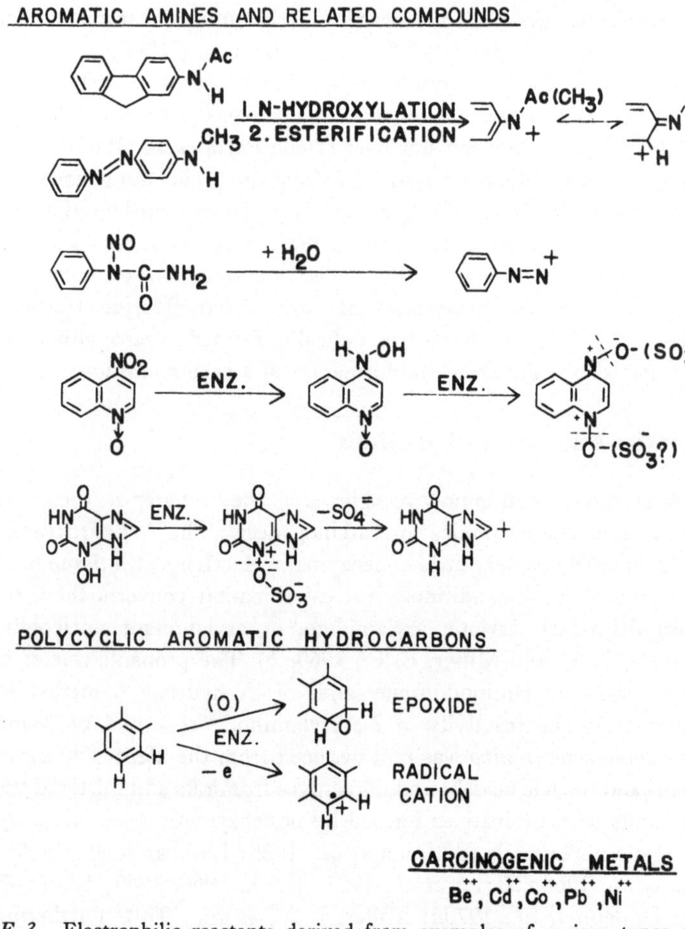

AROMATIC AMINES AND RELATED COMPOUNDS

POLYCYCLIC AROMATIC HYDROCARBONS

CARCINOGENIC METALS

$Be^{++}, Cd^{++}, Co^{++}, Pb^{++}, Ni^{++}$

FIGURE 3. Electrophilic reactants derived from examples of various types of aromatic chemical carcinogens and carcinogenic metals.

(DeBaun et al., 1970a). The sulfuric acid ester also appears to be a major reactive form of N-hydroxy-2-acetylaminofluorene in vivo, since the levels of protein-, DNA-, and RNA-bound derivatives of N-hydroxy-2-acetyl-aminofluorene in rat liver were decreased when the amount of sulfate was made limiting by administration of p-hydroxyacetanilide to the rats (DeBaun et al., 1970b). The inhibition of 2-acetylaminofluorene-induced hepato-carcinogenesis by the simultaneous administration of acetanilide (Yamamoto et al., 1968) and the partial prevention of this inhibition by dietary sodium sulfate (Weisburger et al., 1970) provide circumstantial evidence for a role of the sulfuric acid ester in hepatic carcinogenesis; through conversion to *p*-hydroxyacetanilide and conjugation of the latter metabolite, acetanilide could deplete the hepatic reservoir of 3'-phosphoadenosine-5'-phosphosulfate (Büch *et al.*, 1968). Direct tests on the carcinogenicity of 2-acetylamino-fluorene-*N*-sulfate have been disappointing, since only a few tumors have been obtained long after repeated administration of the ester to the skin or subcutaneous tissue of the rat (Miller, J. A., and Miller, E. C., 1969*a*). Apparently, the short half-life of this ester (less than 1 min in water) and its ionic character prevent appreciable amounts of the compound from penetrating to the critical site in the cell for the induction of neoplasia prior to its reaction with extracellular or noncritical cellular nucleophils.

The possible roles of other esters of *N*-hydroxy-2-acetylaminofluorene, esters of *N*-hydroxy-2-aminofluorene, and the glucuronides of these two compounds in hepatic carcinogenesis require further elucidation. The data available indicate that the hepatic levels of phosphotransferase and acetyltransferase for *N*-hydroxy-2-acetylaminofluorene in rat liver, if present, are low compared to those for sulfotransferase activity for *N*-hydroxy-2-acetylaminofluorene (DeBaun *et al.*, 1970*a*). However, these or other esters, if formed *in vivo*, would be expected to be reactive. The glucuronide of *N*-hydroxy-2-acetylaminofluorene, a major metabolite of *N*-hydroxy-2-acetylaminofluorene in a number of species (Cramer *et al.*, 1960; Miller, E. C., *et al.*, 1964; Irving, 1962, 1965; Irving *et al.*, 1967), reacts with nucleophils in the same manner as esters of *N*-hydroxy-2-acetylaminofluorene, but at a very much slower rate (Miller, E. C., *et al.*, 1968). From the lack of correlation between the amounts of this glucuronide which are formed *in vivo* and the susceptibility of the animals to hepatic carcinogenesis by *N*-hydroxy-2-acetylaminofluorene, this glucuronide does not appear to play a major role in hepatic carcinogenesis by *N*-hydroxy-2-acetylaminofluorene (Miller, E. C., *et al.*, 1970). The glucuronide of *N*-hydroxy-2-aminofluorene is very reactive *in vitro* under physiological conditions (Irving and Russell, 1970), but there is, as yet, no evidence that appreciable amounts are formed *in vivo*. However, the possible importance of derivatives of *N*-hydroxy-2-aminofluorene in the formation of DNA-bound derivatives *in vivo* is apparent from the observation that the major share of the DNA-bound derivatives of *N*-hydroxy-2-acetylaminofluorene in rat liver do not retain the *N*-acetyl group (Kriek, 1968, 1969; Irving and Veazey, 1969; Irving *et al.*, 1969*a,b*). The possible roles of all of the derivatives just discussed have been considered only with respect to hepatic carcinogenesis, and no information is available on the reactive forms of *N*-hydroxy-2-acetylaminofluorene in other tissues.

By analogy with the metabolic activation of *N*-hydroxy-2-acetylamino-fluorene, esterification has also been suggested as the most likely activation reaction for other *N*-hydroxyamines and -amides (Miller, J. A., and Miller, E. C., 1969*a*). While the *N*-acetoxy derivatives of 4-acetylaminostilbene,

4-acetylaminobiphenyl, 2-acetylaminophenanthrene, 1-acetylaminonaphtha-
lene, and 2-acetylaminonaphthalene differ greatly among themselves with
respect to their reactivity with nucleophilic components at neutrality, the
reactivities of the esters are, in each case, greater than those of the corresponding hydroxamic acids (Miller, J. A., and Miller, E. C., 1969a; Scribner
et al., 1970). Furthermore, each of the above N-acetoxy derivatives has
shown greater carcinogenic activity at the site of subcutaneous injection in
rats than the corresponding hydroxamic acid. Except for N-hydroxy-1-
naphthylamine, each of the aryl hydroxylamines in this series is less carcinogenic at the site of injection than the corresponding hydroxamic acid or its
acetic acid ester; N-hydroxy-1-naphthylamine, however, is much more
carcinogenic than its two derivatives (Miller, J. A., and Miller, E. C., 1969a;
Scribner et al., unpublished).

D. 4-Nitroquinoline-1-oxide

The demonstration that 4-nitroquinoline-1-oxide (Fig. 3) is metabolized
to 4-hydroxyaminoquinoline-1-oxide (Sugimura et al., 1965; Matsushima
et al., 1968) and the greater carcinogenic activity of the latter compound
(Endo and Kume, 1965; Shirasu, 1965) suggest that the nitro compound is
carcinogenic as a consequence of metabolic reduction to the hydroxylamine
and, possibly, subsequent esterification of the latter compound. Thus, the
diacetyl derivative of 4-hydroxyaminoquinoline-1-oxide reacts with a number of nucleophils at neutrality (Enomoto et al., 1968). Furthermore, the
DNA- and RNA-bound derivatives formed by reaction with the diacetyl
derivatives in vitro have similar fluorescent properties to those of the nucleic
acid derivatives isolated from ascites hepatomas of rats treated with 4-
hydroxyaminoquinoline-1-oxide (Matsushima et al., 1967; Tada et al.,
1967). However, in view of the electrophilic reactivity of 4-nitroquinoline-
1-oxide, as exemplified by its reaction with sulfhydryl groups (Nakahara et
al., 1957; Endo, 1958), the possibility that this compound is an ultimate
carcinogen per se can not be dismissed.

E. N-Hydroxypurines

Brown and his coworkers (Brown et al., 1965; Sugiura et al., 1970)
have found several N-hydroxypurines (tautomeric to purine N-oxides),
notably 3-hydroxyxanthine and guanine 3-N-oxide, to have considerable
carcinogenic activity in the subcutaneous tissue of the rat. These compounds
yield potent electrophils upon esterification in vitro (Wölcke et al., 1969)
and considerable evidence indicates that in vivo they yield metabolites,
probably including 3-N-sulfates, which upon decomposition yield purine

intermediates with a strong electrophilic center at C-8 (Stöhrer and Brown, 1970; Stöhrer et al., 1970).

F. Polycyclic Aromatic Hydrocarbons

The carcinogenic polycyclic aromatic hydrocarbons form the only major group of chemical carcinogens for which the nature of the reactive forms is still relatively obscure. These carcinogens can intercalate, or at least associate physically, with bases in DNA in vitro (Boyland, 1964; Ts'o et al., 1969), but the derivatives bound covalently in vivo to nucleic acids and proteins are probably derived from metabolically activated forms (Brookes and Lawley, 1964; Heidelberger, 1964; Grover and Sims, 1969; Gelboin, 1969). The hydrocarbons have relatively low ionization potentials,and it has been suggested that radical cations (Fig. 3) arising from reactions with oxidants in the cells might play a primary role in hydrocarbon carcinogenesis (Fried and Schumm, 1967; Morreal et al., 1969; Wilk and Girke, 1969; Wilk and Hoppe, 1969). On the other hand, Boyland and Sims (1962, 1964, 1965b,c) have presented evidence that a number of epoxides are formed in the metabolism of phenanthrene, benz(a)anthracene, and dibenz(a,h) anthracene and, by analogy, epoxides are probably formed in the metabolism of many carcinogenic hydrocarbons. The recent demonstration of Grover and Sims (1970) of the reactivity of K-region epoxides of phenanthrene and dibenz(a,h)anthracene with DNA, RNA, and histones in vitro suggests that these metabolites may be ultimate reactive (and carcinogenic?) derivatives of the hydrocarbons in vivo. However, the K-region epoxides have shown limited or no carcinogenic activity when administered to the skin of mice or subcutaneously to rats or mice (Boyland and Sims, 1967; Miller, E. C., and Miller, J. A., 1967; Van Duuren et al., 1967).

Hydroxymethyl derivatives (or their esters) are also possible ultimate carcinogenic forms of the potent carcinogen 7,12-dimethylbenz(a)anthracene and of other hydrocarbons with methyl groups. Oxidative attack of these methyl groups in vivo has been demonstrated by Boyland and Sims (1965a), but both 7-hydroxymethyl-12-methylbenz(a)anthracene and 7-methyl-12-hydroxymethylbenz(a)anthracene are less potent carcinogens than the parent hydrocarbon (Boyland et al., 1965; Boyland, 1969). In model studies 7-bromomethylbenz(a)anthracene and 7-bromomethyl-12-methylbenz(a)-anthracene have shown electrophilic reactivity with DNA and other cellular constituents (Brookes and Dipple, 1969), and an ester of a benzylic alcohol might be expected to have similar reactivity.

G. Urethan

The identities of the ultimate carcinogenic forms of urethan and

N-hydroxy urethan, which have similar tumorigenic activities for mouse lung and skin and which also induce tumors at other sites (Mirvish, 1968), have not been elucidated. However, Boyland and his associates (Boyland and Nery, 1965; Boyland and Williams, 1969) found carboxyethylation of sulfhydryl groups *in vitro* and of the 5-carbon of cytosine in lung and liver RNA after administration of urethan to mice. Furthermore, Nery (1969) has shown that *N*-hydroxyurethan and, to a much greater extent, its esters nonenzymatically carboxyethylate the amino group of cytosine under physiological conditions.

H. *N*-Nitroso-*N*-phenylurea

This aromatic nitrosamide decomposes quickly in water to form the electrophilic phenyl diazonium ion (Fig. 3). This reaction product may participate in the carcinogenic activity of *N*-nitroso-*N*-phenylurea in the subcutaneous tissue of the rat (Preussmann *et al.*, 1968).

I. Metal Ions

The carcinogenic metals are, in their ionic forms, electrophils *per se*, and the reaction of metal ions with guanine is of interest in this regard (Shapiro, 1968). Furthermore, many of the carcinogenic metal ions (Be^{2+}, Cd^{2+}, Co^{2+}, Pb^{2+}, Ni^{2+}) (Clayson, 1962; Hueper and Conway, 1964) form very insoluble phosphates and could conceivably interact with the tertiary phosphate ions in the nucleic acids.

J. Conclusion

It is now apparent that the structures of the reactive forms of the chemical carcinogens share a unity which is not apparent from a consideration of the structures of the parent compounds. Thus, while only a few classes of carcinogens, especially the conventional alkylating agents, can be considered electrophilic reactants as administered, most, if not all, other chemical carcinogens appear to be converted to electrophilic reactants *in vivo*. As electrophils or potential electrophils the carcinogens have common nucleophilic targets which include the guanine, adenine, and cytosine bases and the tertiary phosphate groups of the nucleic acids, amino acids such as methionine, cysteine, tyrosine, and histidine in proteins, and probably nucleophilic centers in tissue components of smaller molecular weights. While the absolute importance of any of these reactions to the carcinogenic process is not known for any chemical carcinogen, the ability to yield strong electrophilic reactants *in vivo* does appear to be an essential feature of the known chemical carcinogens.

III. MUTAGENIC ACTIVITY OF CARCINOGENIC CHEMICALS

A. Assay Systems

As detailed elsewhere in this volume, a wide variety of tests are available for assaying chemicals for mutagenic activity. Assays in which isolated nucleic acids are treated with the test compound and then assayed for mutations in a biological system in the absence of the test compound are the most direct tests, if the nature of the ultimate reactive form of the carcinogen is known and if this compound is available for test. The most widely used assay of this type is the Freese and Strack (1962) assay for mutation of transforming DNA from *Bacillus subtilis*. Because this is a nonmetabolizing system, negative results may mean only that a nonultimate form of the carcinogen was assayed. Further, the system may have limited usefulness when the active forms are radical-producing agents, since they may cause too great an inactivation of the DNA to permit adequate mutagenicity assays (Freese and Freese, 1966). Mutation of phage (such as bacteriophage T4) can be similarly used for assay of ultimate carcinogenic forms. when the phage is treated extracellularly. However, the DNA of the phage may be less accessible to the test material than is the isolated nucleic acid in the Freese and Strack assay.

Most other mutagenicity systems provide an opportunity for metabolism of the test compound; this metabolism may be either an advantage or a disadvantage depending on whether the chemical is converted to reactive forms or is metabolized predominantly to nonmutagenic forms. Furthermore, it will be important to determine whether the active form in the mutagenicity test is the same as the active form in the carcinogenicity test. These assay systems have most frequently utilized bacteria or fungi, although more complex organisms such as various plants and *Drosophila* have also been widely used. Use of mammalian cell culture systems has been limited by the number of useful markers.

The most widely used mammalian assay is the dominant lethal test, in which the chemical is administered to male mice. The mutations are scored as the frequency of early fetal deaths after the treated males are bred with successive groups of normal females (Epstein, this book). The usefulness of this assay is limited by the requirement that a mutagenic derivative of the chemical reach or be formed in the sperm or precursor cells in the testes.

The advantages of the bacterial and fungal assays have been combined with the metabolic capabilities of mammalian hosts in the development of host-mediated bacterial and fungal assays (Legator and Malling, this book). Since the microorganisms can be grown in various tissues of the host and the test chemical can be administered by a variety of routes, these host-mediated assays appear to be among the most suitable for assaying the potential mutagenic capacity of a carcinogen for which the ultimate reactive forms are not known.

B. Problems in Interpretation

Delineation of the extent of correlation of mutagenic and carcinogenic activities of chemicals is limited by the types and amount of data available. Some of the better-known mutagenic agents have received only limited tests for carcinogenic activity. More importantly for this discussion, many of the chemical carcinogens have received only limited tests for mutagenic activity. Furthermore, while a variety of assays are useful in the determination of each type of activity, the results from different carcinogenicity or different mutagenicity assays can not always be readily pooled. Thus, those carcinogens which are active only at sites distant from the port of entry may be assumed to require activation to an ultimate carcinogenic form. However, the converse is not true; those compounds which are carcinogenic at the site of administration may be in the active form as administered *or* activation may be possible in the cells at the site of treatment. Similarly, mutagens capable of mutagenizing transforming DNA *in vitro* may be assumed to have interacted with the DNA in the form administered or through the mediation of some nonenzymatically formed product, while those which are mutagenic in cellular systems may have acted directly or through the mediation of enzymatic or nonenzymatic reactions. Thus, possible complications due to lack of penetration to critical cellular sites, lack of activation, and too facile enzymatic or nonenzymatic decomposition must all be considered in assessing each result in either mutagenicity or carcinogenicity tests. Further, the specificity of some mutagenicity assays for detection of only particular types of mutants requires caution in the interpretation of negative data.

In the discussion which follows and in the accompanying Tables 1–3, both carcinogenic and mutagenic activities are rated only as either positive or negative. Attempts to rate the degrees of these activities did not seem profitable, in view of the dependence of the degree of activity on the types of assays employed.

C. Alkylating Agents

As a class the alkylating agents have generally proved to have potent mutagenic activity, and the mechanisms by which they cause mutations have been studied in detail (Ross, 1962; Freese, 1963; Krieg, 1963; Lawley, 1966; Orgel, 1965; Singer and Fraenkel-Conrat, 1969; Loveless, 1969;

Lawley and Thatcher, 1970). A few alkylating agents have proved to be potent carcinogens (Abell *et al.*, 1965; Druckrey *et al.*, 1968; Swann and Magee, 1969), but most of the alkylating agents which have been tested for carcinogenic activity or as initiators for mouse skin papilloma formation have been less active than was anticipated from their mutagenicity (Roe, 1957; Clayson, 1962; Trainin *et al.*, 1964; Koller, 1969; Van Duuren and Sivak, 1968; Van Duuren, 1969). However, it has been generally appreciated that the high reactivity of these electrophilic reactants with water and various other nucleophilic extracellular tissue components reduces the amounts which penetrate to critical intracellular sites in mammalian tissues and thus prejudices their activity as carcinogens. Furthermore, many alkylating agents have received only limited tests for carcinogenic activity. The importance of the alkylating agents in studies on mutagenesis and carcinogenesis is that they are all electrophilic reactants. The electrophilic property of these structurally diverse compounds appears to be responsible for both the mutagenic and carcinogenic (even if weak) properties of those alkylating agents which possess both of these biological properties. The association of these two biological properties with the electrophilic nature of alkylating agents has become more important since it has become evident, as discussed above, that the reactive forms of chemical carcinogens also appear to be electrophils.

D. Potential Alkylating Agents

1. *Nitrosamides and Nitrosamines*

The comprehensive review of Magee and Barnes (1967) on the carcinogenicity, reactivity, and mutagenicity of the nitrosamides and nitrosamines obviates the need for a detailed examination of these compounds here. From their review and the subsequent literature, it is evident that the high carcinogenic activity of the nitrosamides is paralleled, where it has been tested, by potent mutagenic activity in nearly all of the cellular systems examined. These compounds, as exemplified by N-methyl-N-nitroso-N'-nitroguanidine and N-nitrosoethylurethan, require only reaction with sulfhydryl groups for conversion to very reactive carbonium ions or alkyl diazonium ions.

The strongly carcinogenic nitrosamines, on the other hand, have shown a much more limited range of mutagenic activity. In the extensive data compiled by Magee and Barnes (1967) the nitrosamines were found to be mutagenic only in tests with *Drosophila;* no mutagenic activity was noted in a number of assays which utilized fungi or bacteria. That the lack of mutagenicity in these systems was a consequence of the lack of metabolism of the nitrosamines is apparent from more recent studies. Thus, Malling (1966) showed that diethylnitrosamine and dimethylnitrosamine were each mutagenic for *Neurospora* if the conidia were suspended during treatment with the chemicals in the nonenzymatic hydroxylating system of Udenfriend *et al.* (1954). These conditions presumably caused oxidative dealkylation of some of the dialkylnitrosamine to the unstable monoalkylnitrosamine. Similarly, the mutagenicity of dimethylnitrosamine has been demonstrated in the host-mediated assay of Gabridge and Legator (1969) in which the test compound was administered to mice infected with *Salmonella typhimurium*, which was subsequently reisolated for assay of induced mutations.

2. Pyrrolizidine Alkaloids

Among the large group of pyrrolizidine alkaloids those with an allylic ester and, preferably, a second ester grouping are generally hepatotoxic to rats and other species (Schoental, 1963, 1968a,b; Bull *et al.*, 1968). The high toxicity of these compounds has made assays for carcinogenic activity difficult, but retrorsine, isatidine, monocrotaline, lasiocarpine, heliotrine, and seniciphylline have all induced liver carcinomas. Heliotrine and lasiocarpine are both mutagenic for *Aspergillus nidulans* (Alderson and Clark, 1966), and a number of the toxic pyrrolizidine alkaloids have also proved to be mutagenic for *D. melanogaster* (Clark, 1960, 1963; Brink, 1966, 1969). Thus, monocrotaline, lasiocarpine, and heliotrine have very strong mutagenic activity; each of their *N*-oxides is also mutagenic, although less active than the parent compounds. Other pyrrolizidine alkaloids with mutagenic activities in *Drosophila* similar to those of the latter three *N*-oxides are echinatine, echimidine, senecionine, supinine, jacobine, and platyphylline. Heliotric acid and heliotridine, hydrolysis products of heliotrine, exhibited little or no mutagenic activity. According to Culvenor *et al.* (1969), dehydroheliotrine and dehydrolasiocarpine also have very weak mutagenic activity for *Drosophila*. On the other hand, 7-hydroxy-1-hydroxymethyldihydro-5H-pyrrolizine, a pyrrole metabolite of heliotrine and lasiocarpine in the rat, showed mutagenic activity in both *Drosophila* and *Aspergillus*. Thus, while it is not possible to correlate mutagenic and carcinogenic activity for individual compounds of the pyrrolizidine alkaloid group, the structural requirements for these two activities, as well as for general toxicity, are very similar.

3. Cycasin

As discussed above, the carcinogenicity of cycasin, the β-methyl glucoside of methylazoxymethanol, is dependent on the hydrolysis of the glucosidic linkage to yield the proximate form, methylazoxymethanol, which decomposes readily to yield a carbonium ion. Except for the skin and subcutis of newborn rats (Spatz, 1968) rat tissues do not have detectable quantities of glucosidase activity, so that the carcinogenic activity of cycasin for rats depends on the β-glucosidase activity of the intestinal bacteria. Thus, cycasin is a potent carcinogen when administered orally to conventional rats, but not when administered parenterally to conventional rats or orally to germ-free rats, while methylazoxymethanol is carcinogenic under the latter conditions (Laqueur and Spatz, 1968). The mutagenicity data parallel the carcinogenicity data. Thus, when tested in *S. typhimurium* methylazoxymethanol, but not cycasin, was mutagenic (Smith, 1966). However, when *S. typhimurium* was used in the host-mediated assay, both cycasin and methylazoxymethanol were mutagenic when conventional mice were used as hosts and the compounds were administered orally; with germ-free mice or with the compounds administered parenterally only methylazoxymethanol was mutagenic (Gabridge and Legator, 1969; Legator and Epstein, in press).

E. Aromatic Amines and Amides

The data on the mutagenicity of the carcinogenic aromatic amines and amides and their derivatives are presented in Table 1. Despite the fact that many of the compounds listed show strong carcinogenic activity in one or more systems, the majority of the mutagenicity assays have given negative results.

Many of the positive mutagenicity results listed in Table 1 were obtained with esters of *N*-hydroxy amides or of *N*-hydroxy amines; these compounds are electrophilic reactants and appear to be at least prototypes of the ultimate carcinogenic forms *in vivo*. The majority of the positive results with these *N*-hydroxy esters were obtained with *B. subtilis* transforming DNA treated *in vitro*, but some of the esters were also mutagenic in systems utilizing T4 phage, *E. coli*, and *Drosophila*. *N*-Hydroxy-2-aminofluorene, *N*-hydroxy-1-aminonaphthalene, and *N*-hydroxy-2-aminonaphthalene were each mutagenic for *E. coli*, but none of the hydroxylamines tested showed mutagenic activity when reacted with transforming DNA *in vitro*. In general, the parent amines and amides and the *N*-hydroxyamides have not displayed much, if any, mutagenic activity in any system. The lack of mutagenic activity of these "precarcinogens" is in accord with the lack of reactivity of the amines and amides and very weak reactivity of the *N*-hydroxyamides and *N*-hydroxyamines with DNA under physiological conditions (Miller, J. A., and Miller, E. C., 1969*a,b*). Apparently, inadequate amounts of these precursors are metabolized to forms reactive with DNA in the organisms which have been used in the mutagenicity asays.

Two strongly mutagenic compounds (the sulfuric acid esters of *N*-hydroxy-2-acetylaminofluorene and of *N*-hydroxy-4-acetylaminobiphenyl) have not induced tumors when administered to rats, even though the former compound appears to be at least one of the ultimate carcinogenic metabolites of 2-acetylaminofluorene and its *N*-hydroxy metabolite in the rat liver (DeBaun *et al.*, 1970*a,b*; Weisburger *et al.*, 1970). In these cases the high reactivity (short half-lives) and ionic forms of the sulfuric acid

TABLE 1. The Mutagenic Activity of Derivatives of Carcinogenic Aromatic Amines and Amides and Related Compounds

Compound	Carcinogenicity		Mutagenicity		
	Activity	Reference[a]	Test system	Activity	References[b]
2-Acetylaminofluorene and related compounds					
2-Acetylaminofluorene	+	1	Drosophila	—	1
			D. melanogaster—sex-linked lethal mutations	—	2
			N. crassa—forward mutations	—	3
			E. coli—reverse mutations	—	4
			B. subtilis transforming DNA—forward mutations	—	5
2-Aminofluorene	+	1	B. subtilis transforming DNA—forward mutations	—	5
N-Hydroxy-2-aminofluorene	+	2	E. coli—reverse mutations	+	4
			B. subtilis transforming DNA—forward mutations	—	5
N-Hydroxy-2-acetylamino-fluorene	+	3, 4	E. coli—reverse mutations	—	4
			B. subtilis transforming DNA—forward mutations	—	5
N-Acetoxy-2-acetylamino-fluorene	+	2	D. melanogaster—sex-linked lethals and induction of viable X-chromosome fragments	—	2
			D. melanogaster—small deletions	+	2
			E. coli—reverse mutations	+	4
			B. subtilis transforming DNA—forward mutations	+	5
N-Benzoyloxy-2-acetylamino-fluorene	+	2	T4 phage in E. coli—forward mutations	+	6
			B. subtilis transforming DNA—forward mutations	+	5

Compound	Result	Ref	Test system	Result	Ref
2-Acetylaminofluorene-N-sulfate	−[c]	2	B. subtilis transforming DNA—forward mutations	+	5
7-Fluoro-N-acetoxy-2-acetylaminofluorene	n.t.[d]		T4 phage in E. coli—forward mutations	+	6
Glucuronide of N-hydroxy-2-acetylaminofluorene	±	5	T4 phage in E. coli—forward mutations	−	6
2-Nitrosofluorene	+	2	B. subtilis transforming DNA—forward mutations	−	5
			B. subtilis transforming DNA—forward mutations	−	5
1-Aminonaphthalene and related compounds					
1-Aminonaphthalene	+	6	*Drosophila*		1
			E. coli—reverse mutations	−	4, 7
N-Hydroxy-1-aminonaphthalene	+	7, 8, 9	E. coli—reverse mutations	+	4, 7, 8
			T4 phage in E. coli—forward mutations	−	6
N-Acetoxy-1-acetylaminonaphthalene	+	8	E. coli—reverse mutations	−	4
1-Amino-2-naphthol	+, −	6, 10	E. coli—reverse mutations	−	8
1-Amino-2-naphthyl sulfate	−	6	E. coli—reverse mutations	−	8
2-Aminonaphthalene and related compounds					
2-Aminonaphthalene	+	6	*Drosophila*		1
			D. melanogaster—sex-linked lethal mutations; viable X-chromosome fragments	−	2
			D. melanogaster—small deletions	+	2
			E. coli—reverse mutations	−	4, 7, 8
N-Hydroxy-2-aminonaphthalene	+	7, 8, 9, 11, 12	E. coli—reverse mutations	+	4, 7, 8
			T4 phage in E. coli—forward mutations	−	6
N-Acetoxy-2-acetylaminonaphthalene	+	8	E. coli—reverse mutations	−	4
2-Amino-1-naphthol	+, −	6, 12	E. coli—reverse mutations	−	8

TABLE 1. (Continued)

Compound	Carcinogenicity		Mutagenicity		
	Activity	Reference[a]	Test system	Activity	Reference[b]
2-Aminonaphthalene and related compounds					
Di-(2-amino-1-naphthyl)-hydrogen phosphate	+	6	E. coli—reverse mutations	−	8
4-Acetylaminobiphenyl and related compounds					
4-Acetylaminobiphenyl	+	13	B. subtilis transforming DNA—forward mutations	−	9
4-Aminobiphenyl	+	6	B. subtilis transforming DNA—forward mutations	−	9
N-Hydroxy-4-aminobiphenyl	n.t.		B. subtilis transforming DNA—forward mutations	−	9
N-Hydroxy-4-acetylamino-biphenyl	+	14	B. subtilis transforming DNA—forward mutations	−	9
N-Acetoxy-4-acetylamino-biphenyl	+	2	B. subtilis transforming DNA—forward mutations	+	9
4-Acetylaminobiphenyl-N-sulfate	−	2	B. subtilis transforming DNA—forward mutations	+	9
4-Acetylaminostilbene and related compounds					
4-Acetylaminostilbene	+	6, 15	B. subtilis transforming DNA—forward mutations	−	9
4-Dimethylaminostilbene	+	6	N. crassa—forward mutations	−	3
4-Aminostilbene	+	6	B. subtilis transforming DNA—forward mutations	−	9
N-Hydroxy-4-aminostilbene	+	15	B. subtilis transforming DNA—forward mutations	−	9
N-Hydroxy-4-acetylamino-stilbene	+	15, 16	B. subtilis transforming DNA—forward mutations	−	9

Compound	Result	Ref.	Test system	Result	Ref.
N-Acetoxy-4-acetylamino-stilbene	+	2	B. subtilis transforming DNA—forward mutations	−	9
2-Acetylaminophenanthrene and related compounds					
2-Acetylaminophenanthrene	+	17	B. subtilis transforming DNA—forward mutations	−	9
2-Aminophenanthrene	n.t.		B. subtilis transforming DNA—forward mutations	−	9
N-Hydroxy-2-aminophenanthrene	n.t.		B. subtilis transforming DNA—forward mutations	−	9
N-Hydroxy-2-acetylamino-phenanthrene	+	18	B. subtilis transforming DNA—forward mutations	−	9
N-Acetoxy-2-acetylamino-phenanthrene	+	2	B. subtilis transforming DNA—forward mutations	+	9
N,N-Dimethyl-4-aminoazobenzene and related compounds					
N,N-Dimethyl-4-aminoazo-benzene	+	19	Drosophila	−	1
3'-Methyl-N,N-dimethyl-4-aminoazobenzene	+	19	N. crassa—forward mutations	+	3
N-Methyl-4-aminoazobenzene	+	19	B. subtilis transforming DNA—forward mutations	−	5
N-Benzoyloxy-N-methyl-4-aminoazobenzene	+	20	B. subtilis transforming DNA—forward mutations	+	5
2,3'-Dimethyl-4-aminoazo-benzene	+	19	E. coli—forward mutations	+	10
N,N-Diethyl-4-aminoazo-benzene	−	19	D. melanogaster—sex-linked lethal mutations	+	11
4-Aminoazobenzene	−	19	D. melanogaster—sex-linked lethal mutations	−	1
Acridine orange	+	21, 22	D. melanogaster—sex-linked lethal mutations	−	1
			Bacteriophage T4	+	12

TABLE 1. (Continued)

Compound	Carcinogenicity		Mutagenicity		
	Activity	Reference[a]	Test system	Activity	References[b]
4-Nitroquinoline-1-oxide and related compounds					
4-Nitroquinoline-1-oxide	+	23	S. cerevisiae—respiration-deficient mutants	+	13, 14
			Tobacco mosaic virus—plaque variants	+	15
			Tobacco mosaic virus RNA—plaque variants	+	15
4-Hydroxyaminoquinoline-1-oxide	+	24	S. cerevisiae—respiration-deficient mutants	+	14

[a] References to carcinogenicity data: (1) Weisburger, and Weisburger (1958); (2) Miller, J. A., and Miller, E. C. (1969a); (3) Miller, E. C., et al. (1961); (4) Miller, E. C., et al. (1964); (5) Miller, E. C., et al. (1968); (6) Clayson (1962); (7) Belman et al. (1968); (8) Scribner et al. (unpublished data); (9) Boyland et al. (1964); (10) Irving et al. (1963); (11) Boyland et al. (1963); (12) Bryan et al. (1964); (13) Miller, E. C., et al. (1956); (14) Miller, J. A., et al. (1961); (15) Andersen et al. (1964); (16) Baldwin and Smith (1965); (17) Miller. J. A., et al. (1955); (18) Miller E. C., et al. (1966b); (19) Miller, J. A., and Miller, E. C. (1953); (20) Poirier et al. (1967); (21) Munn (1967); (22) Van Duuren et al. (1969); (23) Nakahara et al. (1957); (24) Shirasu (1965).

[b] References to mutagenicity data: (1) Demerc et al. (1949); (2) Fahmy and Fahmy (1970); (3) Barratt and Tatum (1951, 1958); (4) Mukai and Troll (1969); (5) Maher et al. (1968); (6) Corbett et al. (1970); (7) Belman et al. (1968); (8) Perez and Radomski (1965); (9) Maher et al. (1970); (10) Scherr et al. (1954); (11) Demerc (1948); (12) Orgel (1965); (13) Nagai (1969); (14) Epstein and St. Pierre (1969); (15) Endo et al. (1961).

[c] 2-Acetylaminofluorene-N-sulfate appears to be an ultimate carcinogenic metabolite of N-hydroxy-2-acetylaminofluorene in rat liver (DeBaun et al., 1970a,b).

[d] n.t. = not tested.

esters appear to prevent the transport of sufficient amounts of the compounds to critical sites in the cell before they are used up by reaction with
noncritical cellular and extracellular targets.

Acridine orange has potent mutagenic activity for T4 phage and weak
carcinogenic activity (Table 1), but the two activities are probably not related. The mutagenic activity may be dependent on the ability of acridine
to intercalate with DNA (Lerman, 1965). The carcinogenicity of the compound may well depend on the metabolism of the N-dimethylamino groups
to reactive forms in the same manner as for other aromatic amines.

Both 4-nitroquinoline-1-oxide and 4-hydroxyaminoquinoline-1-oxide
are potent carcinogenic and mutagenic compounds (Table 1). Since the nitro
compound is metablized to the hydroxylamine in the rat (Sugimura et al.,
1965; Matsushima et al., 1968)and the latter compound is the stronger
carcinogen (Endo and Kume, 1965; Shirasu, 1965), the carcinogenic activity of 4-nitroquinoline-1-oxide is now generally attributed to this metabolic conversion. However, the reported mutagenic activity of 4-nitroquino-
line-1-oxide for isolated tobacco mosaic virus RNA raises the question of
whether or not this reactive compound may also produce carcinogenic
effects directly.

F. Polycyclic Aromatic Hydrocarbons

While the mutagenicity assays on the carcinogenic hydrocarbons have
usually given negative results, most of these carcinogens did show some
activity in one or more of the test systems (Table 2). In the studies of Demerec
et al. (1949) and Zimmerman (1969a) no conclusion was possible with respect
to correlating mutagenicity and carcinogenicity, since both the carcinogenic
and noncarcinogenic hydrocarbons failed to induce mutations. The recent
studies of Chu et al. (1970) with Chinese hamster cells suggest that there
may be a correlation between these activities. Thus, the carcinogens 7,12-
dimethylbenz(a)anthracene and dibenz(a,h)anthracene were both mutagenic,
while the noncarcinogenic hydrocarbons benz(e)pyrene and dibenz(a,c)-
anthracene were both nonmutagenic. Benz(a)pyrene, which is carcinogenic
but not mutagenic, was the exception to the correlation.

The studies on the polycyclic hydrocarbons are limited by our lack of
knowledge with respect to their ultimate carcinogenic and reactive forms *in
vivo*. Thus, unless one considers that the carcinogenic hydrocarbons may
be mutagenic as a consequence of some physical interaction with DNA, all
of the mutagenicity tests would have required metabolic activation of the
hydrocarbons for a positive result. 7-Bromomethylbenz(a)anthracene, which
has so far been inactive in both mutagenicity and carcinogenicity assays, is
an electrophilic reactant, but its short half-life may reduce its probability of

TABLE 2. The Mutagenicity of Certain Carcinogenic Polycyclic Aromatic Hydrocarbons and Related Compounds

Compound	Carcinogenicity		Mutagenicity		
	Activity	References[a]	Test system	Activity	References[b]
3-Methylcholanthrene	+	1	*Drosophila*	+	1
			Drosophila	−	2, 3
			D. melanogaster—sex-linked lethal mutations	−	4, 5, 6
			N. crassa—forward mutations	+	7, 8
			N. crassa—forward and reverse mutations	−	9
			Neurospora—reverse mutations	−	10
			E. coli	−	11
			E. coli—forward mutations	+	12
			S. cerevisiae—morphological variants	+	13
			Mice	+[c]	14, 15
Dibenz(*a,h*)anthracene	+	1	*Drosophila*	−	2, 16
			Drosophila—sex-linked lethal mutations	−	4
			N. crassa—forward mutations	+	7, 8
			N. crassa—forward and reverse mutations	−	9
			S. cerevisiae—mitotic gene conversion	−	17
			Chinese hamster cells—forward mutations	+	18
Dibenz(*a,h*)anthracene-endosuccinate	+	2	*E. coli*—forward mutations	+	12
Dibenz(*a,c*)anthracene	−	1	Chinese hamster cells—forward mutations	−	18
Benz(*a*)pyrene	+	1	*Drosophila*—sex-linked lethal mutations	−	4
			E. coli—forward mutations·	+	12
			S. cerevisiae—mitotic gene conversion	−	17
			Mice—dominant lethal mutations	+	19
			Chinese hamster cells—forward mutations	−	18

Compound	[a]	Result	Test	[b]
Benz(e)pyrene	1	−	Chinese hamster cells—forward mutations	18
7,12-Dimethylbenz(a)-anthracene	1	+	S. cerevisiae—mitotic gene conversion	17
		+	N. crassa—forward mutations	7, 8
		+	D. melanogaster—small deletions	20
		+	Chinese hamster cells—forward mutations	18
7-Bromomethylbenz(a)-anthracene	3	−	D. melanogaster—sex-linked lethal mutations	6
Benz(a)anthracene	1	±	Drosophila—sex-linked lethal mutations	4
		−	S. cerevisiae—mitotic gene conversion	17
7-Carboxylic acid-12-methyl-benz(a)anthracene	n.t.[d]	−	D. melanogaster—sex-linked lethal mutations; X-chromosome fragments	6
		+	D. melanogaster—small deletions	6
7,12-Bis(hydroxymethyl)-benz(a)anthracene	2	+, −	E. coli—forward mutations	12
Pyrene	1	−	Drosophila—sex-linked lethal mutations	4
		−	S. cerevisiae—mitotic gene conversion	17
Anthracene	1	−	Drosophila—sex-linked lethal mutations	4
		−	E. coli—forward mutations	12
Phenanthrene	1	−	Drosophila—sex-linked lethal mutations	4

[a] References to carcinogenicity data: (1) Clayson (1962); (2) Hartwell (1951); (3) Dipple and Slade (in press).
[b] References to mutagenicity data: (1) Sacharov (1938); (2) Auerbach (1939–40); (3) Bhattacharya (1949); (4) Demerec et al. (1949); (5) Burdette (1952); (6) Fahmy and Fahmy (1970); (7) Barratt and Tatum (1951); (8) Barratt and Tatum (1958); (9) Burdette and Haddox (1954); (10) Jensen et al. (1951); (11) Latarjet et al. (1950); (12) Scherr et al. (1954); (13) Maisin et al. (1953); (14) Strong (1945); (15) Strong (1952); (16) Burdette (1955); (17) Zimmerman (1969a); (18) Chu et al. (1970); (19) Epstein and Shafner (1968); (20) Fahmy and Fahmy (1969).
[c] See criticism and discussion of these data by Burdette (1955).
[d] n.t. = not tested.

causing significant biological effects in cellular systems (Brookes and Dipple, 1967; Dipple and Slade, in press).

G. Urethan (Ethyl Carbamate)

Urethan has shown mutagenic activity in tests with Drosophila and E. coli, but not in studies which employed Neurospora or B. subtilis or in the dominant lethal test in mice (Table 3). However, the significance of these observations in terms of carcinogenesis is not clear, since i-propyl and n-propyl carbamate, which are weaker carcinogens than urethan, and n-butyl and i-amyl carbamate, which are not carcinogenic, had mutagenic activities similar to those of urethan. N-Hydroxy urethan, a metabolite of urethan with a carcinogenic potency similar to that of urethan, and N,O-diacetyl-N-hydroxy urethan, an acylating agent (Nery, 1969) with no known importance in carcinogenesis, were also mutagenic for Drosophila. Urethan is not chemically reactive under physiological conditions, and it seems necessary to expect that some metabolic derivative(s) are involved in both its carcinogenic and its mutagenic activity.

H. Aflatoxins

The potent hepatocarcinogenic aflatoxins (Wogan, 1966) have shown mutagenic activity in two tests. A mixture of aflatoxins B_1 and G_1 caused an eleven-fold increase in mutation frequency in the dominant lethal test in male mice (Epstein and Shafner, 1968). Aflatoxin B_1 was also mutagenic in the in vitro B. subtilis transforming DNA assay (Maher and Summers, 1970). The latter result demonstrated that aflatoxin B_1 can have a direct effect of biological importance on DNA. However, this direct mutagenic effect may not be directly related to its carcinogenic action. Thus, the idea that aflatoxin requires metabolic conversion to an active form can be deduced from the finding that aflatoxin hepatocarcinogenesis in rats is greatly inhibited by hypophysectomy (Goodall and Butler, 1969); hypophysectomy does not inhibit dimethylnitrosamine-induced hepatocarcinogenesis and thus apparently does not interfere with the initiation and growth of liver tumors (Lee and Goodall, 1968).

I. Mitomycin C

Mitomycin C is reduced in vivo to a bifunctional alkylating agent (Szybalski and Iyer, 1967). This antibiotic is active at low dosage as a carcinogen in the subcutaneous tissue of mice (Ikegami et al., 1967) and as a mutagen in E. coli (Iijima and Hagiwara, 1960) and in Habrobracon sperm (Smith, 1969).

TABLE 3. The Mutagenicity of Urethan and N-Hydroxyurethan and Related Compounds

Compound	Carcinogenicity[a]	Mutagenicity		
		Test system	Activity	References[b]
Urethan (ethyl carbamate)	+	D. melanogaster—sex-linked lethal mutations	+	1, 2
		E. coli	+	3
		E. coli—reverse mutations	+	4
		Neurospora—reverse mutations	−	5
		B. subtilis—reverse mutations	−	6
		Mice—dominant lethal mutations	−	7
Methyl carbamate	−	E. coli—reverse mutations	−	4
		B. subtilis—reverse mutations	−	6
Isopropyl carbamate	+	E. coli	+	3
		B. subtilis—reverse mutations	−	6
n-Propyl carbamate	+	E. coli	+	3
		E. coli—reverse mutations	+	4
		B. subtilis—reverse mutations	−	6
n-Butyl carbamate	−	E. coli	−	3
		E. coli—reverse mutations	+	4
Iso-amyl carbamate	−	E. coli—reverse mutations	+	4
N-Hydroxy urethan	+	D. melanogaster—sex-linked lethal mutations	+	2
		B. subtilis—reverse mutations	−	6
N,O-Diacetyl-N-hydroxy urethan	n.t.[c]	D. melanogaster—sex-linked lethal mutations; small chromosome deletions	+	2
		D. melanogaster—induction of viable X-chromosome fragments	−	2

[a] Mirvish (1968).
[b] References to mutagenicity data: (1) Vogt (1948); (2) Fahmy and Fahmy (1970); (3) Latarjet et al. (1950); (4) Demerec et al. (1951); (5) Jensen et al. (1951); (6) Giovanni-Donnelly et al. (1967); (7) Bateman (1967).
[c] n.t. = not tested.

J. Hydrazine

Hydrazine in relatively large doses causes the development of leukemia, reticulum cell sarcoma, and lung adenomas in mice (Juhasz, 1967). This compound, apparently through attack on pyrimidine bases, is also a weak mutagen for T4 phage (Freese *et al.*, 1961). Hydrazine is a strong nucleophil. It is not known if its activity *in vivo* involves this property or its conversion to a strong electrophil.

K. Metals

Data on the mutagenic activity of metals and their ions are too limited to permit much attempt at correlation between mutagenic and carcinogenic activities. Mn^{2+}, which appears not to have been tested for carcinogenic activity, has strong mutagenic activity for *E. coli* under some conditions (Demerec and Hanson, 1951). Fe^{2+} has mutagenic activity for *E. coli* (Demerec *et al.*, 1949); iron-containing compounds have induced tumors only under special conditions (e.g., often-repeated doses of iron dextran), and the role of iron in the induction of these tumors is not clear (Haddow and Horning, 1960; Grasso and Golberg, 1966). Salts of Co^{2+} and Pb^{2+} and calcium chromate, each of which is carcinogenic for rats (Clayson, 1962; Hueper and Conway, 1964), were not mutagenic when assayed in T4 phage grown in *E. coli* (Corbett *et al.*, 1970).

IV. CONCLUSIONS

Consideration of the foregoing data has suggested a general qualitative correlation between electrophilic reactivity, carcinogenicity, and mutagenicity wherever enough information is available on the nature of the probable carcinogenic metabolites. Thus, many alkylating agents possess all three of these properties, and it appears likely that the lack of demonstrable carcinogenic activity in a given test system for some alkylating agents is a consequence of their reaction with other nucleophils before they reach the critical target(s) for induction of neoplasia. The potential alkylating agents have generally shown mutagenic and carcinogenic activity under conditions which appear to permit their conversion to alkylating agents, and their lack of activity in other situations appears to be explicable in terms of a lack of opportunity for such conversion. Examples of this are found in the studies on the nitrosamides, nitrosamines, cycasin, and the pyrrolizidine alkaloids. Similarly, among the aromatic amine derivatives the esters of hydroxamic

acids have generally shown the greatest mutagenic activity; these compounds are electrophilic reactants and appear to be prototypes of the ultimate
carcinogenic metabolites *in vivo*. Some aromatic hydroxylamines are also
both carcinogenic and mutagenic. The natures of the reactive mutagenic
and carcinogenic metabolites of urethan and the polycylic aromatic hydrocarbons are not certain.

The numerous examples of carcinogenic compounds which have not
shown mutagenic activity have probably resulted in most cases from the
failure of these compounds to be metabolized in the mutagenicity systems
to reactive eletrophils. However, lack of demonstrable mutagenicity could
also result from an efficient "repair" of altered DNA, so that the premutational damage caused by reaction with the mutagen was eliminated before
the DNA was replicated (Kimball, 1968; Cleaver, 1969; Wolff and Scott,
1969). To the extent that any instance of chemical carcinogenesis results
from alteration of DNA, the efficiency of "repair" of the altered DNA could
also markedly affect the likelihood of tumor development. It should also be
noted that in those cases in which nonreactive compounds are carcinogenic
and mutagenic in different cellular systems there is no certainty that the
same electrophilic metabolite(s) are involved in both activities. Proof of the
similarity of the metabolites critical to the carcinogenic and mutagenic
activities would depend on careful metabolic studies under both conditions.

There are several mutagens (Freese, 1963; Orgel, 1965) which do not
appear to depend on electrophilic reactivity for mutagenic activity and
which, in the limited tests available, have not proven to be carcinogenic.
This category includes several base analogs and their nucleosides; 5-bromo-
and 5-iododeoxyuridine, 2-aminopurine, and 2,6-diaminopurine did not
initiate skin tumors in the mouse (Salaman and Roe, 1956; Trainin *et al.*,
1964). 5-Fluorouracil, 5-fluorouridine, 5-fluorodeoxyuridine, 5-bromo-
deoxyuridine, and 5-iodouridine did not induce tumors on injection into
newborn mice (Poirier, 1965). There are no data on whether or not these
base analogs were incorporated into the nucleic acids of the mouse tissues
under the conditions of these experiments. Hydroxylamine and nitrous acid
are each mutagenic for transforming DNA through attack on certain bases
(Freese *et al.*, 1961; Freese and Freese, 1966). Druckrey *et al.* (1963) found
no increase in tumor incidence in rats given large amounts of sodium nitrite
in the drinking water for long periods of time, but the stomach may have
been the only tissue of sufficient acidity to make the conditions at all comparable to those under which nitrite is mutagenic (Freese and Freese, 1966).
Hydroxylamine did not prove to be carcinogenic on injection into newborn
mice (Poirier, 1965).

The factors responsible for the development of "spontaneous" tumors
in animals not known to have been subjected to carcinogenic agents have
been a matter of conjecture. Endogenous and exogenous chemicals, latent or

integrated virus genomes, and radiations may all play a role. The recent finding of Malling and Cosgrove (in press) that the conidia of *Neurospora crassa* show a ten-fold increase in mutation frequency after 36 hr in untreated rats or mice may be circumstantial evidence for the importance of reactive endogenous mutagens. Thus, a reasonable explanation for the enhanced mutation frequency of conidia maintained *in vivo* is interaction with electrophilic reactants which arise either in metabolism or as a function of some phase of the environment.

The interpretation of any correlation between mutagenic and carcinogenic activity in terms of the mechanisms involved in chemical carcinogenesis is a much more difficult problem. Even a one-to-one correlation between these activities could not be taken as *a priori* evidence for the concept that carcinogenesis involves a mutagenic event. Thus, the many mutagenic and carcinogenic chemicals which appear to be biologically active via electrophilic intermediates can react with any available and suitably nucleophilic center. Such nucleophilic sites occur in proteins, RNA, and numerous low molecular weight molecules, as well as in DNA. Indeed, studies on the reactions of carcinogens with macromolecules in the target tissues have invariably shown carcinogen-binding to each of the macromolecules, and in most cases the same electrophilic intermediates from a given carcinogen appear to bind to each type of macromolecule. Nevertheless, the concept that the primary carcinogenic reaction is a mutagenic event remains an attractive one (see recent data and discussions of Malling and de Serres, 1969; Zimmerman, 1969*b*; and Fahmy and Fahmy, 1970). However, gross correlations between mutagenic activity and carcinogenic activity cannot define the possible relationship between these activities. The key role played by certain repressors in preventing the expression of the major share of integrated phage genomes (Szybalski *et al.*, 1969) and the great diversity of expression of a given genotype in the various tissues of higher plants and animals emphasize the important role which alterations in the expression of the host or viral genomes play in determining the character of a cell and its progeny. Determination of the locus of the change(s) involved in the origin of neoplasia remains a key problem.

In summary, it appears that many, and perhaps all, chemical carcinogens are potential mutagens. Similarly, many, but possibly not all, mutagens are potential carcinogens. In many instances, reactivity, metabolic, and permeability factors appear to have obscured correlations of mutagenic and carcinogenic activities. These potentialities and complications are consequences of the electrophilic nature of the ultimate carcinogenic and mutagenic forms of many carcinogenic and mutagenic chemicals. From this viewpoint, the ability of these electrophils to react with nucleophils in many cell constituents other than DNA makes it impossible to support a somatic mutation theory of carcinogenesis by a chemical *only* by gross correlations between carcinogenesis and mutagenesis. From a practical point of view, priorities for testing technologically and medically important chemicals for carcinogenicity in the present necessary, but time-consuming, mammalian assays might be usefully determined in part from the results of short and relevant mutagenicity tests such as the mammalian host-mediated assays. The latter assays may eventually prove of value in predicting the carcinogenicity of chemicals.

V. REFERENCES

Abell, C. W., Falk, H. L., Shimkin, M. B., Weisburger, E. K., Weisburger, J. H., and Gubareff, N. (1965), *Science 147*, 1443.

Alderson, T., and Clark, A. M. (1966), *Nature 210*, 593.

Andersen, R. A., Enomoto, M., Miller, E. C., and Miller, J. A. (1964), *Cancer Res. 24*, 128.

Auerbach, C. (1939–40), *Proc. Roy. Soc. Edinburgh, Ser. B. 60*, 164 (quoted by Burdette, W. J. (1955), *Cancer Res. 15*, 201).

Auerbach, C. (1967), *Science 158*, 1141.

Auerbach, C., and Robson, J. M. (1944), *Nature 154*, 81.

Auerbach, C., and Robson, J. M. (1946), *Nature 157*, 302.

Bauer, K. H. (1928), "Mutationstheorie der Geschwulst-Enstehung. Übergang von Körperzellen in Geschwulstzellen durch Gen-Änderung," Springer-Verlag, Berlin.

Bauer, K. H. (1963) "Das Krebsproblem," Springer-Verlag, Berlin.

Baldwin, R. W., and Smith, W. R. D. (1965), *Brit. J. Cancer 19*, 433.

Barratt, R. W., and Tatum, E. L. (1951), *Cancer Res. 11*, 234.

Barratt, R. W., and Tatum, E. L. (1958), *Ann. N.Y. Acad. Sci. 71*, 1072.

Bateman, A. J. (1967), *Mutation Res. 4*, 710.

Belman, S., Troll, W., Teebor, G., and Mukai, F. (1968), *Cancer Res. 28*, 535.

Bhattacharya, S. A. (1949), *Proc. Zool. Soc. Bengal 2*, 187 (quoted by Burdette, W. J. (1955), *Cancer Res. 15*, 201).

Boutwell, R. K., Colburn, N. H., and Muckerman, C. C. (1969), *Ann. N.Y. Acad. Sci. 163*, 751.

Boveri, T. (1929), "The Origin of Malignant Tumors" (first published Jena, 1914), Williams and Wilkins Co., Baltimore.

Boyland, E. (1964), *Brit. Med. Bull. 20*, 121.

Boyland, E. (1969), *in* "The Jerusalem Symposia on Quantum Chemistry and Biochemistry. Vol. I. Physico-Chemical Mechanisms of Carcinogenesis" (E. D. Bergmann and B. Pullman, eds.) p. 25, Israel Academy of Sciences and Humanities, Jerusalem.

Boyland, E., and Nery, R. (1965), *Biochem. J. 94*, 198.

Boyland, E., and Sims, P. (1962), *Biochem. J. 84*, 571.

Boyland, E., and Sims, P. (1964), *Biochem. J. 91*, 493.

Boyland, E., and Sims, P. (1965a), *Biochem. J. 95*, 780.

Boyland, E., and Sims, P. (1965b), *Biochem. J. 95*, 788.

Boyland, E., and Sims, P. (1965c), *Biochem. J. 97*, 7.

Boyland, E., and Sims, P. (1967), *Internat. J. Cancer 2*, 500.

Boyland, E., and Williams, K. (1969), *Biochem. J. 111*, 121.

Boyland, E., Dukes, C. E., and Grover, P. L. (1963), *Brit. J. Cancer 18*, 575.
Boyland, E., Busby, E. R., Dukes, C. E., Grover, P. L., and Manson, D. (1964), *Brit. J. Cancer 19*, 575.
Boyland, E., Sims, P., and Huggins, C. (1965), *Nature 207*, 816.
Brink, N. G. (1966), *Mutation Res. 3*, 66.
Brink, N. G. (1969), *Mutation Res. 8*, 139.
Brookes, P., and Dipple, A. (1967), "British Empire Cancer Campaign 45th Annual Report," Part II, p. 20.
Brookes, P., and Dipple, A. (1969), *in* "The Jerusalem Symposia on Quantum Chemistry and Biochemistry. Vol. I. Physico-Chemical Mechansisms of Carcinogenesis" (E. D. Bergmann and B. Pullman, eds.) p. 139, Israel Academy of Sciences and Humanities, Jerusalem.
Brookes, P., and Lawley, P. D. (1960), *Biochem. J. 77*, 478.
Brookes, P., and Lawley, P. D. (1964), *J. Cell. Comp. Physiol. 64* (Suppl. 1), 111.
Brown, G. B., Sugiura, K., and Cresswell, R. M. (1965), *Cancer Res. 25*, 986.
Bryan, G. T., Brown, R. R., and Price, J. M. (1964), *Cancer Res. 24*, 596.
Büch, H., Rummel, W., Pfleger, K., Eschrich, C., and Texter, N. (1968), *Naunyn-Schmiedebergs Arch. Pharm. Exp. Pathol. 259*, 276.
Bull, L. B., Culvenor, C. C. J., and Dick, A. T. (1968), "The Pyrrolizidine Alkaloids," North Holland Publishing Co., Amsterdam.
Burdette, W. J. (1952), *Cancer Res. 12*, 201.
Burdette, W. J. (1955), *Cancer Res. 15*, 201.
Burdette, W. J., and Haddox, C. H., Jr. (1954), *Cancer Res. 14*, 163.
Butler, T. C. (1961), *J. Pharmacol. Exp. Therap. 134*, 311.
Cessi, C., Colombini, C., and Mameli, L. (1966), *Biochem. J. 101*, 46c.
Chu, E. H. Y., Bailiff, E. E., and Malling, H. V. (1970), "Abstracts Tenth International Cancer Congress," p. 62, Houston.
Clark, A. M. (1960), *Z. Vererb. 91*, 74.
Clark, A. M. (1963), *Z. Vererb. 94*, 115.
Clayson, D. B. (1962), "Chemical Carcinogenesis," Little, Brown Co., Boston.
Cleaver, J. E. (1969), *Radiation Res. 37*, 334.
Colburn, N. H., and Boutwell, R. K. (1968a), *Cancer Res. 28*, 642.
Colburn, N. H., and Boutwell, R. K. (1968b), *Cancer Res. 28*, 653.
Corbett, T. H., Dove, W. F., and Heidelberger, C. (1970), *Mol. Pharmacol.*, in press.
Costa, A., Weber, G., and St. Omer, F. B. (1963), *Arch. de Vecchi Anat. Path. 39*, 357.
Cramer, J. W., Miller, J. A., and Miller, E. C. (1960), *J. Biol. Chem. 235*, 250.
Culvenor, C. C. J., Dann, A. T., and Dick, A. T. (1962), *Nature 195*, 570.
Culvenor, C. C. J., Downing, D. T., Edgard, J. A., and Jago, M. V. (1969), *Ann. Ann. N.Y. Acad. Sci. 163*, 837.
DeBaun, J. R., Miller, E. C., and Miller, J. A. (1970a), *Cancer Res.*, *30*, 577.
DeBaun, J. R., Smith, J. Y. R., Miller, E. C., and Miller, J. A. (1970b), *Science 167*, 184.
Della Porta, G., Terracini, B., and Shubik, P. (1961), *J. Nat. Cancer Inst. 26*, 855.
Demerec, M. (1948), "Proceedings of the Eighth International Congress on Genetics," p. 201, *Hereditas* Supplement.
Demerec, M., and Hanson, J. (1951), *Cold Spring Harbor Symp. Quant. Biol. 16*, 215.
Demerec, M., Wallace, B., Witkin, E. M., and Bertani, G. (1949), *Carnegie Inst. Wash. Year Book 48*, 156.
Demerec, M., Bertani, G., and Flint, J. (1951), *Am. Nat. 85*, 119.
Dipple, A., and Slade, T. A. *Europ. J. Cancer* (in press).

Druckrey, H., Steinhoff, D., Beuthner, H., Schneider, H., and Klärner, P. (1963), *Arzneimittel-Forsch. 13*, 320.

Druckrey, H., Preussmann, R., Ivankovic, S., Schmähl, D., Afkham, J., Blum, G., Mennel, H. D., Muller, M., Petropoulos, P., and Schneider, H. (1967), *Z. Krebsforsch. 69*, 103.

Druckrey, H., Kruse, H., and Preussmann, R. (1968), *Naturwissenschaften 55*, 449.

Druckrey, H., Preussmann, R., and Ivankovic, S. (1969), *Ann. N.Y. Acad. Sci. 163*, 676.

Endo, H. (1958), *Gann 49*, 151.

Endo, H., and Kume, F. (1965), *Gann 56*, 261.

Endo, H., Wada, A., Miura, K. I., Hidaka, Z., and Hiruki, C. (1961), *Nature 190*, 833.

Enomoto, M., Sato, K., Miller, E. C., and Miller, J. A. (1968), *Life Sci. 7*, 1025.

Epstein, S. S. (1970), *in* "Environmental Chemical Mutagens—Principles and Methods for Their Detection" (A. Hollander, ed.) Plenum Press, New York.

Epstein, S. S., and St. Pierre, J. A. (1969), *Toxicol. Appl. Pharmacol. 15*, 451.

Epstein, S. S., and Shafner, H. (1968) *Nature 219*, 385.

Fahmy, O. G., and Fahmy, M. J. (1969), *Nature 224*, 1328.

Fahmy, O. G., and Fahmy, M. J. (1970), *Cancer Res. 30*, 195.

Farber, E. (1963), *Adv. Cancer Res. 7*, 383.

Farber, E., McConomy, J., Franzen, B., Marroquin, F., Stewart, G. A., and Magee, P. N. (1967), *Cancer Res. 27*, 1761.

Fowler, J. S. L. (1969), *Brit. J. Pharmacol. 36*, 181P.

Freese, E. (1963), *in* "Molecular Genetics" (J. H. Taylor, ed.) Part I, Chapter V, Academic Press, New York.

Freese, E. (1970), *in* "Environmental Chemical Mutagens—Principles and Methods for Their Detection" (A. Hollander, ed.) Plenum Press, New York.

Freese, E., and Freese, E. B. (1966), *Radiation Res. 6* (Suppl.), 97.

Freese, E., and Strack, H. B. (1962). *Proc. Nat. Acad. Sci., 48*, 1796.

Freese, E., Bautz, F., and Freese, E. B. (1961), *Proc. Nat. Acad. Sci. 47*, 845.

Fried, J., and Schumm, D. E. (1967), *J. Am. Chem. Soc. 89*, 5508.

Gabridge, M. G., and Legator, M. S. (1969), *Proc. Soc. Exp. Biol. Med. 130*, 831.

Gelboin, H. V. (1967), *Adv. Cancer Res. 10*, 1.

Gelboin, H. V. (1969), *Cancer Res. 29*, 1272.

Giovanni-Donnelly, R. de, Kolbye, S. M., and DiPaolo, J. A. (1967), *Mutation Res. 4*, 543.

Goldschmidt, B. M., Blazej, T. P., and Van Duuren, B. L. (1968), *Tetrahedron Letters*, p. 1583.

Goodall, C. M., and Butler, W. H. (1969), *Internat. J. Cancer 4*, 422.

Grasso, P., and Golberg, L. (1966), *Food Cosmet. Toxicol. 4*, 297.

Grover, P. L., and Sims, P. (1969), *Biochem. J. 110*, 159.

Grover, P. L., and Sims, P. (1970), *Biochem. Pharmacol. 19*, 2251.

Haddow, A., and Horning, E. S. (1960), *J. Nat. Cancer Inst. 24*, 109.

Hartwell, J. L. (1951), "Survey of Compounds Which Have Been Tested for Carcinogenic Activity," U.S. Public Health Service Publication No. 149.

Heidelberger, C. (1964), *J. Cell. Comp. Physiol. 64* (Suppl. 1), 129.

Hueper, W. C., and Conway, W. D. (1964), "Chemical Carcinogenesis and Cancers," Charles C Thomas, Springfield, Ill.

Iijima, T., and Hagiwara, A. (1960), *Nature 185*, 395.

Ikegami, R., Akamatsu, Y., and Haruta, M (1967), *Acta Pathol. Japan 17*, 495.

Innes, J. R. M., Ulland, B. M., Valerno, M. G., Petrucelli, L., Fishbein, L., Hart, E. R., Pallotta, A. J., Bates, R. R., Falk, H. L., Gart, J. J., Klein, M., Mitchell, I., and Peters, J. (1969), *J. Nat. Cancer Inst. 42*, 1101.

Irving, C. C. (1962), *Cancer Res. 22*, 867.

Irving, C. C. (1965), *J. Biol. Chem. 240*, 1011.

Irving, C. C., and Russell, L. T. (1970), *Biochemistry 9*, 2471.

Irving, C. C., and Veazey, R. A. (1969), *Cancer Res. 29*, 1799.

Irving, C. C., Gutmann, H. R., and Larson, D. M. (1963), *Cancer Res. 23*, 1782.

Irving, C. C., Wiseman, R., Jr., and Hill, J. T. (1967), *Cancer Res. 27*, 2309.

Irving, C. C., Veazey, R. A., and Hill, J. T. (1969a), *Blochim. Biophys. Acta 179*, 189.

Irving, C. C., Veaszey, R. A., and Russell, L. T. (1969b), *Chemico-Biol. Interactions 1*, 19.

Jensen, K. A., Kirk, I., Kolmark, G., and Westergaard, M. (1951), *Cold Spring Harbor Symp. Quant. Biol. 16*, 245.

Juhasz, J. (1967), *in* "Potential Carcinogenic Hazards from Drugs" (R. Truhaut, ed.), U.I.€.C. Monograph, Vol. 7, p. 180 Springer-Verlag, Berlin.

Kimball, R. F. (1968), *Photochem. Photobiol. 8*, 515.

Koller, P. C. (1969), *Mutation Res. 8*, 199.

Krieg, D. R. (1963), *Prog. Nucleic Acid Res. 2*, 125.

Kriek, E. (1968), *Biochim. Biophys. Acta 161*, 273.

Kriek, E. (1969), *Chemico-Biol. Interactions 1*, 3.

Kriek, E., Miller, J. A., Juhl, U., and Miller, E. C. (1967), *Biochemistry 6*, 117.

Krüger, F. W., Preussmann, R., and Niepelt, N. (1970), "Abstracts, Tenth International Cancer Congress," p. 4, Houston.

Laqueur, G. L., and Matsumoto, H. (1966), *J. Nat. Cancer Inst. 37*, 217.

Laqueur, G. L., and Spatz, M. (1968), *Cancer Res. 28*, 2262.

Laqueur, G. L., McDaniel, E. G., and Matsumoto, H. (1967), *J. Nat. Cancer Inst. 39*, 355.

Latarjet, R., Buu-Hoi, N. P., and Elias, C. A. (1950), *Publ. Staz. Zool. Napoli 22*, 78 (quoted by Burdette, W. J. (1955), *Cancer Res. 15*, 201).

Lawley, P. D. (1966), *Prog. Nucleic Acid Res. 5*, 89.

Lawley, P. D., and Brookes, P. (1963), *Biochem. J. 89*, 127.

Lawley, P. D., and Thatcher, C. J. (1970), *Biochem. J. 116*, 693.

Lee, K. Y., and Goodall, C. M. (1968), *Biochem. J. 106*, 767.

Legator, M., and Epstein, S. S. (in press), *in* "Non-psychiatric Hazards of Drugs of Abuse" (S. S. Epstein and J. Lederberg, eds.) National Institutes of Mental Health, U.S. Government Printing Office.

Legator, M. S., and Malling, H. V. (1971), *in* "Chemical Mutagens—Principles and Methods for Their Detection (A. Hollaender, ed.) Plenum Press, New York.

Lerman, L. S. (1965), *J. Cell. Comp. Physiol. 64* (Suppl. 1), 1.

Lijinsky, W., and Ross, A. E. (1969), *J. Nat. Cancer Inst. 42*, 1095.

Lin, J.-K., Miller, J. A., and Miller, E. C. (1968), *Biochemistry 7*, 1889.

Lin, J.-K., Miller, J. A., and Miller, E. C. (1969), *Biochemistry 8*, 1573.

Lotlikar, P. D., Scribner, J. D., Miller, J. A., and Miller, E. C. (1966), *Life Sci. 5*, 1263.

Loveless, A. (1969), *Nature 223*, 206.

Magee, P. N., and Barnes, J. M. (1967), *Adv. Cancer Res. 10*, 163.

Magee, P. N., and Schoental, R. (1964), *Brit. Med. Bull. 20*, 102.

Maher, V. M., and Summers, W. C. (1970), *Nature 225*, 68.

Maher, V. M., Miller, E. C., Miller, J. A., and Szybalski, W. (1968), *Mol. Pharmacol. 4*, 411.

Maher, V. M., Miller, E. C., Miller, J. A., and Summers, W. C. (1970), *Cancer Res., 30*, 1473.

Maisin, J. H., Lambert, G., and Van Duyse, E. (1953), *Acta Unio Internat. Contra Cancrum 9*, 693.

Malling, H. V. (1966), *Mutation Res. 3*, 537.

Malling, H. V., and Cosgrove, G. E., "Proc. Symposium: Chemische Mutagenese bei Sauger und Mensch. Test Systeme und Ergebnisse," Mainz, Germany (October 5–8, 1969) (in press).

Malling, H. V., and de Serres, F. J. (1969), *Ann. N.Y. Acad. Sci. 163*, 788.

Matsumoto, H., and Higa, H. H. (1966), *Biochem. J. 98*, 20C.

Matsushima, T., Kobuna, I., and Sugimura, T. (1967), *Nature 216*, 508.

Matsushima, T., Kobuna, I., Fukuoka, F., and Sugimura, T. (1968), *Gann 59*, 247.

Mattocks, A. R. (1968), *Nature 217*, 723.

Miller, E. C., and Miller, J. A. (1966), *Pharmacol. Rev. 18*, 805.

Miller, E. C., and Miller, J. A. (1967), *Proc. Soc. Exp. Biol. Med. 124*, 915.

Miller, E. C., Sandin, R. B., Miller, J. A., and Rusch, H. P. (1956), *Cancer Res. 16*, 525.

Miller, E. C., Miller, J. A., and Hartmann, H. A. (1961), *Cancer Res. 21*, 815.

Miller, E. C., Miller, J. A., and Enomoto, M. (1964), *Cancer Res. 24*, 2018.

Miller, E. C., Juhl, U., and Miller, J. A. (1966a), *Science 153*, 1125.

Miller, E. C., Lotlikar, P. D., Pitot, H. C., Fletcher, T. L., and Miller, J. A. (1966b), *Cancer Res. 26*, 2239.

Miller, E. C., Lotlikar, P. D., Miller, J. A. Butler, B. W., Irving, C. C., and Hill, J. T. (1968), *Mol. Pharmacol. 4*, 147.

Miller, E. C., Smith, J. Y., and Miller, J. A. (1970), *Proc. Am. Assoc. Cancer Res. 11*, 56.

Miller, J. A. (1970), *Cancer Res., 30*, 559.

Miller, J. A., and Miller, E. C. (1953), *Adv. Cancer Res. 1*, 339.

Miller, J. A., and Miller, E. C. (1969a), *Prog. Exp. Tumor Res. 11*, 273.

Miller, J. A., and Miller, E. C. (1969b), *in* "The Jerusalem Symposia on Quantum Chemistry and Biochemistry. Vol. I. Physico-Chemical Mechanisms of Carcinogenesis" (E. D. Bergman and B. Pullman, eds.) p. 237, The Israel Academy of Sciences and Humanities, Jerusalem.

Miller, J. A., Sandin, R. B., Miller, E. C., and Rusch, H. P. (1955), *Cancer Res. 15*, 188.

Miller, J. A., Wyatt, C. S., Miller, E. C., and Hartmann, H. A. (1961), *Cancer Res. 21*, 1465.

Miller, J. A., Lin, J.-K., and Miller, E. C. (1970), *Proc. Am. Assoc. Cancer Res. 11*, 56.

Mirvish, S. S. (1968), *Adv. Cancer Res. 11*, 1.

Morreal, C. E., Dao, T. L., Eskins, K., King, C. L., and Dienstag, J. (1969), *Biochim. Biophys. Acta 169*, 224.

Mukai, F., and Troll, W. (1969), *Ann. N.Y. Acad. Sci. 163*, 828.

Munn, A. (1967), *in* "Bladder Cancer, A Symposium" (W. B. Deichmann, K. F. Lampe, R. A. Penalver, A. Soto, and J. L. Radomski, eds.) p. 187, Aesculapius Publishing Co., Birmingham, Ala.

Nagai, S. (1969), *Mutation Res. 7*, 333.

Nagata, Y., and Matsumoto, H. (1969), *Proc. Soc. Exp. Biol. Med. 132*, 383.

Nakahara, W., Fukuoka, F., and Sugimura, T. (1957), *Gann 48*, 129.

Nery, R. (1969), *J. Chem. Soc.*, p. 1860.

Orenstein, J. M., and Marsh, W. H. (1968), *Biochem. J. 109*, 697.

Orgel, L. E. (1965), *Adv. Enzymol. 27*, 289.

Ortwerth, B. J., and Novelli, G. D. (1969), *Cancer Res. 29*, 380.

Perez, G., and Radomski, J. L. (1965), *Ind. Med. Surg. 34*, 714.

Pitot, H. C., and Heidelberger, C. (1963), *Cancer Res. 23*, 1694.

Poirier, L. A. (1965), Ph.D. thesis, University of Wisconsim, Madison, Wis.

Poirier, L. A., Miller, J. A., Miller, E. C., and Sato, K. (1967), *Cancer Res. 27*, 1600.

Preussman, R., Druckrey, H., and Bücheler, J. (1968), *Z. Krebsforsch. 71*, 63.

Preussman, R., Druckrey, H., Ivankovic, S., and Von Hodenberg, A. (1969a), *Ann. N.Y. Acad. Sci. 163*, 697.

Preussman, R., Von Hodenberg, A., and Hengy, H. (1969b), *Biochem. Pharmacol. 18*, 1.

Price, C. C., Gaucher, G. M., Koneru, P., Shibakawa, R., Sowa, J. R., and Yamaguchi, M. (1968), *Biochim. Biophys. Acta 166*, 327.

Rao, K. S., and Recknagel, R. O. (1969), *Exp. Mol. Pathol. 10*, 219.

Reuber, M. D., and Glover, E. L. (1970), *J. Nat. Cancer Inst. 44*, 419.

Reynolds, E. S. (1967), *J. Pharmacol. Exp. Therap. 155*, 117.

Roe, F. J. C. (1957), *Cancer Res. 17*, 64.

Rosen, L. (1968), *Biochem. Biophys. Res. Commun. 33*, 546.

Ross, W. C. J. (1962), "Biological Alkylating Agents," Butterworth and Co., London.

Sacharov, V. V. (1938), *Biol. Zh. 7*, 595 (quoted by Burdette, W. J. (1955), *Cancer Res. 15*, 201).

Salaman, M. H., and Roe, F. J. C. (1956), *Brit. J. Cancer 10*, 363.

Scherr, G. H., Fishman, M., and Weaver, R. H. (1954), *Genetics 39*, 141.

Schoental, R. (1963), *Bull. World Health Org. 29*, 823.

Schoental, R. (1968a), *Cancer Res. 28*, 2237.

Schoental, R. (1968b), *Israel J. Med. Sci. 4*, 1133.

Scribner, J. D., Miller, E. C., and Miller, J. A., Unpublished data.

Scribner, J. D., Miller, J. A., and Miller, E. C. (1965), *Biochem. Biophys. Res. Commun. 20*, 560.

Scribner, J. D., Miller, J. A., and Miller, E. C. (1970), *Cancer Res., 30*, 1570.

Shapiro, R. (1968), *Prog. Nucleic Acid Res. Mol. Biol. 8*, 73.

Shapiro, R. (1969), *Ann. N.Y. Acad. Sci. 163*, 624.

Shirasu, Y. (1965), *Proc. Soc. Exp. Biol. Med. 118*, 812.

Singer, B., and Fraenkel-Conrat, H. (1969), *Prog. Nucleic Acid Res. Mol. Biol. 9*, 1.

Smith, D. W. E. (1966), *Science 152*, 1273.

Smith, R. H. (1969), *Mutation Res. 7*, 231.

Spatz, M. (1968), *Proc. Soc. Exp. Biol. Med. 128*, 1005.

Spatz, M. (1969), *Ann. N.Y. Acad. Sci. 163*, 848.

Stekol, J. A. (1963), *Adv. Enzymol. 25*, 369.

Stekol, J. A. (1965), *in* "Transmethylation and Methionine Biosynthesis" (S. K. Shapiro and F. Schlenk, eds.) p. 231, University of Chicago Press, Chicago.

Stöhrer, G., and Brown, G. B. (1970), *Science 167*, 1622.

Stöhrer, G., Corbin, E., and Brown, G. B. (1970), *Proc. Am. Assoc. Cancer Res. 11*, 76.

Strong, L. C. (1945), *Proc. Nat. Acad. Sci. 31*, 290.

Strong, L. C. (1952), *Cancer Res. 12*, 300.

Sugimura, T., Okabe, K., and Endo, H. (1965), *Gann 56*, 489.

Sugiura, K., Teller, M. N., Parham, J. C., and Brown, G. B. (1970), *Cancer Res. 30*, 184.

Swann, P. F., and Magee, P. N. (1968), *Biochem. J. 110*, 39.

Swann, P. F., and Magee, P. N. (1969), *Nature 223*, 947.

Szybalski, W., and Iyer, V. N. (1967), *in* "Antibiotics" (D. Gottlieb and P. D. Shaw, eds.) Vol. I, p. 211, Springer-Verlag, Berlin.

Szybalski, W., Bovre, K., Fiandt, M., Guha, A., Hradecna, Z., Kumar, S., Lozeron, H. A., Maher, V. M., Nijkamp, H. J. J., Summers, W. C., and Taylor, K. (1969), *J. Cell. Physiol. 74* (Suppl. 1), 33.

Tada, M., Tada, M., and Takahashi, T. (1967), *Biochem. Biophys. Res. Commun.* *29*, 469.

Trainin, N., Kaye, A. M., and Berenblum, I. (1964), *Biochem. Pharmacol. 13*, 263.

Ts'o, P. O. P., Lesko, S. A., and Umans, R. S. (1969), *in* "The Jerusalem Symposia on Quantum Chemistry and Biochemistry. Vol. I. Physico-Chemical Mechanisms of Carcinogenesis" (E. D. Bergman and B. Pullman, eds.) p. 106, Israel Academy of Sciences and Humanities, Jerusalem.

Udenfriend, S., Clark, C. T., Axelrod, J., and Brodie, B. B. (1954), *J. Biol. Chem.* *208*, 731.

Van Duuren (1969), *Ann. N.Y. Acad. Sci. 163*, 633.

Van Duuren, B. L., and Sivak, A. (1968), *Cancer Res. 28*, 2349.

Van Duuren, B. L., Langseth, L., Goldschmidt, B. M., and Orris, L. (1967), *J. Nat. Cancer Inst. 39*, 1217.

Van Duuren, B. L., Sivak, A., Katz, C., and Melchionne, S. (1969), *Brit. J. Cancer* *23*, 587.

Vogt, M. (1948), *Experientia 4*, 68.

Weinstein, I. B. (1969), *in* "Genetic Concepts and Neoplasia, Twenty-third Annual Symposium on Fundamental Cancer Research," p. 380, Williams and Wilkins, Baltimore.

Weisburger, E. K., and Weisburger, J. H. (1958), *Adv. Cancer Res. 5*, 331.

Weisburger, J. H., Yamamoto, R. S., Grantham, P. H., and Weisburger, E. K. (1970), *Proc. Am. Assoc. Cancer Res. 11*, 82.

Wilk, M., and Girke (1969), *in* "The Jerusalem Symposia on Quantum Chemistry and Biochemistry. Vol. I. Physico-Chemical Mechanisms of Carcinogenesis" (E. D. Bergmann and B. Pullman, eds.) p. 91, Israel Academy of Sciences and Humanities, Israel.

Wilk, M., and Hoppe, U. (1970), *Liebigs Ann. Chem. 727*, 81.

Wogan, G. N. (1966), *Bacteriol. Rev. 30*, 460.

Wölcke, U., Birdsall, N. J. M., and Brown, G. B. (1969), *Tetrahedron Letters 10*, 785.

Wolff, S., and Scott, D. (1969), *Exp. Cell Res. 55*, 9.

Yamamoto, R. S., Glass, R. M., Fankel, H. H., Weisburger, E. K., and Weisburger, J. H. (1968), *Toxicol. Appl. Pharmacol. 13*, 108.

Zimmerman, F. K. (1969a), *Z. Krebsforsch. 72*, 65.

Zimmerman, F. K. (1969b), *Arzneimittel-Forsch. 7*, 1046.

CHAPTER 4

Effects on DNA: Chemical Methods

P. Brookes and
P. D. Lawley

Chester Beatty Research Institute
Institute of Cancer Research: Royal Cancer Hospital
London, England

I. INTRODUCTION

The general acceptance of the role of DNA as the primary genetic material, and the knowledge of its structure and mechanism of replication following from the Crick and Watson model, leads to the postulate that any chemical modification of DNA is potentially mutagenic.

In this chapter we summarize the methods used for DNA isolation and for the subsequent identification of the chemical changes induced by mutagens.

II. ISOLATION OF DNA

Many workers have described the reactions of mutagens with DNA *in vitro*, but it is obviously necessary to show that such reactions can result from treatment of biological material suitable for the detection of genetic changes. The first essential of such an approach is to isolate the DNA from the treated organism or tissue. In general, the procedures can be considered as three stages: the rupture of cell walls or membranes; the release of DNA from its complex with proteins; and the final purification of the DNA of likely impurities such as RNA, protein, and carbohydrate.

With bacteriophages, release of nucleic acid is readily achieved by solubilization of the protein coat with phenol. This property of phenol to dissolve most proteins provides the basis of several widely used procedures, as discussed and reviewed by Kirby (1964, 1967).

Marmur (1961) described a general method for isolation of DNA from bacteria, in which, following lysis of the cells, DNA is released by detergent, and protein is rendered insoluble by chloroform-octanol denaturation. The principles involved in this method are essentially the same as used and described in detail by Zamenhof (1958) for isolation of DNA from mammalian tissues. Marmur discusses the problem of Gram-positive bacteria which are not lysed by detergent. In these cases, enzymes (e.g., lysozyme) or, failing this, mechanical methods (e.g., sonication or use of a pressure cell) must be employed. These and other related problems have been reviewed by Tittensor and Walker (1967).

A particular problem arose in the isolation of DNA from sperm (bull or human) when an initial treatment with mercaptoethanol proved necessary before the nucleic acid could be released by phenol or detergent (Borenfreund et al., 1961).

In general, choice of a suitable method can ensure the isolation of DNA of high purity from most sources, but in very few cases can it be said with certainty that the last trace of impurity, particularly protein, has been eliminated.

Isopycnic centrifugation in cesium chloride has frequently been used for final purification of isolated DNA, and in some cases also for the direct isolation of DNA. For example, Dingman and Sporn (1967) used this technique to obtain DNA from nuclei of the livers of rats fed diets containing azo dye.

Because of the difficulty of obtaining absolutely pure DNA, and since treatment with a mutagen *in vivo* will normally result in only a very small extent of DNA reaction, it is always desirable and sometimes essential to isolate and characterize the product of the reaction by comparison with an authentic specimen. This eliminates the possibility that the product derives from reaction with a trace of impurity.

III. DEGRADATION OF DNA

A. Enzymatic Methods

1. Nucleotides

Treatment of DNA with deoxyribonuclease in the presence of

magnesium ions yields a mixture of oligonucleotides. Subsequent treatment with snake venom phosphodiesterase yields the four 5′-mononucleotides. For example, 1 mg of DNA in 1.0 ml of 0.015 M magnesium acetate, pH 7, was treated with 0.05 ml of a solution of DNase (1 mg/ml in water) at 37°C for 3 hr. The pH was raised to approximately 8.5 by the addition of ammonia solution, before adding 0.1 ml of a solution of snake venom phosphodisterase (1 mg/ml in water). Incubation was continued at 37°C for 18 hr to yield a mixture of the mononucleoside-5′-phosphates.

2. Nucleosides

The addition of bacterial alkaline phosphatase (50 μg/ml) to the mixture of mononucleoside-5′-phosphates prepared as described above yields the four nucleosides. Further details are given by Hall (1967), who applied to DNA the method he had earlier used to degrade RNA to its constituent nucleosides.

For the degradation of chemically modified DNA, the control of pH may be of importance, since some products may be partially or completely unstable under acid or alkaline conditions, and even prolonged incubation at neutral pH may lead to significant modifications. For these reasons, the enzymic degradation of DNA in nonbuffered solutions with the pH maintained by the intermittent addition of alkali should be avoided.

There is evidence from several laboratories that chemically modified DNA may be partially resistant to degradation by enzymes, leading to the production of some oligonucleotide fragments containing the altered base (Brookes and Heidelberger, 1969; Kriek, 1968).

B. Chemical Methods

1. Purines

The most generally used method for the degradation of chemically modified DNA is some form of acid hydrolysis (Loring, 1955; Wyatt, 1955). The glycosidic bond in purine deoxyribonucleotides is the bond in DNA most susceptible to acid hydrolysis, and in some instances alkylated purines are liberated by hydrolysis at an appreciable rate even at pH 7 (Lawley and Brookes, 1963, 1964). Thus, 3- or 7-alkyladenines and 7-alkylguanines are released from alkylated DNA by heating a solution of the DNA in 1 mM sodium phosphate buffer, pH 7, at 100°C for 20 min.

The complete liberation of the purines from DNA requires somewhat more drastic conditions than would suffice to hydrolyze isolated purine deoxyribonucleotides, due to the insolubility of the initially produced polynucleotides in dilute acid. Suitable methods include incubation of the DNA solutions at pH 1, 37°C, for 16 hr or heating at pH 1, 70°C, for 30 min. In each case, a clear solution should result and the time required to achieve this is normally shortened if the DNA is first dissolved in neutral solution with heating and the pH then adjusted. Such relatively mild hydrolysis leaves the pyrimidine bases as oligonucleotides of the type Py_np_{n+1} as a consequence of the partial hydrolysis of apurinic acid by the β-elimination mechanism (Brown and Todd, 1955).

Interest in the precise conditions used for this type of DNA hydrolysis has recently been stimulated following the suggestion by Loveless (1969) that O-6-alkyguanines could be products of DNA alkylation. These are known to be unstable to acid hydrolysis with loss of the alkyl group. It has been shown, however, that 2-amino-6-methoxypurine is stable at 70°C, pH 1, for at least 1 hr, but is dealkylated at a measurable rate at pH 1, 100°C (Lawley and Thatcher, 1970).

2. Pyrimidines

As discussed above, mild acid hydrolysis of DNA yields pyrimidine oligonucleotides. In order to obtain pyrimidine bases, more drastic conditions are required; for example, either perchloric acid (70% w/v) at 100°C for 1 hr, or 98% formic acid at 170°C (sealed tube) for 1 hr. Perchloric acid digestion results in some destruction of thymine and the formation of a carbon like precipitate, but has the advantage of simplicity.

An alternative method for hydrolysis of alkylated pyrimidine nucleotides consists in refluxing with 50% HBr for 3 hr (Price et al., 1968).

IV. METHODS OF SEPARATION OF PRODUCTS

A. Chromatography

The most widely used method of separation of DNA hydrolysis products is chromatography on paper. This technique requires only readily available apparatus, is cheap and reasonably rapid, and can separate most nonvolatile products which are then capable of detection by simple techniques. Disadvantages are that only comparatively small amounts of material can be applied to each paper, and in some cases quite different products have very similar R_f values even in many different solvent systems. These problems may be partially overcome by two-dimensional chromatography or by the elution of products from the paper and their rechromatography under different conditions (Heppel, 1967).

1. Paper Chromatography

Effective separation of nucleotides usually requires solvents containing salts, e.g., saturated aqueous ammonium sulfate–0.1 M phosphate buffer, pH 6–propan-2-ol (79:19:2 by volume); or isobutyric acid–0.5 N ammonia (10:6 by volume, pH 3.6) (Wyatt, 1955). Use of salts can be avoided by prolonged chromatography with propan-2-ol–water (70:30 by volume) or propan-2-ol–5 N ammonia (65:35 by volume) (Burton, 1962). Nucleotides can be effectively fractionated on ion-exchange paper (e.g., DEAE-paper) using aqueous buffers as solvent (see review, Grav, 1967).

For nucleosides, Hall (1967) has quoted R_f values for 30 commonly encountered products in five solvent systems. The R_f values of a deoxy-nucleoside are generally similar to the corresponding nucleoside, but acid solvents will normally convert the purine deoxynucleosides to the free base.

The separation on paper of the bases derived from nucleic acids generally requires the use of strongly acid or alkaline solvents, since only in such solvents will guanine and its derivatives be sufficiently soluble to have positive R_f values. Commonly used solvent systems are propan-2-ol–conc. HCl–H_2O, 170:41:39 by volume; methanol–conc. HCl–H_2O, 7:2:1 by volume; and butan-1-ol–conc. NH_3–H_2O, 86:2:12 by volume. Furthermore, some pairs of alkylated bases are separated inadequately even after chromatography in two dimensions. For example, for the case of methylated DNA it was found advantageous to partially hydrolyze the DNA by heating a neutral solution at 100°C, thus liberating 7-methylguanine, 3-methyladenine, and 7-methyladenine. These bases could then be separated from the residual polynucleotides by chromatography in an alkaline solvent in which the latter have zero R_f values (Lawley and Brookes, 1964). Values of R_f for the normal bases, nucleosides, and nucleotides in several solvent systems are given by Wyatt (1955) and by Grav (1967). However, no comprehensive tables of R_f values for mutagen-induced products are available and therefore an attempt has been made to give such data in Table 1.

In general, an increase in size of the alkyl chain leads to an increased Rf value. However, the addition of aromatic groups may have the reverse effect, due to the enhanced adsorption of such moieties to paper. Cross-linked bases have R_f values much lower than those of the individual bases.

2. Column Chromatography

a. *Nucleotides.* Ion-exchange methods are clearly suited to separation of both mono- and oligonucleotides. Cohn (1955) discussed the use of the anion exchange resins, eluted with either decreasing pH or increasing salt concentration at constant pH, for the separation of mononucleotides. For deoxynucleotides, pH values below about 4 must be avoided, but elution from Dowex-1 can be effected with a gradient of acetate buffer pH 4.4 (Anderson et al., 1963). In this and similar systems, 3- and 7-alkylated adenines and 7-alkylguanines would be obtained as bases.

Fractionation of nucleotides at near neutral pH can be achieved on

TABLE 1. R_f Values and Spectral Properties of DNA Bases and Some Derivatives

	R_f value relative to adenine in solvent[a]			
	A	B	C	pH
A. Adenine derivatives				
Adenine	1.0	1.0	1.0	2
				12
1-Methyladenine	1.3	0.6	1.6	4
				13
2'-Deoxy-1-methyladenosine	—	—	—	ca. 6
				11
1-Ethyladenine	1.5	1.2	—	2
				12
1-Benzyladenine	—	—	—	3
				13
3-Methyladenine	1.5	0.95	1.2	2
				13
3-Ethyladenine	1.8	1.5	—	2
				12
3-(2-Diethylaminoethyl)adenine	—	—	—	2
				11
6-Methylaminopurine	1.4	1.9	—	0
				12
2'-Deoxy-N^6-methyladenosine	—	—	—	ca. 6
				11
6-Ethylaminopurine	—	2.7	—	1
				11
7-Methyladenine	1.2	0.8	—	1
				12
7-Ethyladenine	1.5	1.3	—	2
				12
3,7-Dimethyladenine	1.8	0.3	2.1	4
				13
5-Aminoimidazole-4-N'-methyl-carboxamidine	1.3	0.5	1.8	4
				12
5-Aminoimidazole-4-N'-(2-diethyl-aminoethyl)carboxamidine	—	0.5	—	2
				11

[a] See pp. 132–133.

λ_{max},mμ	10^{-3} ε_{max}	$\dfrac{A_{260}\ m\mu}{A_{280}\ m\mu}$	pK_a	References
262.5	13.2	0.38	4.1	Beaven *et al.* (1955)
269	12.3	0.60	9.8	
259	11.7	0.23	7.2	Brookes and Lawley (1960)
270	14.4	0.85	11.0	
257.5	15.1	—	—	Jones and Robins (1963)
257.5	15.1	—	—	
260	—	—	6.9	Pal (1962); Ludlum (1969)
271	—	—	11.4	
259	11.8	—	7.0	Brookes *et al.* (1968)
270	14.4	—	11.3	
274	15.9	1.26	6.1	Brookes and Lawley (1960)
273	12.8	1.48	—	
274	18.4	1.32	ca. 5.4	Denayer (1962)
273	13.4	1.33	—	
275	—	1.50	—	Price *et al.* (1968)
274	—	1.50	—	
267	14.9	0.73	4.2	Elion *et al.* (1952)
273	15.9	1.20	10.0	
265	15.4	—	—	Jones and Robins (1963)
265	15.4	—	—	
270	16.3	—	—	Elion *et al.* (1952); Rosen (1968)
273	17.0	—	—	
272	13.8	—	ca. 3.5	Denayer (1962); Lawley and
270	10.5	—	—	Brookes (1964)
272	12.4	1.03	—	Lawley and Brookes (1964)
270	9.5	0.76	—	
276	15.9	1.50	—	Brookes and Lawley (1960);
279	14.1	2.00	11.0	Broom *et al.* (1964)
281	12.8	1.80	—	Brookes and Lawley (1960)
290	15.4	2.60	9.5	
286	—	1.90	—	Price *et al.* (1968)
289	—	1.80	—	

TABLE 1. (*Continued*)

	R_f values in solvent[a]					
	A	B	D	E	F	G
B. Guanine derivatives						
Guanine	—	0.15	—	0.45	0.22	0.30
1-Methylguanine	—	—	—	0.50	0.32	—
2'-Deoxy-1-methylguanosine	—	—	—	—	—	—
1-Ethylguanine	—	1.4[b]	—	—	—	—
N^2-methylguanine (6-hydroxy-2-methyl-aminopurine)	—	0.18	—	—	0.50	—
N^2-ethylguanine (2-ethylamino-6-hydroxypurine)	—	1.6[b]	—	—	0.61	—
3-Methylguanine	—	—	—	0.23	—	—
O^6-methylguanine (2-amino-6-methoxypurine)	0.55	0.50	—	—	—	0.62
2'-Deoxy-O^6-methylguanosine	—	—	—	—	—	0.78
O^6-ethylguanine (2-amino-6-ethoxypurine)	—	—	—	—	—	0.75
7-Methylguanine	0.27	0.12	0.24	—	—	0.35
7-Ethylguanine	0.33	0.30	0.45	—	—	0.61
7-(2-Carboxyethyl) guanine	0.46	0.0	0.45	—	—	—
Di-(2-guanin-7-ylethyl) methylamine	0.05	0.0	0.0	—	—	—
2,4-Diamino-6-hydroxy-5-N-methyl formamidopyrimidine	—	—	—	—	0.42	—
8-(N-2-fluorenylamino) guanine	—	—	—	—	—	—
8-(N-2-fluorenylacetamido) guanine	—	—	—	—	—	—

[a] See pp. 132–133.   [b] R_f relative to adenine.

pH	λ_{max},mμ	$10^{-3}\,\varepsilon_{max}$	$\dfrac{A_{280\,m\mu}}{A_{260\,m\mu}}$	pK_a	References
1	249, 276	11.4, 7.4	0.84	0	Beaven et al. (1955)
7	246, 276	10.7, 8.2	1.04	3.2	
11	246, 274	6.3, 8.0	1.14	9.6	
				ca. 12.5	
1	250, 272	10.2, 7.1	0.81	3.1	Broom et al. (1964);
7	248, 272	10.0, 7.9	0.93	10.5	Smith and Dunn (1959);
13	277	8.7	1.19		Pfleiderer (1961)
11	254	13.6	—	—	Broom et al. (1964)
—	—	—	—	—	Rosen (1968)
1	252, 275	12.3, 6.9	—	3.3	Smith and Dunn (1959);
7	249, 277	—	—	8.9	Shapiro and Gordon
14	255, 277	—	—	12.8	(1964)
1	253	—	—	—	Rosen (1968); Smith and Dunn (1959)
1	244, 264	8.3, 11.2	—	—	Townsend and Robins
11	273	13.7	—		(1962)
1	286	11.2	—	—	Balsiger and
7	240, 280	7.4, 7.9			Montgomery (1960)
13	246, 284	4.5, 7.9			
ca. 6	244, 276	—	—	—	Friedman et al. (1965)
1	286	11.7	—	—	Balsiger and
7	240, 281	8.2, 8.3			Montgomery (1960)
13	246, 284	4.1, 8.2			
1	250, 270	10.6, 6.9	0.79	3.5	Brookes and Lawley
7	248, 283	5.7, 7.4	1.8	10.0	(1961b); Pfleiderer
12	281	7.3	1.9		(1961)
1	250, 274	11.1, 7.0	0.81	—	Brookes and Lawley
7	245, 284	5.9, 7.7	1.9		(1961b)
12	280	7.4	1.8		
0	250	10.1	0.73	—	Roberts and
7	283	7.2	1.8		Warwick (1963);
13	280	7.4	1.8		Colburn and Boutwell (1966)
0.4	252	20.7	0.55	—	Brookes and Lawley
7	284	12.5	1.8		(1961b)
12	281	14.2	1.8		
1	262	17.8	—	3.8	Haines et al. (1962)
7	264	13.8			
13	261	9.8			
EtOH	282, 332	—	—	—	Miller et al. (1966)
PrOH: H₂O (3:2)	268, 300	29, 20	—	—	Kriek et al. (1967)

TABLE 1. (*Continued*)

| | \multicolumn{7}{c}{R_f in solvent[a]} | | | | | | |
	A	B	D	E	F	H	J
C. Cytosine derivatives							
Cytosine	—	0.28	—	—	0.44	—	0.22
3-Methylcytosine	0.67	0.33	0.67	—	—	—	—
5-Methylcytosine	—	0.36	—	—	0.52	—	0.29
2'-Deoxy-3-methylcytidine	—	—	—	—	—	—	0.10
2'-Deoxy-5-methylcytidine	—	—	—	—	—	—	—
5-Hydroxymethyl-3-methylcytosine	0.66	0.20	—	—	—	—	—
N^4-hydroxycytosine	—	—	—	—	0.44	—	—
N^4-hydroxy-5-hydroxymethyl-cytosine	—	—	—	—	—	—	—
N^4-hydroxy-5-methylcytosine	—	—	—	—	—	0.65	—
2'-Deoxy-N^4-hydroxycytidine	—	—	—	—	—	0.77	—
2'-Deoxy-N^4-hydroxy-5-methyl-cytidine	—	—	—	—	—	—	—
N^4-aminocytosine	—	—	—	—	—	—	—
O^2-methylcytosine (2-methoxy-4-aminopyrimidine)	—	—	—	—	—	—	—
2'-Deoxy-N^4-amino-5-methyl-cytidine	—	—	—	—	—	—	—
N^4-hydroxy-5,6-dihydro-6-hydroxylaminocytosine	—	—	—	—	—	—	0.07
N^4-hydroxy-5,6-dihydro-5-hydroxymethyl-6-hydroxylaminocytosine	—	—	—	—	—	0.72	—
2'-Deoxy-N^4-hydroxy-5,6-dihydro-6-hydroxylaminocytidine	—	—	—	—	—	—	0.18

[a] See pp. 132–133.

pH	λ_{max}, mμ	$10^{-3}\,\varepsilon_{max}$	$\dfrac{A_{280}\ m\mu}{A_{260}\ m\mu}$	pK$_a$	References
2	276	10.0	1.53	4.5	Beaven *et al.* (1955)
7	267	6.1	0.58	12.2	
14	282	7.9	3.28		
4	274	9.4	1.83	7.4	Brookes and Lawley
12	294	11.4	5.7		(1962)
2	284	9.8	2.7	4.8	Beaven *et al.* (1955)
7	274	6.2	1.2	12.4	
14	290	8.1	3.75		
0	278	12.0	1.94	—	Brookes and Lawley (unpublished)
1	287	12.4	—	4.4	
7	277	8.5	—	13	Fox *et al.* (1959)
2	278	—	1.9	7.1	Lawley and Brookes
12	296	—	3.5		(1963)
1	216, 276	8.7, 11.8	—	2.25	Brown and Schell
7	233, 273	10.6, 5.3	—		(1965)
1	217, 279	9.4, 10.3	—	—	Brown and Schell
7	230, 273	9.3, 5.6	—		(1965)
7	232, 272	8.8, 6.7	—	2.6	Brown and Schell (1965)
1	219, 278	8.5, 13.0	—	—	Brown and Schell
7	234, 272	11.1, 6.4			(1965)
0	285	13.5	2.4	2.3	Fox *et al.* (1959)
7.4	236, 269	11.6, 8.7	0.9	11.1	
13	254	11.7	0.76		
1	276	11.0			Brown (1967)
ca. 6	268	7.6			
2	230, 261	8.8, 9.5	—	5.3	Shugar and Fox (1952)
7	225, 271	8.0, 7.2			
1	218, 283	10.0, 13.3	—	—	Fox *et al.* (1959)
7	278	10.9	—	—	
7	220	10.8	—	—	Brown and Schell (1965)
7	220	9.2	—	—	Brown and Schell (1965)
7	222	12.0	—	—	Brown and Schell (1965)

TABLE 1. (Continued)

	R_f in solvent[a]					
	B	E	F	G	J	K
D. Thymine derivatives						
Thymine	0.50	—	0.76	—	0.52	—
3-Methylthymine	—	—	—	—	—	0.9
3-Methylthymidine	—	—	—	0.87	—	0.88
cis-syn-Thymine dimer	—	0.29	0.60	—	0.18	—

[a] Solvent: A. Methanol–conc. HCl–H$_2$O (7:2:1).
B. Butan-1-ol–conc. NH$_3$–H$_2$O (85:2:13, or variants as given in the references).
C. Propan-2-ol–5% aq. (NH$_4$)$_2$SO$_4$ (1:19).
D. Ethanol–conc. NH$_3$–H$_2$O (80:2:18).
E. Butan-1-ol–glacial acetic acid–H$_2$O (2:1:1).
F. Propan-2-ol–conc. HCl–H$_2$O (170:41:39).

columns of DEAE-cellulose, using a gradient of 0.01–0.13 M ammonium bicarbonate (Staehelin, 1961).

Chromatography on DEAE-cellulose in the presence of 7 M urea was used by Tomlinson and Tener (1963) to fractionate oligonucleotides into groups having the same number of negative charges; the urea inhibits the adsorption forces between the column material and the bases which otherwise cause separation of the individual nucleotides. Brookes and Heidelberger (1969) employed this same system to fractionate the products derived by enzymic degradation of DNA with bound carcinogenic hydrocarbon. In the absence of urea the hydrocarbon derivative was bound irreversibly to the DEAE-cellulose.

DEAE-Sephadex has also been used in the separation of oligonucleotides, again in the presence of urea (Rushizky et al., 1964).

b. Nucleosides. Separation of the deoxynucleosides derived from from DNA was initially achieved on the cation exchange resin Dowex-50 (ammonium form) eluted with 0.1 M ammonium acetate, pH 3.9 (Reichard and Estborn, 1950). More recently, Hall (1962, 1967) introduced the technique of two-phase partition chromatography on Celite columns, using organic solvents containing formic acid. Both the above methods would lead to conversion of the more acid-labile alkylated purine nucleosides to bases.

Hall has applied his methods extensively to separate the minor nucleosides of RNA, but no application to the problems of mutagen action has

pH	λ_{max}, mμ	$10^{-3}\,\varepsilon_{max}$	$\dfrac{A_{280\,m\mu}}{A_{260\,m\mu}}$	pK$_a$	References
7	264.5	7.9	0.53	ca. 0	Beaven *et al.* (1955)
				9.9	
12	291	5.4	1.31	13	
1	266	—	0.56	—	Friedman *et al.* (1965)
7	264	—	0.50		
13	289	—	3.50		
1	265	—	0.66	—	Haines *et al.* (1964);
7	266	9.1		—	Friedman *et al.* (1965)
12	267	—	0.70		
—	—	—	—	—	Blackburn and Davies (1965); Günther and Prusoff (1967)

G. Propan-2-ol–conc. NH$_3$–H$_2$O (7:1:2).
H. Water.
J. Butan-1-ol–H$_2$O (86:14).
K. Propan-2-ol–H$_2$O (7:3).
All compositions are given by volume.

appeared, although both Celite and Dowex-50 columns are clearly of potential value.

Sephadex G-10, although designed for exclusion chromatography, separates nucleosides on a basis not solely dependent on molecular size, the purines being adsorbed more strongly than the pyrimidines (Braun, 1967). Its use presents the advantage that elution with water or dilute buffers of volatile salts can be used.

The recent introduction of Sephadex LH-20, which can be used with organic solvents, should prove of particular value for the separation of modified nucleosides containing aromatic residues, since such compounds are not readily eluted by aqueous solvents. Kriek (1969) has used this technique to isolate the products from the *in vivo* reaction of 2-acetylamino-fluorene with nucleic acids after degradation to the nucleosides with snake venom diesterase and bacterial alkaline phosphatase.

c. *Bases.* The separation of nucleic acid bases on Dowex-50 H$^+$-form eluted with 2 N hydrochloric acid was introduced by Cohn (1949). This method has been used extensively with alkylated nucleic acids (Brookes and Lawley, 1961a; Craddock and Magee, 1966). Figure 1 shows an example of the application of this technique to a DNA methylated with di-[C^{14}]methyl sulfate. DNA from salmon sperm was treated with di-[C^{14}]methyl sulfate at 37°C, pH 7, to give an extent of reaction of 2.9 mmole methyl per mole DNA-P, isolated by precipitation with ethanol and

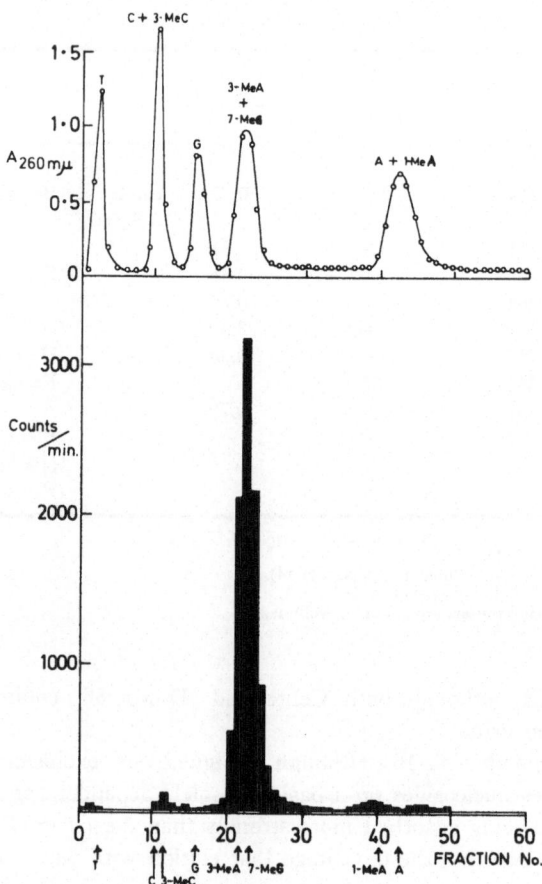

FIGURE 1. Cation-exchange chromatography of an acid hydrolysate of ^{14}C methylated DNA.

hydrolyzed with $HClO_4$ (72% w/v) for 1 hr at 100°C. The hydrolysate was applied together with added unlabeled marker methylated bases to a column of Dowex-50, H^+ form, 13 by 1 cm, eluted with a linear gradient 0.75–2 N-HCl, 10 ml fractions, 0.5 ml/min. Bases and added unlabeled marker methylated bases are denoted by their initial letters, and were detected by measurements of the UV absorption of the fractions; ^{14}C-methylated bases were detected by evaporation of aliquots of the fractions and scintillation counting of the residues. The major methylation products, 3-methyladenine and 7-methylguanine, are not separated, eluting together between guanine and adenine, while 1-methyladenine elutes with adenine. Any O-6 methylated guanine present in the DNA might be at least partially demethylated by this method and would certainly be degraded if the fractions were evaporated without prior neutralization.

Anion exchange resins can also be used for fractionation of nucleic acid bases as discussed by Cohn (1949). In this case, the disadvantages are that the alkaline solvent used for elution might modify certain alkylated bases, and the salts used would complicate reisolation of the products.

As with nucleosides, Sephadex G-10 can be used for separation of bases. Lawley (1968) found that 3-methyladenine absorbed quite strongly to this material in water, and dilute ammonia was required for satisfactory elution.

3. Thin Layer Chromatography

This technique has developed considerably during the past 5 years, and may be considered as essentially a development from paper chromatography. In general, separations which can be obtained on paper can be reproduced with the same solvents on cellulose thin layers. The advantages of thin layer methods over paper chromatography are speed (often ten times faster than paper), sharpness of resolution, and ease of reisolation of products. For a general outline of techniques, the reader is referred to Randerath (1966).

Although of potential value for nucleotide and nucleoside separations, few examples are available of the application of this method to mutagen-treated DNA.

Olson (1968) has separated methylated bases, derived from DNA treated with N-methyl-N-nitroso-p-toluenesulfonamide, on cellulose thin layers, using solvents previously found suitable for this separation on paper, e.g., methanol–conc. HCl–H_2O, 7:2:1 by volume, and butan-1-ol–conc. NH_3–H_2O, 88:2:10 by volume.

4. Gas Chromatography

Several authors have considered the application of gas chromatography to analysis of nucleic acid products. The difficulty lies in the conversion of the nucleosides or bases to suitably volatile derivatives. Trimethysilylation has been used for this purpose (Sasaki and Hashizume, 1966; Gehrke et al., 1967) and also exhaustive methylation (MacGee, 1966). With chemically modified DNA, the need to modify further before being able to apply gas chromatography is an obvious disadvantage, and no publications are available using this method for mutagen studies. An alternative approach might be the separation and identification of pyrolysis products from treated DNA (Jennings and Dimick, 1962), but again no applications to the present problem appear to have been made.

B. Electrophoresis

The principle, technique, and application of electrophoresis to the separation of bases, nucleosides, and nucleotides have been recently reviewed in detail by Zweig and Whitaker (1967). In general, Whatman 3 MM paper is employed, with a voltage of about 70 V/cm for 1 hr in 0.05 M borate, pH 9.5, or 2–3 hr in 0.05 M ammonium formate, pH 3.5. Again, one must note that some modified nucleotides or nucleosides would be unstable under these conditions. Cellulose acetate and thin layer supports have tended to supersede paper because of the greater speed and higher capacity, but no applications to mutagen-treated DNA have been reported.

The mobilities are of course mainly dependent on the net charge of the molecules, but in addition those of the purine derivatives are generally less than those of pyrimidine derivatives at the same net charge, and nucleosides have higher mobilities than their parent bases. For example, at pH 9.5, the following mobilities were observed (given in cm/hr at 78 V/cm): adenine, 3.5; cytosine, 2.6; guanine, 8.1; adenosine, 15.6; cytidine, 19.2; guanosine, 25.0. For nucleotides, at pH 3.5, 20 V/cm, the mobilities relative to that of UMP were: AMP, 0.5; CMP, 0.41; GMP, 0.87 (TMP would be expected to resemble UMP).

V. METHODS OF DETECTION AND IDENTIFICATION OF PRODUCTS

A. Spectroscopy

1. Ultraviolet Absorption

By far the most widely used method for detection and identification of DNA purine and pyrimidine derivatives is ultraviolet absorption spectroscopy.

Spots on paper or thin layer plates can be visualized under a standard low-pressure mercury arc lamp suitably filtered to give 254 mμ radiation, when the pyrimidine and adenine compounds absorb, while guanine and its derivatives generally give a blue fluorescence.

After elution from paper, thin layers, or columns, the individual products can often be characterized by their UV absorption spectra in the range 210–350 mμ. The molar extinction coefficients at the maxima are generally of the order of 10^4, so that 10 μg of a base in 3 ml of solution in a conventional cuvette of 1 cm light path would give an absorption of about 0.2.

A most valuable criterion for the identification of isolated products of this type is the variation of spectra with pH, indicating the presence of basic or acidic ionizations. Furthermore, the occurrence of isosbestic points for spectra determined at pH values over the range in which such ionizations occur, or during the slow conversion reactions of such products, is a criterion valuable both for identification and as evidence of purity.

It can generally be assumed that the substitution of simple groups such as alkyl at a given position in the purine or pyrimidine ring will always result in a characteristic type of spectrum, with only minor variations resulting from the nature of the group involved. However, if the attached group contains an aromatic or heterocyclic residue, then a significant effect on the spectra would be predicted. If the aromatic group is conjugated with the nucleic acid base residue, then an entirely new spectrum will result, while if the two residues are separated by one or more saturated carbon atoms the resulting spectrum will approximate to the sum of that of the aromatic moiety plus that of the appropriate substituted DNA base.

Table 1 gives spectral and pK$_a$ values of a number of compounds which are of interest in this field, some of which have been shown to result from the action of mutagens on DNA.

When identifying a product by spectral comparison with an authentic sample it is essential that the same solvent is used for both. For example, if the product has been obtained by elution from paper or by evaporation of a chromatographic fraction, its spectrum should be determined using an appropriate blank solution and compared with an authentic sample similarly isolated. Spectral comparisons are much more reliable at the higher wavelengths, since, at values much below the usual maxima of around 260 mμ, absorption by trace impurities is more likely to become significant.

In general, it is desirable when identifying a product by comparison with a known compound that the two should have identical R_f values in three solvent systems and identical spectra at two pH values at which different forms of the compound will predominate, e.g., base, acid, or neutral molecule.

2. Fluorescence Spectra

As mentioned above, the fluorescence of guanine and its derivatives when exposed to UV light assists in their location on paper or thin layer chromatograms. However, it should be noted that minor traces of highly fluorescent materials are sometimes found in DNA hydrolysates which after elution give no measurable absorption. Although some substituted purines do show fluorescence spectra (e.g., 7-alkylguanines), few if any applications to the identification of mutagen-induced products have been made. Leng *et al.* (1968) have studied the fluorescence of 7-methylguanosine, and Börresen (1967) has reported spectra for 1-methyladenosine and 7-methyladenine.

3. Infrared Spectra

The measurement of IR spectra is a standard method in organic chemistry for identification of unknown products both by comparison with authentic specimens and by direct interpretation of the position and intensity of the spectral bands. However, generally relatively large amounts of material are required and the need to use nonaqueous solvents or D_2O makes IR methods of little value for the problems considered here.

The interpretation of IR spectra of nucleic acid derivatives is complex, but the technique has proved valuable in determining their tautomeric forms.

B. Use of Radioactive Materials

1. Labeled Mutagens

The scope of spectroscopic methods for detection and identification is, as indicated above, limited by the amounts of minor products available. Greater sensitivity can be achieved by the use of isotopically labeled mutagens; this approach has been used extensively and is often the method of choice.

The limit of detection of a labeled product will clearly depend on the specific radioactivity of the mutagen and on the efficiency of detection of the particular isotope used.

For carbon-14 the theoretically possible specific radioactivity is 62.4 mC/m-atom (Bayly and Evans, 1968), but in practice values of the order of 40 mC/mmole are available for ^{14}C compounds. Liquid scintillation counting is obviously the method of choice, and for the problem under consideration it is generally necessary to count aqueous solutions. A scintillation fluid based on toluene:dioxan:ethanol (385:385:230 by volume) will take up to 0.5 ml of water per 10 ml of phosphor, and for ^{14}C an efficiency of about 70% is obtainable with a background of 20–30 counts/min. When estimating radioactivity in DNA it is essential to degrade the DNA to prevent its precipitation in the scintillation fluid. This is readily achieved with either deoxyribonuclease at pH 7 or by hydrolysis in 5% (w/v) trichloroacetic acid at 100°C to a clear solution (a few minutes). The amount of DNA being assayed is best estimated from the UV absorption of the digest after suitable dilution. The ε_p for degraded DNA can be taken as 8750 under these conditions (Brookes and Lawley, 1964). Thus, if 1 mg of DNA is available, a convenient procedure would be to dissolve this in 0.9 ml, degrade by the addition of 0.1 ml of deoxyribonuclease solution (1 mg/ml in 0.1 M magnesium acetate), count aliquots of 0.1–0.4 ml, and dilute 0.1 ml to 2.5 ml with water for measurement of UV absorption.

With the introduction of liquid scintillation counting, tritium assay became greatly simplified, and the use of this isotope correspondingly increased. The theoretical limit of specific activity for tritium, 29.2 C/m-atom, is much greater than for ^{14}C, and frequently several atoms per molecule can be introduced. Tritium compounds are often available at specific activities of the order of 1 C/mmole, and thus despite the lower efficiency of counting about 25%, the detectable amount of a reaction product is about ten times lower than for ^{14}C-mutagens.

A disadvantage of tritium labeling is the problem of radiolysis encountered with the more highly radioactive compounds. The factors governing this process are complex, and the extent of breakdown depends not only only the tritium content, but also on the condition of storage, e.g., whether as a solid or in solution, and in which solvent. The problem is discussed in detail by Evans (1966), who gives suitable methods of storage for many compounds. It is advisable whenever possible to purify ^{3}H-labeled compounds before use after prolonged storage.

Labeling with sulfur-35, although applicable to only a few mutagens, presents the greatest advantage with respect to sensitivity of detection of products, since the theoretical limit of specific activity (1.5×10^{3} C/m-atom) is higher than for tritium and the efficiency of counting is as great as for ^{14}C.

Radioactive products can be detected and estimated on paper chromatograms or thin layer plates by direct scanning, using commercially available apparatus fitted with windowless gas-flow detectors. With such apparatus about 0.1 mμC of ^{14}C or 1.0 mμC of ^{3}H can be detected. Greater sensitivity can be achieved by cutting up the paper chromatograms into segments, or by scraping off the thin layer support, and counting directly in scintillation vials. Alternatively, the products can be eluted and then counted, but this usually involves evaporation of the solvent used. The sensitivity achieved in this way is of the order of 0.005 mμC for ^{14}C or ^{35}S and perhaps 0.015 mμC for ^{3}H, assuming that a value 50% above background is significant. With some counters it is important when counting ^{3}H on paper segments to ensure that the orientation of the paper in the bottles is uniform throughout a series.

Visualization of radioactive products on paper or thin layer chromatograms by radioautography using x-ray sensitive films is valuable when a permanent visible record is required. Exposure times of one month or more can be used and 1 mμC of ^{14}C or ^{35}S is detectable, but for ^{3}H about 100 mμC is required.

Radioactive material eluted from columns can be assayed by evaporation of the solvent in the liquid scintillation vials prior to the addition of phosphor, assuming that the product is nonvolatile. It may be necessary, however, to check the effect of any salts or other nonvolatile residue from the solvent on counting efficiency.

TABLE 2. Limits of Detection of Extent of Reaction of Labeled Mutagens with DNA[a]

Label	Specific radioactivity of mutagen, mC/mmole	Lower detection limit, mole DNA-P per mole mutagen
^{14}C	40	10^7
^{3}H	1000	8×10^7
^{35}S	1000	3×10^8

[a] Assume that 1 mg DNA (as usually isolated, 2.5 μmole DNA-P) is counted with an efficiency of 70% (^{14}C or ^{35}S) or 30% (^{3}H) and that one-half of background (15 counts/min ^{14}C or ^{35}S, or 20 counts/min ^{3}H) is significant.

Table 2 shows the limits of detection of the extent of reaction of a mutagen with DNA of which 1 mg is available for assay. These data are based on the assumption that the reagent was isotopically labeled to the highest extent commercially available and that a count 50% above background is significant. It may be noted that in no case does the sensitivity permit detection of a single reaction event in the whole DNA of a mammlian cell comprising about 10^{10} nucleotide units. However, about 100 such events could be detected. If more than 1 mg of DNA were available, the sensitivity would be improved, but with liquid scintillation there is a limit to the amount of material that can be assayed before adversely affecting the efficiency of counting.

2. Use of Labeled DNA

Some mutagens, for example, nitrous acid (and of course radiation), cannot be studied by mutagen labeling. In these cases, use can be made of DNA prelabeled *in vivo*. Craddock (1969b) applied this method in studies on the methylation of rat liver DNA by N-methyl-N'-nitro-N-nitrosoguanidine (MNNG). Newborn animals were injected with high specific activity ^{14}C formate to obtain generalized labeling of DNA purine bases, while the label in other cellular constituents decreased rapidly as the animals grew. By this method, Craddock deduced that the extent of deamination of DNA bases of liver of rats following administration of MNNG could not be greater than 0.1%, and the formation of 7-methylguanine in DNA was readily detected. Earlier, Craddock and Magee (1966) had used the same principle of show that dimethylnitrosamine administration resulted in methylation of 0.5% of the guanine moieties of DNA without detectable deamination or conversion of cytosine residues to N^4-hydroxycytosine.

Related to the present discussion is the problem of incorporation of radioactivity from the ^{3}H or ^{14}C methyl labeled mutagens into normal DNA bases, following metabolic degradation of the mutagen to labeled formaldehyde. The work of Magee and his group indicated the extent to which this can occur in DNA isolated at relatively long times after administration of [^{14}C]dimethylnitrosamine to the rat (Craddock, 1969a). It is apparent that ^{3}H can also be incorporated into normal nucleic acid bases from [^{3}H]-dimethylnitrosamine, despite the assumed ease of exchange of tritium in formate with hydrogen derived from the cellular media (Lee et al., 1964).

Use of isotopic phosphorus has not been reported for in vivo chemical studies of mutagen action. The isotope ^{33}P has a somewhat longer half-life (25 days) than the more commonly used ^{32}P (14.3 days) and also has the advantage of a softer β-radiation. In experiments using phosphorus labeling, it would be essential to have an adequate separation of DNA from RNA since both would be labeled. Various physicochemical studies have been reported of the degradation of isotopically prelabeled DNA following mutagen treatment, but these are outside the scope of the present chapter.

C. Mass Spectroscopy

The application of this powerful technique to the problems of mutagen action are at present very limited, but it is likely to become of increasing importance (Foster, 1969).

Methylated purines, particularly guanine derivatives, and pyrimidines are readily characterized as molecular ions (Rice et al., 1965; Rice and Dudek, 1967). Provided a method of isolation is available (perhaps by gas chromatography), the sensitivity of detection of mutatgen-induced products could be of the order of nanograms, which would perhaps exceed that achieved by ^{3}H-labeling. McCloskey et al. (1968) have reported mass spectra of trimethylsilyl derivatives of nucleic acid components which have been used for gas chromatographic separations. The mass spectral method is of course potentially capable of giving valuable information about any compound detected. From the accurately determined mass of the molecular ion can be derived the atomic composition of the unknown, and a detailed study of the fragmentation pattern might enable a structure to be asigned.

The most significant application of mass spectra so far has been by Lijinsky et al. (1968). Dimethylnitrosamine labeled with deuterium in the methyl group was injected into rats, and the 7-methylguanine produced in DNA of the liver of the treated animals was shown to contain the CD_3 group intact. In these experiments the ^{3}H-labeled mutagen was injected together with the deuterated analog to provide a chromatographic marker. The 7-alkylguanine was isolated from the DNA hydrolysate by Dowex-50 column chromatography and shown by mass spectra to give a molecular ion having m/e—168, corresponding to 7-CD_3-guanine. This provides an elegant proof that diazomethane could not have been the methylating intermediate derived from dimethylnitrosamine.

The major factor limiting the extension of this potentially powerful technique is the problem of isolating the degradation products of DNA in a form suitable for introduction into the mass spectrometer. The value of gas chromatography for this purpose remains for further investigation.

VI. REFERENCES

Anderson, N. G., Green, J. G., Barber, M. L., and Ladd, F. C. (1963), *Anal. Biochem. 6*, 153.

Balsiger, R. W., and Montgomery, J. A. (1960), *J. Org. Chem. 25*, 1573.

Bayly, R. J., and Evans, E. A. (1968), "Storage and Stability of Compounds Labelled with Radioisotopes," The Radiochemical Centre, Amersham, England.

Beaven, G. H., Holiday, E. R., and Johnson, E. A. (1955), *in* "The Nucleic Acids," Vol. I, Academic Press, New York, p. 502.

Blackburn, G. M., and Davies, R. J. H. (1965), *Chem. Commun. 11*, 215.

Borenfreund, E., Fitt, E., and Bendich, A. (1961), *Nature 191*, 1375.

Börresen, H. C. (1967), *Acta Chem. Scand. 21*, 2463.

Braun, R. (1967), *Biochim. Biophys. Acta 142*, 267.

Brookes, P., and Heidelberger, C. (1969), *Cancer Res. 29*, 157.

Brookes, P., and Lawley, P. D. (1960), *J. Chem. Soc.*, p. 539.

Brookes, P., and Lawley, P. D. (1961*a*), *Biochem. J. 80*, 496.

Brookes, P., and Lawley, P. D. (1961*b*), *J. Chem. Soc.*, p. 3923.

Brookes, P., and Lawley, P. D. (1962), *J. Chem. Soc.*, p. 1348.

Brookes, P., and Lawley, P. D. (1964), *Nature 202*, 781.

Brookes, P., Dipple, A., and Lawley, P. D. (1968), *J. Chem. Soc.*, p. 2026.

Broom, A. D., Townsend, L. B., Jones, J. W., and Robins, R. K. (1964), *Biochemistry 3*, 494.

Brown, D. M. (1967), *in* "Methods in Enzymology" (L. Grossman and K. Moldave, eds.) Vol. XII, p. 31, Academic Press, New York.

Brown, D. M., and Schell, P. (1965), *J. Chem. Soc.*, p. 208.

Brown, D. M., and Todd, A. R. (1955), *in* "The Nucleic Acids" (E. Chargaff and J. N. Davidson, eds.) Vol. I, p. 409, Academic Press, New York.

Burton, K. (1962), *Biochim. Biophys. Acta 55*, 412.

Cohn, W. E. (1949), *Science 109*, 377.

Cohn, W. E. (1955), *in* "The Nucleic Acids" (W. E. Cohn and J. N. Davidson, eds.) Part I, p. 211, Academic Press, New York.

Colburn, N. H., and Boutwell, R. K. (1966), *Cancer Res. 26*, 1701.

Craddock, V. M. (1969*a*), *Biochem. J. 111*, 497.

Craddock, V. M. (1969*b*), *Biochem. J. 111*, 615.

Craddock, V. M., and Magee, P. N. (1966), *Biochem. J. 100*, 724.

Denayer, R. (1962), *Bull. Soc. Chim. France*, p. 1358.

Dingman, C. W., and Sporn, M. B. (1967), *Cancer Res. 27*, 938.

Elion, G. B., Burgi, E., and Hitchings, G. H. (1952), *J. Am. Chem. Soc. 74*, 411.

Evans, E. A. (1966), "Tritium and its Compounds," p. 316, Butterworths, London.

Foster, A. B. (1969), *Lab. Practice 18*, 743.

Fox, J. J., Van Praag, D., Wempen, I., Doerr, I. L., Cheong, L., Knoll, J. E., Eidinoff, M. L., Bendich, A., and Brown, G. B. (1959), *J. Am. Chem. Soc. 81*, 178.

Friedman, O. M., Mahapatra, G. N., Dash, B., and Stevenson, R. (1965), *Biochim. Biophys. Acta 103*, 286.

Gehrke, C. W., Stalling, D. L., and Ruyle, C. D. (1967), *Biochem. Biophys. Res. Commnu. 28*, 869.

Grav, H. J. (1967), *in* "Methods in Cancer Research" (H. Busch, ed.) Vol. III, p. 300, Academic Press, New York.

Günther, H. L., and Prusoff, W. H. (1967), *in* "Methods in Enzymology" (L. Grossman and K. Moldave, eds.) Vol. XII, p. 19, Academic Press, New York.

Haines, J. A., Reese, C. B., and Todd, A. R. (1962), *J. Chem. Soc.*, p. 5281.

Haines, J. A., Reese, C. B., and Todd, A. R. (1964), *J. Chem. Soc.*, p. 1406.

Hall, R. H. (1962), *J. Biol. Chem. 237*, 2283.

Hall, R. H. (1967), *in* "Methods in Enzymology" (L. Grossman and K. Moldave, eds.) Vol. XII, p. 305, Academic Press, New York.

Heppel, L. A. (1967), *in* "Methods in Enzymology" (L. Grossman and K. Moldave, eds.) Vol. XII, p. 316, Academic Press, New York.

Jennings, E. C., and Dimick, K. P. (1962), *Anal. Chem. 34*, 1543.

Jones, J. W., and Robins, R. K. (1963), *J. Am. Chem. Soc. 85*, 193.

Kirby, K. S. (1964), *in* "Progress in Nucleic Acid Research and Molecular Biology" (J. N. Davidson and W. E. Cohn, eds.) Vol. 3, p. 1, Academic Press, New York.

Kirby, K. S. (1967), *in* "Methods in Cancer Research" (H. Busch, ed.) Vol. 3, p. 1, Academic Press, New York.

Kriek, E. (1968), *Biochim. Biophys. Acta 161*, 273.

Kriek, E. (1969), *Chem. Biol. Interactions 1*, 3.

Kriek, E., Miller, J. A., Juhl, U., and Miller, E. C. (1967), *Biochemistry 6*, 177.

Lawley, P. D. (1968), *Nature 218*, 580.

Lawley, P. D., and Brookes, P. (1963), *Biochem. J. 89*, 127.

Lawley, P. D., and Brookes, P. (1964), *Biochem. J. 92*, 19c.

Lawley, P. D., and Thatcher, C. J. (1970), *Biochem. J. 116*, 693.

Lee, K. Y., Lijinsky, W., and Magee, P. N. (1964), *J. Nat. Cancer Inst. 32*, 65.

Leng, M., Pochon, F., and Michelson, A. M. (1968), *Biochim. Biophys. Acta 169*, 338.

Lijinsky, W., Loo, J., and Ross, A. E. (1968), *Nature 218*, 1174.

Loring, H. S. (1955), *in* "The Nucleic Acids" (E. Chargaff and J. N. Davidson, eds.) Vol. I, p. 191, Academic Press, New York.

Loveless, A. (1969), *Nature, 223*, 206.

Ludlum, D. B. (1969), *Biochim. Biophys. Acta 174*, 773.

MacGee, J. (1966), *Anal. Biochem. 14*, 305.

Marmur, J. (1961), *J. Mol. Biol. 3*, 261.

McCloskey, J. A., Lawson, A. M., Tsuboyama, K., Krueger, P. M., and Stillwell, R. N. (1968), *J. Am. Chem. Soc. 90*, 4182.

Miller, E. C., Juhl, V., and Miller, J. A. (1966), *Science, 153*, 1125.

Olson, A. C. (1968), *J. Chromatog. 35*, 292.

Pal, B. C. (1962), *Biochemistry 1*, 558.

Pfleiderer, W. (1961), *Ann. 647*, 167.

Price, C. C., Gaucher, G. M., Koneru, P., Shibakawa, R., Sowa, J. R., and Yamaguchi, M. (1968), *Biochim. Biophys. Acta 166*, 327.

Randerath, K. (1966), "Thin-Layer Chromatography," Academic Press, New York.

Reichard, P., and Estborn, G. (1950), *Acta Chem. Scand. 4*, 1047.

Rice, J. M., and Dudek, G. O. (1967), *J. Am. Chem. Soc. 89*, 2719.

Rice, J. M., Dudek, G. O., and Barber, M. (1965), *J. Am. Chem. Soc. 87*, 4569.

Roberts, J. J., and Warwick, G. P. (1963), *Biochem. Pharmacol. 12*, 1441.

Rosen, L. (1968), *Biochem. Biophys. Res. Commun. 33*, 546.

Rushizky, G. W., Bartos, E. M., and Sober, H. A. (1964), *Biochemistry 3*, 626.

Sasaki, Y., and Hashizume, T. (1966), *Anal. Biochem. 16*, 1.

Shapiro, R., and Gordon, C. N. (1964), *Biochem. Biophys. Res. Commun. 17*, 160.

Shugar, D., and Fox, J. J. (1952), *Biochim. Biophys. Acta 9*, 152.

Smith, J. D., and Dunn, D. B. (1959), *Biochem. J. 72*, 294.

Staehelin, M. (1961), *Biochim. Biophys. Acta 49*, 11.

Tittensor, J. R., and Walker, R. T. (1967), *Europ. Polymer J. 3*, 691.

Tomlinson, R. V., and Tener, G. M. (1963), *Biochemistry 2*, 697.

Townsend, L. B., and Robins, R. K. (1962), *J. Am. Chem. Soc. 84*, 3008.

Wyatt, G. R. (1955), *in* "The Nucleic Acids" (E. Chargaff and J. N. Davidson, eds.) Vol. I, p. 243, Academic Press, New York.

Zamenhof, S. (1958), *in* "Biochemical Preparations" (C. S. Vestling, ed.) Vol. 6, p. 8, Wiley, New York.

Zweig, G., and Whitaker, J. R. (1967), "Paper Chromatography and Electrophoresis. Vol. I. Electrophoresis in Stabilizing Media," Academic Press, New York.

CHAPTER 5

Physical-Chemical Methods for the Detection of the Effect of Mutagens on DNA*

Bernard S. Strauss

Department of Microbiology
The University of Chicago
Chicago, Illinois

I. INTRODUCTION

Mutation is a hereditible change in the DNA of an organism. All mutagens must therefore alter the structure of DNA either directly or indirectly. However, there are imposing barriers to the study of the mutagenic effectiveness of particular compounds by determination of the physical changes they produce in DNA. The major problem comes from the amplification factor inherent in gene action. The DNA of an organism such as the bacterium *Escherichia coli*, with a total molecular weight of about 4.7×10^9, will contain about 1.6×10^7 nucleotides (McQuillen, 1965). A change in any one of these nucleotides may produce a recognizable mutation. Yet most standard chemical methods are unable to detect changes of one part per thousand, let alone one per 20,000,000. Furthermore, the mutagenic reaction is not necessarily the only, or even the major, change produced by reaction of DNA with a mutagen. For example, nitrous acid deaminates bases (Schuster,

* The work reported from this laboratory was supported by grants from the National Institutes of Health (USPHS 2 R01 GM 07816), the National Science Foundation (NSF GB 8514), and the Atomic Energy Commission (AEC AT (11-1) 2040)).

1960) and should lead to transition mutations, but it also produces crosslinks (Becker *et al.*, 1964) by an unknown mechanism and leads to large deletions of genetic material (Tessman, 1962). In many cases, the putative mutagen may not even be the actual one and the compound may be transformed *in vivo* into the actual mutagen (Legàtor *et al.*, 1969; Leahy *et al.*, 1967; Kojima and Ichibagase, 1966). In such cases, *in vitro* mixture of the original, nontransformed compound and DNA will give no meaningful result.

Freese and Freese (1966) have made a useful distinction between mutagenic and inactivating changes in DNA: the inactivating changes prevent replication of the DNA; the mutagenic changes in general do not prevent replication, are much less drastic, and hence are harder to detect. It is often possible to convert, *in vitro*, a DNA alteration into an inactivating change which is more easily detected by physical methods, i.e., into a single-strand break. However, only a biological test can tell whether any particular DNA alteration is indeed mutagenic. Physical tests are useful and may give indications of a possible mutagenic action but only the direct demonstration of mutagenicity in a biological system, preferably as close to the system of interest as possible, can be considered final.

Some idea of the problem can be gained by considering the reaction of ethyl methanesulfonate with bacteriophage T2. Loveless (1959) first reported that EMS* was mutagenic for mature bacteriophage treated *in vitro*. In his experiments, incubation of phage for 40 min with 0.2 M EMS led to a mutation frequency of about $7 \times 10^{-3} r$ mutants per phage. Such r mutants are produced as a result of changes within at least 1 percent of the total bacteriophage DNA (Mosig, 1968). Phage T2 DNA has a molecular weight of about 130×10^6 (MacHattie and Thomas, 1968) and therefore has a genome of about 4×10^5 nucleotides. Change in any one of $4 \times 10^5 \times 0.01$ = 4000 nucleotides might result in detectable r mutation. Brookes and Lawley (1963) measured the actual number of ethyl groups added to T2 DNA by reaction with EMS under conditions similar to Loveless' mutation studies. They reported 1050 ethyl groups fixed per mole of phage DNA after 4 hr reaction with 0.2 M EMS. Assuming the alkylation to be linear with time, there would be about 175 ethyl groups fixed per bacteriophage particle in 40 min or about two ethylated bases produced in 40 min within the region of DNA that could give an r mutation. At this dose the observed mutation frequency was 7×10^{-3}. In order to increase the r mutation frequency to 1, that is, an r mutation in each phage, there would have to be $(175/7 \times 10^{-3}) = 25 \times 10^3$ ethyl groups per phage or about 250 ethyl groups added per 4000 nucleotides. One out of every 16 nucleotides, or one out of every eight adenine and guanine residues (which are the major reaction sites for this mutagen) would have to be ethylated. A better way of looking at the

---

* Abbreviations: EMS: ethyl methanesulfonate; MMS: methyl methanesulfonate.

TABLE 1. Representative Mutagens and the Changes They Produce

Chemical change	Mutagenic change	Representative mutagen	References
Base alteration—chemical change in the purine or pyrimidine bases of DNA as a result of the direct chemical action of a mutagen[a]	Transition	Hydroxylamine	Freese et al. (1961); Phillips and Brown (1967)
		Ethyl methanesulfonate	Krieg (1963)
		N-methyl N'-nitro-N-nitrosoguanidine	
		Nitrous acid	Vielmetter and Schuster (1960); Orgel (1965)
Base substitution—incorporation of base analogs	Transition	Bromouracil	Freese (1959); Lawley and Brookes (1962)
Base removal—formation of apurinic sites as a result of chemical action	Transversion (?)	Ethyl methanesulfonate	Krieg (1963)
Intercalation into DNA—insertion of materials into the DNA so that the dimensions or properties of the helix are altered	Frame shift	Proflavin	Streisinger et al. (1966)
		ICR 170	Ames and Whitfield (1966)
Crosslinking of DNA strands	Deletions; frameshifts	HNO_2	Tessman (1962); Becker et al. (1964)
		Nitrogen and sulfur mustard	Kihlman (1966)

[a] Nothing is known of the possible genetic effect of substitutions on other portions of the DNA molecule—for example, additions to the deoxyribose.

data is as follows: If the frequency of r mutations is about 7×10^{-3} and this represents a change in about 1 percent of the genome, the overall mutation frequency for all phage genes must be 0.7. That is, $175/0.7 = 250$ ethyl groups fixed to the whole DNA are sufficient to produce, on the average, one mutation per bacteriophage. On the other hand, 900 to 1000 ethylations are required to produce a single lethal hit* in the virus. Lethality is the result of either depurination or breaks occurring as a result of processes secondary to the initial alkylation (see below); these inactivating lesions are relatively simple to detect by physical means in contrast to the mutagenic lesion.

The contrast between the number of lesions required to produce mutation and the number required to inactivate illustrates the problem in the detection of potential mutagens by physical means: the most effective mutagens need not be particularly lethal and need not produce immediate and drastic physical changes in the DNA. Their mutagenic effects may be produced at levels of reaction only 20 percent or less of that required for inactivation. In addition, there is evidence that some of the efficient mutagens produce mutation and inactivate by quite different mechanisms (Cerdá-Olmedo and Hanawalt, 1967). Methods are therefore required which will detect on the order of one or two alterations per 4000 nucleotides. In addition, we require means by which the subtle mutagenic changes can be converted into readily recognizable physical effects. Those chemical changes often associated with mutation are indicated in Table 1.

II. DNA SUBSTRATES FOR PHYSICAL STUDIES

Discussion of the most efficient means of detecting mutagenic change requires a brief discussion of the major features of the structure of DNA as it is found in free living organisms, in viruses, and in the cytoplasmic particles of bacteria. The double helical structure of DNA is made up of a series of four nucleotides linked together by phosphodiester bonds to form two independent chains which are wound around each other. At each level of the helix, two nucleotide bases, one in each chain, form a complementary pair: adenine with thymine, guanine with cytosine. The structure which results has measurable crystalline parameters (Table 2; Davies, 1967), and these

---

* A lethal hit is defined as the dose required to reduce survival to a fraction $N/N_0 = 0.37$ where N_0 is an initial number and N the surviving number of organisms. If inactivation is described by the exponential $N/N_0 = e^{-kD}$, one lethal hit is defined as that dose, D, at which $kD = 1$. A lethal hit may also be used to define the fraction of molecules surviving a given dose; i.e., one lethal hit of X-rays has been administered to a solution of DNA when only 37 percent (e^{-1}) of the molecules are intact with no single-strand breaks.

TABLE 2. Structural Parameters for DNA (from Davies, 1967)[a]

DNA	Pitch	Residues per turn	Translation per residue	Rotation per residue	Angle between perpendicular to helix axis and bases	Dihedral angle between base planes	Furanose[b] out of plane atoms	φCN[b]
A form, Na salt, 75% humidity	28.15 ± 0.16	11	2.55	32.7°	20°	16°	C-2' = −0.13 C-3' = +0.53	−14.1°
B form, Na salt, 92% humidity	34.6	10	3.46	36°				
B form, Li salt, 66% humidity	33.7 ± 0.1	10	3.37	36°	2°	5°	C-2' = +0.19[c] C-3' = 0.10	−86.7°
C form, Li salt, 66% humidity	31.0	9.3	3.32	39°	6°	10°	C-2' = +0.41 C-3' = −0.05	−74.6°
DNA–RNA hybrid, Na salt, 75% humidity	28.8 ± 0.5	11	2.62	32.7°	~20°			
dAT, B form, Li salt, 66% humidity	33.4 ± 0.2	10	3.34	36°				
dABrU, B form, Li salt, 66% humidity	33.4 ± 0.2	10	3.34	36°				

[a] Reproduced, by permission, from *Annual Reviews of Biochemistry*.

[b] The values were computed by Langridge and MacEwan. They are the displacements from the plane containing C-1'. O-1' and C-4'. The plus sign indicates that the displacement is on the same side as the C-5' (endo).

[c] Haschemeyer and Rich (personal communication) calculate C-2' to be 0.26 A endo displaced from the least-squares plane through the remaining four atoms.

parameters are changed when dyestuffs such as acridine or ange, ethidium bromide, or proflavin are intercalated into the structure (Luzzati *et al.*, 1961; Neville and Davies, 1966). Dyestuffs which can intercalate may be mutagenic in a variety of systems (Lerman, 1964), although the exact mechanism of mutagenicity can still be argued.

DNA structures in organisms have higher levels of organization, and a discussion of this organization is pertinent because of the practical importance of choosing DNA molecules with convenient physical parameters for investigation. Many DNA molecules occurring in nature are circular. The more common circular DNAs include the DNA of viruses such as polyoma and $\phi \times 174$ (see Thomas and MacHattie, 1967), the DNA of eukaryotic plasmids such as mitochondria (see Kirschner *et al.*, 1968), and the DNA of bacterial cytoplasmic factors such as the R factors (Novick,

TABLE 3. Some Homogeneous DNA Sources

| Type and Source | Form | Size[a] | | References to method of preparation |
		Daltons ($\times 10^{-6}$)	Length (μ)	
Plasmid DNA (*E. coli*)	Circular	1.45	0.96	Cozzarelli *et al.* (1968)
$\phi174$—single stand (*E. coli*)	Circular	1.7	1.77	Sinsheimer (1966)
$\phi174$—double strand, replicative form (*E. coli*)	Circular	3.4	1.64 (2.26)	Sinsheimer (1966)
Colicin E2, E3 (*E. coli*)	Circular	5	—	Bazaral and Helinski (1968)
Bacteriophage λ, b2b5c (*E. coli*)	Circular	26		Wang (1969)
λ, i.e., λcl	Circular	33		Bode and Kaiser (1965); Young and Sinsheimer (1967)
Bacteriophage 17 (*E. coli*)	Linear	25	12.5	Richardson *et al.* (1964)
Bacteriophage λ (*E. coli*)	Linear	32	17.2	Thomas and Abelson (1966); Miyazawa and Thomas (1965); Wu and Kaiser (1967)
F factor (*E. coli*)	Circular	45		Freifelder and Freifelder (1968); Freifelder (1968*b,c*)
R factor (R15) (*Proteus mirabilis*)	Circular	35	18	Nisioka *et al.* (1969)
Polyoma virus	Circular	2.9–3.4	1.56	Winocour (1963); Weil (1963); Vinograd *et al.* (1968)

[a] Data on size from MacHattie and Thomas (1968).

1969), the fertility or F factors (Freifelder, 1968), and the colicinogenic factors (Bazaral and Helinski, 1968). Although not circular in the mature virus, the DNA of the bacteriophage λ is transformed into a circular form when reproducing in its bacterial host (Bode and Kaiser, 1965).

Circular double-stranded DNA is useful for studies on the physical detection of the effects of mutagenic agents because it provides a readily prepared source of molecularly homogeneous DNA in which single events, particularly breaks or the formation of apurinic sites, can be readily detected (Table 3). Closed circular DNA molecules sediment in characteristic fashion; as a result of the circularity "the resistance to denaturation, the sedimentation velocity in neutral and alkaline solution and the buoyant density are all enhanced" (Radloff et al., 1967). A single single-strand break drastically alters these properties, making detection straightforward. This difference in the properties of closed circles and those with single-strand breaks has been employed in the measurement of the breaks induced in DNA by X-irradiation and by treatment with alkylating agents (Freifelder, 1968a; Boyce and Tepper, 1968; Boyce and Farley, 1968). Circular DNA also interacts with compounds such as proflavin and ethidium bromide, which intercalate into the DNA structure so that the density of the molecule is changed. The circularity results in a differential binding of intercalating dyes as compared to the binding by molecules with single-strand breaks because of the untwisting associated with the binding. The interactions of such dyes with circular DNA can provide data by which the binding of the dyes to DNA can be measured (Bauer and Vinograd, 1968; Vinograd et al., 1968).

Each of the chemical changes produced by the mutagens listed in Table 1 can be detected with greater or less sensitivity by physical means. The sensitivity of the physical methods, however, is a function of the homogeneity of the DNA used as a substrate for the test. If a DNA preparation containing a heterogeneous mixture of molecular sizes is used, the introduction of a few single-strand breaks will make the DNA even more heterogeneous. Although the weight average molecular weight will of course be reduced by additional breaks, the effect of a single additional break per molecule is unlikely to be spectacular. On the other hand, if a homogeneous DNA preparation such as can be isolated from phage T7 is used, the introduction of even a single break per molecule will be clearly evident by sedimentation velocity methods (Davidson et al., 1964). If the initial preparation is both homogeneous and circular (see above), the difference between native and degraded DNA can be greatly magnified. Clearly, the choice of an appropriate DNA substrate is as important for the physical investigation of the in vitro effect of a putative mutagen as is the physical methodology itself (Table 3). Unfortunately, it is often necessary to be interested in less elegant substrates, and one may need to study the attack of mutagens on DNA molecules still contained in cellular organelles such as

TABLE 4. Methods for the Isolation and Purification of DNA

Method	Principle	References
Chloroform—isoamyl alcohol deproteinization; isopropanol precipitation of DNA	Bacteria are lysed with lysozyme and/or detergent. The lysates are deproteinized by repeated treatment with chloroform–isoamyl alcohol. Extracts are treated with ribonuclease and DNA is precipitated with isopropanol. $S_{20,w}$ of up to 29 is obtained.	Marmur (1961)
Phenol extraction—pH 9	Lysates are extracted with phenol at pH 9. This procedure extracts DNA but a minimum of RNA into the aqueous layer.	Saito and Miura (1963); Miura (1967)
Pronase–phenol	Cells are lysed and treated with pronase in a $ZnCl_2$-treated dialysis bag which permits dialysis of peptides. Purification of the DNA is carried out with hot phenol. $S_{20,w}$ of up to 80 can be obtained.	Massie and Zimm (1965a,b)
Pronase–phenol–hydroxy-apatite	Modification of the phenol–pronase method to give higher molecular weight. Hydroxyapatite is used to separate RNA and DNA in the preparation.	Thomas et al. (1966); Thomas and Abelson (1966); Bernardi (1965)
Dextran–polyethylene glycol partition	Protein, single-stranded DNA, and RNA can be separated from native DNA by partition in the dextran–poly-ethylene glycol two-phase system.	Alberts (1967); Rudin and Albertsson (1967); Favre and Pettijohn (1967)
Phenol–amino salicylate	Nucleic acids are more readily separated from protein with phenol–amino salicylate mixture. DNA is separated from RNA by precipitation with ethoxyethanol.	Kirby (1957, 1964); Hastings and Kirby (1966)
Cetyltrimethyl ammonium bromide (CTAB) precipitation	CTAB precipitates DNA free of contaminating polysac-charides. Nucleic acids may be solubilized from the CTAB precipitate by treatment with sodium acetate.	Bellamy and Ralph (1968); Ralph and Bellamy (1964)
Sodium perchlorate	Bacteriophages are dissociated into protein and nucleic acid by treatment with concentrated sodium perchlorate. The nucleic acid can then be separated or used directly for physical tests.	Freifelder (1966, 1967)

Viral DNA—See Table 3 for preparation of particular viral DNAs.

chromosomes. No available method will then be completely satisfactory. However, it is my opinion that many very elegant physical studies have been performed with poorly characterized DNA substrates. Nevertheless, a commercial uncharacterized "salmon sperm DNA" may not be the appropriate material for a detailed physical investigation. Methods for the preparation of DNA of high molecular weight, free of protein, RNA, and polysaccharide, are listed in Table 4.

III. PHYSICAL METHODS

Since mutagenic effects may represent very rare chemical changes, an ideal physical method should be able to detect one or a very few changes in a molecule. Of all the possible changes in a DNA molecule, single-strand breaks are detectable by the greatest variety of methods and with the highest sensitivity (Table 5). Such breaks are inactivating changes per se. However, other more subtle changes can be converted to single-strand breaks by chemical or enzymatic treatment (see below). The enumeration of single-strand breaks is really made by comparison of the molecular weights of treated and untreated DNA samples; hence any physical property which depends on molecular weight can, theoretically, be utilized. Such methods include light scattering, viscosity, sedimentation, etc. The details of the use of the various methods can be found in a variety of manuals of physical techniques (Moore, 1968, 1969; Grossman and Moldave, 1967, 1968; Cantoni and Davies, 1966). The presence of breaks may be measured by the decrease in the difference of irreversible and fast reversible denaturation temperatures (Geiduschek, 1962; Hirschman and Felsenfeld, 1966; Wahl-Synek, 1967). A number of quantitative treatments permit calibration of the average number of breaks from reduction in the weight average molecular weight (Charlesby, 1954; Bernardi and Bach, 1968; Rupp and Howard-Flanders, 1968). However, as pointed out above, the results are much simpler to interpret if circular DNA is employed.

Analytical and/or preparative ultracentrifugation requires a large investment in equipment. A good indication of reaction between DNA and a putative mutagen can be obtained by viscosity measurements in a sensitive but not expensive apparatus (Zimm and Crothers, 1962), and development of methods for gel electrophoresis similar to those now available for RNA (Bishop *et al.*, 1967) should provide useful information. Nonetheless, for most purposes a combination of methods employing the analytical and preparative centrifuges is desirable. The electron microscope may shortly become the most useful instrument for studies on DNA, since direct visualization of DNA is possible by the technique of Kleinschmidt (1968). A direct demonstration of changes in DNA is possible; for example, areas

TABLE 5. A List of Some Physical Methods for Detecting Changes in the Macromolecular Structure of DNA

Physical effect	Method of detection	Description and comment	References
Chain breaks	Ultraviolet spectrophotometry		
	1. Increase in absorption, hyperchromic effect	Absorption at 260 nm increases as DNA is degraded. Useful for endonuclease assay but of low sensitivity.	Laskowski (1966)
	2. Decrease in temperature of irreversible denaturation	Quick cooling of partially denatured DNA results in fast renaturation. As the number of breaks increases, the renaturable portion decreases.	Geiduschek (1962); Strauss and Wahl (1964)
	Viscosity	Double-stranded molecules with single-strand breaks have increased flexibility and their solutions have lowered viscosity. Use of the rotating cylinder viscometer should permit high sensitivity.	Zimm and Crothers (1962); Crothers and Zimm (1965); Massie and Zimm (1965)
	Ultracentrifugation		
	1. Sedimentation velocity	The sedimentation rate of uniformly dispersed DNA decreases as breaks are introduced. Capable of defecting single events.	Davidson et al. (1964)
	2. Band sedimentation	Sedimentation of bands of DNA in the analytical centrifuge permits direct measurement of a concentration-independent sedimentation constant. Requires a special band forming center-piece for the ultracentrifuge cell.	Studier (1965); Vinograd et al. (1963)

3. Sedimentation velocity through sucrose gradients	Treated DNA is layered on a sucrose gradient in a preparative ultracentrifuge. Fractions are collected and analyzed by absorption or radioactivity. The distance traveled by a molecule is a function of its molecular weight and may be determined by comparison with a standard. Single-strand breaks may be determined with an efficiency close to 1, especially when circular DNA is used as a substrate.	Burgi and Hershey (1963); McEwen (1967); Noll (1967); Boyce and Tepper (1968)
Equilibrium density gradient centrifugation in presence of ethidium bromide	Circular DNA molecules with single-strand breaks are separated from intact molecules in the presence of dye since the circular form binds less dye and therefore has a higher density than molecules with breaks.	Radloff et al. (1967); Bauer and Vinograd (1968)
Electron microscopy	The length distribution of molecules of homogeneous DNA preparations can be determined by the Kleinschmidt technique. Single-strand breaks can be detected either directly or by analysis of the length distribution.	Freifelder and Kleinschmidt (1965); Kleinschmidt (1968); Inman and Schnös (1970)
Miscellaneous methods depending on molecular weight		
1. Adherance of denatured DNA to membrane filters	Only high molecular weight denatured DNA is retained by nitrocellulose filters. Accumulation of breaks permits labeled DNA to pass through such filters.	Geiduschek and Daniels (1965)

TABLE 5. *(Continued)*

Physical effect	Method of detection	Description and comment	References
	2. Gel electrophoresis	The movement of charged molecules through a gel under the influence of an electric field is a function of the size and shape of the molecules. Very high resolution is obtained with RNA molecules of different size and the method is also useful for DNA. Single- and double-stranded $\phi \times 174$ DNA are well separated.	Bishop *et al* (1967)
	3. Light scattering	Scattering by single molecules is proportional to the square of the molecular weight. Therefore, contribution of smaller molecules to scattering will be minimal.	Harpst *et al.* (1968*a, b*)
	4. Flow dichroism	Flow dichroism is a linear function of molecular weight. Not used so far for studies of interaction between DNA and mutagens.	Lee and Davidson (1968)
Crosslinking of DNA strands		Denatured DNA will not return to the native state quickly on return to non-denaturing conditions. If the molecule contains crosslinks, however, the native configuration will be quickly regained. Any method which distinguishes native from denatured DNA can therefore be used to estimate crosslinks. The experimental protocol involves denaturation by heat or acid, quick return to non-	

	denaturing conditions, and measurement of the appropriate physical property.	
Spectrophotometry	Denatured DNA has a greater UV absorption—hypochromic effect.	Geiduschek (1961); Felsenfeld (1968); Hirschman and Felsenfeld (1966)
Partition in dextran–carbowax two phase systems	Denatured and native DNAs have very different partition coefficients in this two phase system.	Alberts (1967a,b); Pettijohn (1967)
Equilibrium density gradient centrifugation	Native and denatured DNA differ in density by 0.015 g/cm³.	Becker et al. (1964); Kohn et al. (1966)
Band and boundary sedimentation	DNA with crosslinks has a higher molecular weight and sedimentation constant after denaturation than do molecules in which the strands can separate.	Studier (1965)
Optical rotatory dispersion	The specific rotation of a plane of polarized light depends on helical conformation. Changes in rotation are parallel to those hypochroism.	Adler and Fasman (1968)
Intercalation of dyes and generalized "interaction" between dyes and DNA	The intercalation of planar acridine molecules between successive base pairs changes the spacing from 3.4 to 7 Å (Orgel, 1965). This results in an increase in viscosity and a change in mass per unit length. Any method which can defect this change in mass per unit length will detect the interaction.	

TABLE 5. (*Continued*)

Physical effect	Method of detection	Description and comment	References
	Viscosity enhancement	Intrinsic viscosity depends on contour length and average length of the random coil. Intercalation would extend the helix and diminish the possibility of irregular coiling, increasing viscosity.	Lerman (1961); Drummond *et al.* (1966)
	Low-angle X-ray diffraction	Change in the mass per unit length changes the parameters of the unit cell as measured by crystallographic techniques.	Luzzati *et al.* (1961); Neville and Davies (1966)
	Flow dichroism	If intercalation occurs, the chromophoric ring of the dye will be parallel to the rings of the bases and exhibit an extent of flow dichroism comparable to that of the bases	Lerman (1963); O'Brien (1966); Nagata *et al.* (1966)
	Equilibrium density gradient centrifugation	Intercalation of dyes results in decrease in density as mass per unit length decreases.	Bauer and Vinograd (1968)
	Stopped-flow reaction kinetics	Dye and DNA are mixed and the mixture is perturbed by a temperature jump. The kinetics of return to equilibrium give information about the type of complex.	Li and Crothers (1969)
	Spectrophotometry	Absorption spectra of compounds interacting with DNA shift on addition of DNA. This method was used to demonstrate that actinomycin reacts with guanine in DNA.	Reich and Goldberg (1964); Cerami *et al.* (1967)

Velocity sedimentation	Existence of a complex can be shown by demonstrating that the mixture sediments as a complex.	Krey and Hahn (1969)
Light scattering	The asymptotic expression of the light scattering function permits determination of mass per unit length.	Mauss et al. (1967)
Base alteration		
Spectrophotometric	Mutagens such as hydroxylamine decrease the extinction of the purine and pyrimidine bases by destroying the aromatic character of the ring.	Brown and Phillips (1965); Rhaese and Freese (1968)
Enzymatic	DNA with bases alkylated by alkylating agents or with dimers produced by UV irradiation is a substrate for an endonuclease which produces single-strand breaks detectable by the methods given above. The efficiency may be high enough to detect very few altered bases.	Strauss and Robbins (1968); Friedberg and Goldthwait (1969); Kaplan et al. (1969); Boyce and Farley (1968)
Alteration of melting temperature	Partial destruction of the secondary structure will lead to a decrease in the secondary structure of DNA. Probably requires extensive reaction.	Rhaese and Freese (1968)
Base substitution	Substitution of a significant number of bases or base analogs may change the buoyant density and the melting temperature of DNA. Substitution of an "odd" base, such as hydroxymethyluracil for thymine may change the relationship between melting temperature and buoyant density.	

TABLE 5. (*Continued*)

Physical effect	Method of detection	Description and comment	References
	Increase in thermal denaturation temperature measured spectrophotometrically		Cerami *et al.* (1967)
	Change in density on substitution of bases	Centrifugation in gradients of Cs_2SO_4 is particularly useful, especially when compared to values obtained in CsCl since such determinations permit detection of unusual and altered bases. Mass substitution of bases is probably required.	Szybalski (1968)
	Production of breaks in DNA on irradiation of substituted DNA with visible light—limited to bromouracil-substituted DNA	DNA containing certain base analogs is particularly susceptible to damage by light in the near ultraviolet.	Puck and Kao (1967)

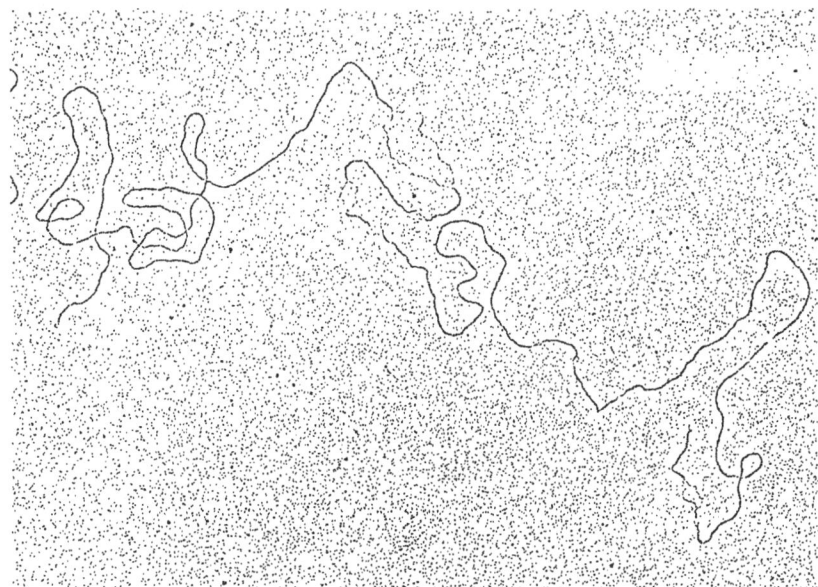

FIGURE 1. Electron Micrograph of partially denatured λ DNA (Inman and Schnös, 1970). The DNA was partially denatured for 11 min at room temperature in 0.02 M Na₂CO₃, 0.005 M Na₂EDTA, and 10 % HCHO adjusted with 1 M NaOH to pH 11.29.

of local denatuation can be seen (Fig. 1. Inman and Bertani, 1969; Inman and Schnös, 1970) and single-strand breaks can be counted (Freifelder and Kleinschmidt, 1965).

Assays for crosslinks are based on the difference in the physical properties of native and denatured DNA. Linear DNA without crosslinks can be separated into its constituent single strands by denaturation with heat or alkali. When even a single covalent bond is present to hold the two strands together, return to an environment in which native DNA is stable results in rapid return of the molecule to the native state since the base pairs are quickly brought into register (Geiduschek, 1961). Native and denatured DNAs have a density difference of 0.015 g/ml in CsCl, which is sufficient to separate the two forms in a CsCl gradient (Marmur *et al.*, 1963). It is therefore possible to estimate the relative amount of crosslinked DNA in a preparation by determining the fractions of native and denatured DNA present after denaturation and quick renaturation (Becker *et al.*, 1964). A new and convenient method for distinguishing native and denatured DNA has been described (Alberts, 1967*a,b*; Pettijohn, 1967), based on the different partition coefficients of these molecules in a two-phase system. Chromatography on hydroxyapatite has also been used to separate native and denatured DNA (Bernardi, 1965).

Methods for the detection of a few altered bases in a DNA molecule or for determining the fact that a few bases have been substituted by analogs would be most desirable for studies of potential mutagens. However, such methods are not generally available. Addition of a few alkyl groups or the substitution of a few DNA bases by the known analogs does not change the properties of DNA in any easily measurable way. In fact, it has been suggested that the most efficient mutagens are those which react with DNA to produce the least modification (Strauss et al., 1968). The ideal mutagenic change will so resemble a normal base that only at replication will there be any recognition that a change has occurred, and the result of that change will merely be a substitution of one base pair for another. Enzymes exist which recognize DNA that has been methylated by alkylating agents (Strauss and Robbins, 1968; Friedberg and Goldthwait, 1969). Such enzymes produce endonucleolytic breaks and might be used to convert molecules with mutagenic changes to molecules with single-strand breaks adjacent to the site of the substituting group. This is only a theoretical possibility at present because the enzymes involved have not been fully characterized or purified nor are their specificities understood. It should be possible to determine the number of pyrimidine dimers produced by ultraviolet radiation by reacting UV-treated DNA with a purified endonuclease with specificity for such intrastrand crosslinks (Kaplan et al., 1969) since one break is made for each dimer. Biological data (see Strauss, 1968) suggest that the UV enzyme should also attack DNA that has intrastrand crosslinks resulting from alkylation with bifunctional alkylating agents such as sulfur mustard. However, no chemical tests have been performed with this enzyme on a chemically crosslinked substrate. This lack of data is due to the difficulty in preparation of a suitable substrate since DNA alkylated with a crosslinking agent will have monofunctional alkylations and breaks as well as crosslinks (see below).

Substitution of large amounts of bromouracil for thymine changes the melting temperature of the DNA as well as causing large changes in the density of the molecule (Inman and Baldwin, 1962; Beers et al., 1967). Substitution of just a few thymine residues with bromouracil, however, should have no obvious effect on the physical properties, and, in fact, this lack of change is the basis for studies on "repair synthesis" (see Strauss, 1968). It might be possible to take advantage of the sensitivity to visible light of DNA containing bromouracil (Puck and Kao, 1967) to convert such substituted sites into detectable single-strand breaks, but such a methodology could not be generally applicable. It is my opinion that the use of analogs of high specific radioactivity is the only satisfactory method of demonstrating small amounts of analog incorporation.

The interaction of intercalating dyestuffs with DNA (Lerman, 1964) is readily measured by a variety of methods (Table 5), from the increase in viscosity owing to the extension of the chain to the change in crystal structure measurable by low-angle X-ray scattering (Neville and Davies, 1966), which occurs for the same reason. When measuring single-strand breaks we are concerned that we should be able to detect a single event due to reaction at a single site. In contrast, measurements of intercalation are almost always done with a large excess of dye; in effect, the magnitude of Avogadro's number (6×10^{23} molecules per mole) ensures that even at low dye molarities, many molecules of dye will be present per molecule of DNA. Since the mutations produced by the intercalating agents are phase shifts, i. e., additions or deletions (Streisinger et al., 1966), and may be the result of alterations in the pairing of DNA molecules in recombination owing to the extension of the molecule, it may be that the inability to detect intercalation of a single dye molecule is not important; that is, it may be necessary to have sufficient dye present to extend the whole molecule before any mutagenic effect can be demonstrated.

IV. CHANGES RESULTING FROM ALKYLATION

The utilization of physical methods for the study of mutagenesis and the limitations of these same methods can be illustrated by consideration of a particular case, reaction of DNA with the mutagenic monofunctional alkylating agents. The monofunctional alkylating agents induce mainly transitions (Krieg, 1963), but other types of change can also be induced (Freese and Freese, 1966). The bifunctional alkylating agents also induce inactivating changes in DNA and in addition to their mutagenic effect produce chromosome aberrations in vivo (Kihlman 1966). The questions I would like to discuss are (a) what changes can be detected and with what efficiency, and (b) does detection of a change by physical means indicate anything about the mutagenicity, in a given system, of the compound(s) giving the effect?

The first reaction of alkylating agents with DNA is, simply, the addition of alkyl groups to the DNA. The great majority of these are added to the bases; any other reaction seems to be quantitatively insignificant (Strauss and Hill, 1970), although there is evidence to support the suggestion that the phosphates are also alkylated (Rhaese and Freese, 1969). Some estimate of the number of alkyl groups added to DNA and to organisms can be made as can a very rough estimate of the number of alkyl groups required to produce a mutation (see above). The number of alkyl groups required to induce a significant increase in the mutation rate of a population need not change the molecular weight, density, denaturability, etc., of the DNA in a way which can be detected. However, very few alkylations are sufficient to sensitize DNA to a nuclease which does produce single-strand breaks at or near the site of alkylation (Strauss and Robbins, 1968). The single-strand breaks produced can then be detected by band centrifugation using either circular (Boyce and Farley, 1968) or linear (Strauss and Robbins, 1968) DNA molecules. The specificity of the endonuclease has not been investigated; there is evidence that it attacks both ethylated and methylated DNA. It is not known whether there are also enzymes which act on DNA with apurinic sites or whether there is an effect of the distribution of alkylated sites within a DNA molecule on sensitivity to enzyme attack (Strauss et al., 1969). The enzyme preparation used by Strauss and Robbins (1968) caused breaks at only about 50 percent of the available methylated sites. MMS and N-methyl-N'-nitro-N nitrosoguanidine (NTG) have been reported to produce the same pattern of alkylated bases (Lawley, 1968), but their relative mutagenic and inactivating effects are different. We do not know whether the alkyl groups added by the methanesulfonates are added to produce the same pattern of alkylated bases as those produced by NTG (or its active product diazomethane) (Cerdá-Olmedo and Hanawalt, 1968) or whether the pattern is important for mutagenesis.

Alkylated DNA spontaneously decomposes to yield DNA with apurinic sites (Lawley, 1966). It takes about 150 hr at 37°C for half of the methylated bases to be lost. This spontaneous reaction is speeded up by heating, and the actual rate of depurination depends on the nature of the alkylated base (Lawley and Brookes, 1963). Polydeoxynucleotide chains with apurinic sites are alkali labile and are broken on incubation at high pH (Strauss et al., 1968). A few apurinic sites in a DNA molecule can be detected by the changed sedimentation of such apurinic DNA through an alkaline sucrose gradient (Fig. 2). DNA with many apurinic sites can be recognized by the liberation of acid-soluble fragments when an apurinic DNA sample is treated with alkali before precipitation with acid (Strauss and Hill, 1970).

The final step in the process of DNA degradation as a result of alkylation is the production of single-strand breaks. Complete characterization of the degradation products requires that breaks be distinguished from apurinic sites, a problem made difficult by the ready conversion of apurinic sites to breaks in alkali. The distinction between breaks and apurinic sites can be made by denaturation in formamide followed by sedimentation in neutral gradients since such denaturaton does not hasten the conversion of apurinic sites into breaks (Strauss et al., 1968).

All of the intermediates following the alkylation of DNA by mutagenic agents can therefore be distinguished. The initial alkylation can be detected by the enzymatic susceptibility of the product or by the conversion of alkylated into apurinic sites, which can be determined by their alkali lability. Single-strand breaks and double-strand scissions can be detected by the ordinary methods of centrifugation. The extent of reaction of DNA bases with any monofunctional alkylating agent can presumably be estimated by

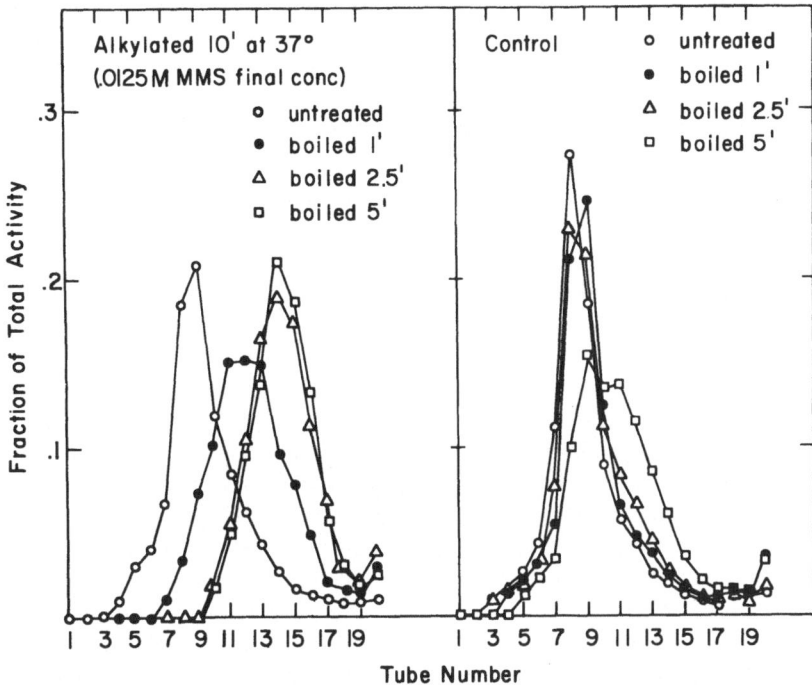

FIGURE 2. Sedimentation of methylated and control T7 DNA through an alkaline sucrose gradient after immersion in a boiling water bath (Strauss *et al.*, 1968). Phage T7 was grown on *E. coli* strain BB as a host. Tritiated DNA was prepared by the addition of ^{3}H-thymidine (Schwarz Bio Research Inc.) 2 min after infection (multiplicity of infection = 5) to a final concentration of 1 μC/ml. DNA was extracted by treating a phage suspension, purified by CsCl density gradient centrifugation, with buffer-saturated phenol. Phage T7 DNA can be cautiously pipetted in serological pipets without shearing. DNA was alkylated as follows: 0.5 ml of T7 DNA (absorbance 3.0 at 260 nm) was mixed with an equal volume of 0.025 M methyl methanesulfonate (0.085 ml methyl methanesulfonate in 40 ml 0.05 M phosphate buffer, pH 7.4), freshly made. After the desired reaction time (10 min in this experiment), the mixture was diluted 1:10 with ice-cold 0.15 M NaCl plus 0.015 M sodium citrate (SSC) and then dialyzed overnight in the cold to remove excess methyl methanesulfonate. Alkaline sucrose gradient centrifugation was performed by layering 0.1 ml of a sample containing approximately 0.6 μg of radioactive DNA onto 4.9 ml of a 5–20 % (w/w) linear sucrose gradient in 0.9 M NaCl + 0.1 M NaOH and centrifuging at 20°C at 35,000 rpm (average centrifugal force 99,972g) in the SW39 (or SW50) rotor of the Beckman L2 centrifuge for 3 hr. Fractions of 16 drops each were collected from the bottom of the tube. DNA samples were precipitated with trichloroacetic acid in the presence of albumin carrier and collected on nitrocellulose filters for counting in a Packard scintillation counter with a scintillation fluid made up of 4.9 g 2.5 diphenyloxazole and 0.1 g 1,4-bis-2-(5-phenyloxazolyl) benzene per liter of toluene.

heating to produce apurinic sites, treating with alkali to hydrolyze the chain at the apurinic sites, and then determining the number of single-strand breaks produced. A comparative estimate of the number of single-strand breaks can be obtained by determination of the acid solubility of the treated DNA (Strauss and Hill, 1970). How can one identify a particular biological effect with a particular chemical change—for example, purine alkylation (Loveless and Hampton, 1968)? It is clear that such correlation requires both chemical and biological studies. Consider, for example, the reaction resulting in the inactivation of transforming DNA after treatment with a chemical mutagen. We know single-strand breaks do inactivate. Bresler *et al.* (1968) make clear that "the additivity of the actions of different in-activating agents shows that the detailed mechanism of DNA damage is irrelevant for the kinetics of inactivation. The important point is the diminution in genetic lengths of DNA molecules because of their damage by any chemical agent." The transforming activity of DNA is a function of the single-strand molecular weight, i.e., is a function of the size of the DNA molecules which can be integrated (Wahl-Synek, 1967; Cato and Guild, 1968), but the fact that DNA can be inactivated by breaks in the chain does not determine whether other alterations, i.e., methylation (alkylation) or de-purination, can also inactivate. It seems unlikely that inactivation occurs as a result of alkylation per se, at least as a result of methylation, since there is good evidence (Prakash and Strauss, 1970) that *B. subtilis* DNA with at least one methylated base per 500 nucleotides can replicate and that DNA with methyl groups added by MMS alkylation is not inactivated for transfor-mation (Strauss and Wahl, 1964). But apurinic sites apparently inactivate both transformation (*loc. cit.*) and bacteriophage development. Phage T7 suffers one inactivating hit for each six to eight depurinations (Lawley *et al.*, 1969), and a single depurination is sufficient to inactivate phage $\varphi \times 174$ DNA treated with the half sulfur mustard, 2-chloroethyl 2-hydroxethyl sulfide (Loveless, 1966). Since the hydrolysis rate of apurinic sites is low compared with the rate of depurination of alkylated sites (Laurence, 1963), the conclusion seems unavoidable that apurinic sites per se can prevent phage DNA replication.

However, there is no way of telling from the physical effects alone whether any change less drastic than chain breakage will be lethal. Acridines increase the viscosity of DNA (Drummond *et al.*, 1966) and decrease the mass per unit length by increasing the distance between the bases (Neville and Davies, 1966), but there is no a priori reason for supposing that *any* substance with these effects will inhibit the replication of extrachromosomal pieces of DNA without affecting the replication of the chromosomes, since the mechanism of the inhibition is unknown (Hohn and Korn, 1969). There is even less possiblity of predicting that a particular compound will have a mutagenic effect in the absence of prior calibration with the biological system of interest. MMS, for example, is highly mutagenic for neurospora (Malling and de Serres, 1970), for bacteria (Strauss, 1961), and transforming DNA (Rhaese and Boetger, 1970) but not for bacteriophage treated in the mature state (Loveless, 1959); EMS is mutagenic for all four. The difference has still not been completely explained (see Loveless and Hampton, 1968). The same type of chemical reaction has presumably occurred in both cases, but the biological responses of the different systems differ.* The physical techniques do provide a methodology for determining whether reaction between DNA and a putative mutagen takes place, and this methodology is certainly useful for determining whether inactivating changes in the DNA could occur if mutagen and DNA came together *in vivo* under conditions approximating the *in vitro* test. Since such inactivating changes often result in the chromosomal aberrations which form a large proportion of the observed mutations in eukaryotic chromosomes, the demonstration that reaction occurs between a compound and DNA is not trivial. However, the actual mutagenic effect of a compound can only be determined by biological means.

V. REFERENCES

Adler, A., and Fasman, G. (1968), Optical rotatory dispersion as a means of determining nucleic acid conformation, *in* "Methods in Enzymology" (L. Grossman and K. Moldave, eds.) Vol. 12b, pp. 268–302, Academic Press, New York.

Alberts, B. (1967a), Fractionation of nucleic acids by dextran–polyethylene glycol two phase systems, *in* "Methods in Enzymology" (L. Grossman and K. Moldave, eds.) Vol. 12a, pp. 566–581, Academic Press, New York.

Alberts, B. (1967b), Efficient separation of single-stranded and double-stranded deoxyribonucleic acid in a dextran–polyethylene glycol two phase system, *Biochemistry 6*, 2527–2532.

Ames, B., and Whitfield, H. (1966), Frameshift mutagenesis in *Salmonella, Cold Spring Harbor Symp. Quant. Biol. 31*, 221–225.

Bauer, W., and Vinograd, J. (1968), The interaction of closed circular DNA with intercalative dyes I. The superhelix density of SV40 DNA in the presence and absence of dye. *J. Mol. Biol. 33*, 141–171.

Bazaral, M., and Helinski, D. (1968), Circular DNA forms of colicinogenic factors E1, E2 and E3 from *Escherichia coli, J. Mol. Biol. 36*, 185–194.

Becker, E., Zimmerman, B., and Geidsuchek, E. (1964), Structure and function of cross linked DNA. I. Reversible denaturation and *Bacillus subtilis* transformation, *J. Mol. Biol. 8*, 377–391.

Beers, W., Cerami, A., and Reich, E. (1967), An experimental model for internal denaturation of linear DNA molecules, *Proc. Nat. Acad. Sci. 58*, 1624–1631.

---

* See note added in proof.

Bellamy, A., and Ralph, R. (1968), Recovery and purification of nucleic acids by means of cetyltrimethyl ammonium bromide, *in* "Methods in Enzymology" (L. Grossman and K. Moldave, eds.) Vol. 12b, pp. 156–160, Academic Press, New York.

Bernardi, G. (1965), Chromatography of nucleic acids on hydroxyapatite, *Nature* 206, 779–783.

Bernardi, G., and Bach, M. (1968), Inactivation of *Haemophilus influenzoe* transforming DNA by spleen acid deoxyribonuclease. Appendix: Estimatation of the ratio of total bonds broken to bonds broken by diplotomic degradation in native DNA digested by spleen acid DNase, *J. Mol. Biol. 37*, 87–98.

Bishop, D., Claybrook, J., and Spiegelman, S. (1967), Electrophoretic separation of viral nucleic acids on polyacrylamide gels., *J. Mol. Biol. 26*, 378–387.

Bode, V. and Kaiser, A. (1965), Changes in the structure and activity of λ DNA in a superinfected immune bacterium, *J. Mol. Biol. 14*, 399–417.

Boyce, R., and Farley, J. (1968), Production of single-strand breaks in covalent circular λ phage DNA in super infected lysogens by monoalkylating agents and the joining of broken DNA strands, *Virology 35*, 601–609.

Boyce, R., and Tepper, M. (1968), X-ray induced single strand breaks and joining of broken strands in superinfecting λ DNA in *E. coli* lysogenic for λ, *Virology 34*, 344–351.

Bresler, S., Kalinin, V., and Perumov, D. (1968), Inactivation and mutagenesis on isolated DNA. III. Additivity of action of different agents on transforming DNA, *Mutation Res. 5*, 209–215.

Brookes, P., and Lawley, P. (1963), Effects of alkylating agents on T2 and T4 bacteriophages, *Biochem. J. 89*, 138–144.

Brown, D., and Phillips, J. (1965), Mechanism of the mutagenic action of hydroxylamine, *J. Mol. Biol. 11*, 663–671.

Burgi, E., and Hershey, A. (1963), Sedimentation rate as a measure of molecular weight of DNA, *Biophys. J. 3*, 309–321.

Cantoni, G., and Davies, D. (1966), "Procedures in Nucleic Acid Research," Harper and Row, New York.

Cato, A., and Guild, W. (1968), Transformation and DNA size. I. Activity of fragments of defined size and a fit to a random double cross-over model, *J. Mol. Biol. 37*, 157–178.

Cerami, A., Reich, E., Ward, D., and Goldberg, I. (1967), The interaction of actinomycin with DNA: Requirement for the 2-amino group of purines, *Proc. Nat. Acad. Sci. 57*, 1036–1042.

Cerdá-Olmedo, E., and Hanawalt, P. (1967), Repair of DNA damaged by N-methyl-N'nitro-N-nitrosoguanidine in *Escherichia coli*, *Mutation Res. 4*, 369–371.

Cerdá-Olmedo, E., and Hanawalt, P. (1968), Diazomethane as the active agent in nitroso guanidine mutagenesis and lethality, *Mol. Gen. Genet. 101*, 191–202.

Charlesby, A. (1954), Molecular weight changes in the degradation of long-chain polymers, *Proc. Roy. Soc. London, A Series, 224*, 120–128.

Cozzarelli, N., Kelly, R., and Kornberg, A. (1968), A minute circular DNA from *Escherichia coli* 15, *Proc. Nat. Acad. Sci. 60*, 992–999.

Crothers, D. M., and Zimm, B. H. (1965), Viscosity and sedimentation of the DNA from bacteriophages T2 and T7 and the relation to molecular weight, *J. Mol. Biol. 12*, 525–536.

Davies, D. R. (1967), X-ray diffraction studies of macromolecules, *Ann. Rev. Biochem. 36*, 321–364.

Davidson, P., Freifelder, D., and Holloway, B. (1964), Interruptions in the polynucleotide strands in bacteriophage DNA, *J. Mol. Biol. 8*, 1–10.

Drummond, D., Pritchard, N., Simpson-Gildemeister, V., and Peacocke, A. (1966), Interaction of amino acridines with deoxyribonucleic acid: Viscosity of the complexes, *Biopolymers 4*, 971–987.

Favre, J., and Pettijohn, D. (1967), A method for extracting purified DNA or protein–DNA complex from *Escherichia coli, Europ. J. Biochem. 3*, 33–41.

Felsenfeld, G. (1968), Ultraviolet spectral analysis of nucleic acid, *in* "Methods in Enzymology" (L. Grossman and K. Moldave, eds.) Vol. 12b, pp. 247–253, Academic Press, New York.

Freese, E. (1959), The specific mutagenic effect of base analogues on phage T4, *J. Mol. Biol. 1*, 87–105.

Freese, E., and Freese, E. B. (1966), Mutagenic and inactivating DNA alterations, *Radiation Res. Suppl. 6*, 97–140.

Freese, E., Freese, E. B., and Bautz, E. (1961), Hydroxylamine as a mutagenic and inactivating agent, *J. Mol. Biol. 3*, 133–143.

Freifelder, A., and Freifelder, D. (1968), Studies on *Escherichia coli* sex factors. I. Specific labeling of F′ lac DNA, *J. Mol. Biol. 32*, 15–23.

Freifelder, D. (1966), Effect of Na₂ ClO₄ on bacteriophage: Release of DNA and evidence for population heterogeneity, *Virology 28*, 742–750.

Freifelder, D. (1967), The use of Na₂ ClO₄ to isolate bacteriophage nucleic acids, *in* "Methods in Enzymology" (L. Grossman and K. Moldave, eds.) Vol. 12a, pp. 550–554, Academic Press, New York.

Freifelder, D. (1968b), Rate of production of single strand breaks in DNA by X-irradiation *in situ, J. Mol. Biol. 35*, 303–310.

Freifelder, D. (1968b), Studies on *Escherichia coli* sex factors. III. Covalently closed F′ lac DNA molecules, *J. Mol. Biol. 34*, 31–38.

Freifelder, D. (1968c), Studies on *Escherichio coli* sex factors. IV. Molecular weights of the DNA of several F′ elements, *J. Mol. Biol. 35*, 95–102.

Freifelder, D., and Kleinschmidt, A. (1965), Single strand breaks in duplex DNA of coliphage T7 as demonstrated by electron microscopy, *J. Mol. Biol. 14*, 271–278.

Friedberg, E., and Goldthwait, D. (1969), Endonuclease II of *E. coli*. I. Isolation and purification, *Proc. Nat. Acad. Sci. 62*, 934–940.

Geiduschek, E. (1961), "Reversible" DNA, *Proc. Nat. Acad. Sci. 47*, 950–955.

Geiduschek, E. (1962), On the factors controlling the reversibility of DNA denaturation, *J. Mol. 4*, 468–487.

Geiduschek, E., and A. Daniels (1965), A simple assay for DNA endonucleases, *Anal. Biochem. 11*, 133–137.

Grossman, L., and Moldave, K., eds. (1967), "Methods in Enzymology. Nucleic Acids," Vol. 12a, Academic Press, New York.

Grossman, L., and Moldave, K., eds. (1968), "Methods in Enzymology. Nucleic Acids," Vol. 12b, Academic Press, New York.

Harpst, J., Krasna, A., and Zimm, B. (1968a), Low-angle light scattering instrument for DNA solutions, *Biopolymers 6*, 585–594.

Harpst, J., Krasna, A. and Zimm, B. (1968b), Molecular weight of T7 and calf thymus DNA by low-angle light scattering, *Biopolymers 6*, 595–603.

Hastings, J.,and Kirby, K. (1966), The nucleic acids of *Drosophila melanogaster, Biochem. J. 100*, 532–539.

Hirschman, S., and Felsenfeld, G. (1966), Determination of DNA composition and concentration by spectral analysis, *J. Mol. Biol. 16*, 347–358.

Hohn, B., and Korn, D. (1969), Cosegregation of a sex factor with the *Escherichia coli* chromosome during curing by acridine orange, *J. Mol. Biol. 45*, 385–395.

Inman, R., and Baldwin, R. (1962), Helix–random coil transitions in synthetic DNA's of alternating sequence, *J. Mol. Biol. 5*, 172–184.

Inman, R., and Bertani, G. (1969), Heat denaturation of P2 bacteriophage DNA: Compositional heterogeneity, *J. Mol. Biol. 44*, 533–530.

Inman, R., and Schnös, M. (1970), Partial denaturation of thymine—and BU-containing λ DNA in alkali, *J. Mol. Biol., 49*, 93–98.

Kaplan, J., Kushner, S., and Grossman, L. (1969), Enzymatic repair of DNA, I. Purification of two enzymes involved in the excision of thymine dimers from ultraviolet-irradiated DNA, *Proc. Nat. Acad. Sci. 63*, 144–151.

Kihlman, B. (1966), "Actions of Chemicals on Dividing Cells," Prentice-Hall, Englewood Cliffs, N.J.

Kirby, K. (1957), A new method for the isolation of deoxyribonucleic acids: Evidence on the nature of bonds between deoxyribonycleic acid and protein., *Biochem. J. 66*, 495–504.

Kirby, K. (1964), Isolatiop and fractionation of nucleic acids, *Prog. Nucleic Acid Res. Mol. Biol. 3*, 1–31.

Kirschner, R., Wolstenholme, D., and Gross, N. (1968), Replicating molecules of circular mitochandrial DNA, *Proc. Nat. Acad. Sci. 60*, 1466–1472.

Kleinschmidt, A. (1968), Monolayer techniques in electron microscopy of nucleic acid molecules, *in* "Methods in Enzymology" (L. Grossman and K. Moldave, eds.) Vol. 12b, pp. 361–377, Academic Press, New York.

Kleinschmidt, A., Lang, D. Jacherts, D., and Zahn, R. (1962), Darstellung und Langenmessungen des gesamten Desoxyribonucleinsäure-Inhaltes von T2-Bakteriophagen, *Biochim. Biophys. Acta 61*, 857–864.

Kohn, K., Spears, C., and Doty, P. (1966), Interstrand crosslinking of DNA by nitrogen mustard, *J. Mol. Biol. 19*, 266–288.

Kojima, S., and Ichibagase, H. (1966), Synthetic sweetening agents. VIII. Cyclohexylamine, a metabolite of sodium cyclamate, *Chem. Pharm. Bull. 14*, 971–974.

Krey, A., and Hahn, F. (1969), Berberine: Complex with DNA, *Science 166*, 755–757.

Krieg, D. (1963), Ethyl methanesulfonate–induced reversion of bacteriophage T4r II mutants, *Genetics 48*, 561–580.

Laskowski, M. (1966), Pancreatic deoxyribonuclease I, *in* "Procedures in Nucleic Acid Research" (Cantoni and Davies, eds.) pp. 85–92. Harper and Row, New York.

Laurence, D. (1963), Chain breakage of deoxyribonucleic acid following treatment with low doses of sulphur mustard, *Proc. Roy. Soc. London, Series A 271*, 520–530.

Lawley, P., and Brookes, P. (1963), Further studies on the alkylation of nucleic acids and their constituent nucleotides, *Biochem. J. 89*, 127–138.

Lawley, P., Lethbridge, J., Edwards, P., and Shooter, K. (1969), Inactivation of bacteriophage T7 by mono- and difunctional sulphur mustards in relation to crosslinking and depurination of bacteriophage DNA, *J. Mol. Biol. 39*, 181–198.

Lawley, P. (1966), Effects of some chemical mutagens and carcinogens on nucleic acids, *Prog. Nucleic Acid Res. Mol. Biiol. 5*, 89–131.

Lawley, P. (1968), Methylation of DNA by *N*-methyl *N*-nitrosourethane and *N*-methyl-*N*-nitroso-*N'*-nitro guanidine, *Nature 218*, 580–581.

Lawley, P., and Brookes, P. (1962), Ionization of DNA bases or base analogues as a possible explanation of mutagenesis with special refernce to 5-bromodeoxyuridine, *J. Mol. Biol. 4*, 216–219.

Leahy, J., Wakefield, M., and Taylor, T. (1967), Urinary excretion of cyclohexylamine following oral administration of sodium cyclamate to man, *Food Cosmet. Toxicol. 5*, 447.

Lee, C., and Davidson, N. (1968), Flow dichroism of deoxyribonucleic acid solutions, *Biopolymers 6*, 531–550.

Legator, M., Palmer, K., Green, S., and Petersen, K. (1969), Cytogenetic studies in rats of cyclohexylamine, a metabolite of cyclamate, *Science 165*, 1139–1140.

Lerman, L. (1961), Structural considerations in the interaction of DNA and acridines, *J. Mol. Biol. 3*, 18–30.

Lerman, L. (1963), The structure of the DNA–acridine complex, *Proc. Nat. Acad. Sci. 49*, 94–102.

Lerman, L. (1964), Acridine mutagens and DNA structure, *J. Cell Comp. Physiol. 64* (Suppl. 1) 1–18.

Li, H., and Crothers, D. (1969), Relaxation studies of the proflavin–DNA complex: The kinetics of an intercalation reaction, *J. Mol. Biol. 39*, 461–478.

Loveless, A. (1959), The influence of radiomimetic substances on deoxyribonucleic acid synthesis and function studied in *Escherichia coli*–phage systems. III. Mutation of T2 bacteriophage as a consequence of alkylation *in vitro*: The uniqueness of ethylation, *Proc. Roy. Soc. London, Series B 150*, 497–508.

Loveless, A. (1966), "Genetic and Allied Effects of Alkylating Agents," Butterworths, London.

Loveless, A., and Hampton, C. (1968), Inactivation and mutation of coliphage T2 by *N*-methyl and *N*-ethyl-*N*-nitrosourea, *Mutation Res. 7*, 1–12.

Luzzati, V., Masson, F., and Lerman, L. (1961), Interaction of DNA and proflavine: A small angle X-ray scattering study, *J. Mol. Biol. 3*, 634–639.

MacHattie, L., and Thomas, C. (1968), Viral DNA molecules, *in* "Handbook of Biochemistry," pp. 113–117, Chemical Rubber Co., Cleveland.

McEwen, C. (1967), Tables for estimating sedimentation through linear concentration gradients of sucrose solution, *Anal. Biochem. 20*, 114–149

McQuillen, K. (1965), The physical organization of nucleic acid and protein synthesis, *Symp. Soc. Gen. Microbiol. 15*, 135–158.

Malling, H. V., and de Serres, F. J. (1970), in press.

Marmur, J. (1961), A procedure for the isolation of deoxyribonucleic acid from microorganisms, *J. Mol. Biol. 3*, 208–218.

Marmur, J., Rownd, R., and Schild (1963), Denaturation and renaturation of deoxyribonucleic acid, *Prog. Nucleic Acid Res. 1*, 231–300.

Massie, H., and Zimm, B. (1965a), Molecular weight of the DNA in the chromosomes of *E. coli* and *B. subtilis*, *Proc. Nat. Acad. Sci. 54*, 1636–1641.

Massie, H., and Zimm, B. (1965b), The use of hot phenol in preparing DNA, *Proc. Nat. Acad. Sci. 54*, 1641–1643.

Mauss, Y., Chambion, J., Duane, M., and Benoit, H. (1967), Etude morphologique par diffusion de la lumiere du complexe forme par le DNA et la proflavine, *J. Mol. Biol. 27*, 579–589.

Miura, K. (1967), Preparation of bacterial DNA by the phenol-pH 9-RNases method, *in* "Methods in Enzymology" (L. Grossman and K. Moldave, eds.) Vol. 12a, pp 543–545, Academic Press, New York.

Miyazawa, Y., and Thomas, C. (1965), Nucleotide composition of short segments of DNA molecules, *J. Mol. Biol. 11*, 223–237.

Moore, D., ed. (1968, 1969), "Physical Techniques in Biological Research," Vol. II, Part A, Physical Chemical Techniques (1968); Vol. II, Part B (1969), Academic Press, New York.

Mosig, G. (1968), A map of distances along the DNA molecule of phage T4, *Genetics 59*, 137–151.

Nagata, C., Kodama, M., Tagashira, Y., and Imamura, A. (1966), Interaction of polynuclear aromatic hydrocarbons, 4-nitroquinoline 1-oxides, and various dues with DNA, *Biopolymers 4*, 409–427.

Neville, D. M., Jr., and Davies, D. (1966), The interaction of acridine dyes with DNA: An X-ray diffraction and optical investigation, *J. Mol. Biol. 17*, 57–74.

Nisioka, T., Mitani, M., and Clowes, R. (1969), Composite circular forms of R factor deoxyribonucleic acid molecules, *J. Bacteriol. 97*, 376–385.

Noll, H. (1967), Characterisation of macromolecules by constant velocity sedimentation, *Nature 215*, 360–363.

Novick, R. (1969), Extrachromosomal inheritance in bacteria, [*Bacteriol. Rev. 33*, 210–263.

O'Brien, R., Allison, J., and Hohn, F. (1966), Evidence for intercalation of chloroquine into DNA, *Biochim. Biophys. Acta 129*, 622–624.

Orgel, L. E. (1965), The chemical basis of mutation. *Ad. Enzymol. 27*, 290–346.

Pettijohn, D. (1967), A study of DNA, partially denatured DNA and protein DNA complexes in the polyethylene glycol–dextran phase system, *Europ. J. Biochem. 3*, 25–32.

Phillips, J., and Brown, D. (1967), The mutagenic action of hydroxylamine, *Prog. Nucleic Acid Res. Mol. Biol. 7*, 349–368.

Prakash, L., and Strauss, B. (1970), Repair of alkylation damage: Stability of methyl groups in *Bacillus subtilis* treated with methyl methanesulfonate, *J. Bacteriol. 102*, 760–766.

Puck, T., and Kao, F. (1967), Genetics of somatic mammalian cells V. Treatment with 5-bromodeoxyuridine and visible light for isolation of nutritionally deficient mutants, *Proc. Nat. Acad. Sci. 58*, 1227–1234.

Radloff, R., Bauer, W., and Vinograd, J. (1967), A dye–buoyant-density method for the detection and isolation of closed circular duplex DNA: The closed circular DNA in HeLa cells, *Proc. Nat. Acad. Sci. 57*, 1514–1521.

Ralph, R., and Bellamy, A. (1964), Isolation and purification of undegraded ribonucleic acids, *Biochim. Biophys. Acta 87*, 9–16.

Reich, E., and Goldberg, I. (1964), Actinomycin and nucleic acid function, *Prog. Nucleic Acid Res. Mol. Biol. 3*, 183–234.

Rhaese and Boetger (1970).

Rhaese, H., and Freese, E. (1969), Chemical analysis of DNA alterations. I. Base liberation and backbone breakage of DNA and oligo-deoxyadenylic acid induced by hydrogen peroxide and hydroxylamine, *Biochim. Biophys. Acta 155*, 476–490.

Richardson, C., Inman, R., and Kornberg, A. (1964), Enzymic synthesis of deoxyribonucleic acid. XVIII. The repair of partially single-stranded DNA templates by DNA polymerase, *J. Mol. Biol. 9*, 46–69.

Rudin, L., and Albertsson, P. (1967), A new method for the isolation of deoxyribonucleic acid from microorganisms, *Biochim. Biophys. Acta 134*, 37–44.

Rupp, W., and Howard-Flanders, P. (1968), Discontinuities in the DNA synthesized in an excision defective strain of *Escherichia coli* following ultraviolet irradiation. Appendix: Theoretical sedimentation pattern of DNA with random breaks, *J. Mol. Biol. 31*, 291–304.

Saito, H., and Miura, K. (1963), Preparation of transforming deoxyribonucleic acid by phenol treatment, *Biochim. Biophys. Acta 22*, 619–629.

Schuster, H. (1960), Die Reaktionsweise der Deoxyribonucleinsäure mit salpetriger Säure, *Z. Naturforsch. 15b*, 298–304.

Sinsheimer, R. (1966), φX 174 DNA, *in* "Procedures in Nucleic Acid Research" (Cantoni and Davies, eds.) pp. 569–576, Harper and Row, New York.

Strauss, B. (1961), Specificity of the mutagenic action of the alkylating agents, *Nature* 191, 730–731.

Strauss, B. (1968), DNA repair mechanisms and their relation to mutation and recombination, *Current Topics Microbiol. Immunol. 44*, 1–85.

Strauss, B., and Hill, T. (1970). The intermediate in the degradation of DNA alkylated with a monofunctional alkylating agent, *Biochim. Biophys. Acta 213*, 14–25.

Strauss, B., and Robbins, M. (1968), DNA methylated *in vitro* by a monofunctional alkylating agent as a substrate for a specific nuclease from *Micrococcus lysodeikticus, Biochim. Biophys. Acta 161*, 68–75.

Strauss, B., and Wahl, R. (1964), The presence of breaks in the deoxyribonucleic acid of *Bacillus subtilis* treated *in vivo* with the alkylating agent, methyl methanesulfonate, *Biochim. Biophys. Acta 80*, 116–126.

Strauss, B., Coyle, M., and Robbins, M. (1968), Alkylation damage and its repair, *Cold Spring Harbor Symp. Quant. Biol. 33*, 277–287.

Strauss, B., Coyle, M., and Robbins, M. (1969), Consequences of alkylation for the behavior of DNA, *Ann. N.Y. Acad. Sci. 163*, 765–787.

Streisinger, G., Okada, Y., Emrich, J., Newton, J., Tsugita, A., Terzaghi, E., and Inouye, M. (1966), Frameshift mutations and the genetic code, *Cold Spring Harbor Symp. Quant. Biol. 31*, 77–84.

Studier, F. (1965), Sedimentation studies of the size and shape of DNA, *J. Mol. Biol. 11*, 373–390.

Szybalski, W. (1968), Use of cesium sulfate for equlibrium density gradient centrifugation, *in* "Methods in Enzymology" (L. Grossman and K. Moldave, eds.) Vol. 12b, pp. 330–360, Academic Press, New York.

Tessman, I. (1962), The induction of large deletions by nitrous acid, *J. Mol. Biol. 5*, 442–445.

Thomas, C., and Abelson, J. (1966), The isolation and characterization of DNA from bacteriophage, *in* "Procedures in Nucleic Acid Research" (Cantoni and Davies, eds.) pp. 553–561, Harper and Row, New York.

Thomas, C., Berns, K., and Kelly, T. (1966), Isolation of high molecular weight DNA from bacteria and cell nuclei, *in* "Procedures in Nucleic Acid Research" (Cantoni and Davies, eds.) pp. 535–540, Harper and Row, New York.

Thomas, C. A., and MacHattie, L. (1967), The anatomy of viral DNA molecules, *Ann. Rev. Biochem. 36*, 485–518.

Vielmetter, W., and Schuster, H. (1960), Die Basenspezifität bei der Induktion von Mutationen durch salpetrige Säure in Phagen T2, *Z. Naturforsch. 15b*, 304–311.

Vinograd, J., Bruner, R., Kent, R., and Weigle, J. (1963), Band-centrifugation of macromolecules and viruses in self-generating density gradients, *Proc. Nat. Acad. Sci. 49*, 902–910.

Vinograd, J., Lebowitz, J., and Watson, R. (1968), Early and late helix-coil transitions in closed circular DNA. The number of superhelical turns in polyoma DNA, *J. Mol. Biol. 33*, 173–197.

Wahl-Synek, R. (1967), Production of single strand breaks as an inactivating effect of the chemical mutagen methyl methanesulfonate on *Bacillus subtilis* DNA, PhD thesis, University of Chicago.

Wang, J. (1969), Degree of superhelicity of covalently closed cyclic DNA from *Escherichia coli, J. Mol. Biol. 43*, 263–272.

Weil, R. (1963), The denaturation and the renaturation of the DNA of polyoma virus, *Proc. Nat. Acad. Sci. 49*, 480–487.

Winocour, E. (1963), Purification of polyoma virus, *Virology 19*, 158–168.

Wu, R., and Kaiser, A. (1967), Mapping the 5'-terminal nucleotides of the DNA of bacteriophage λ and related phages, *Proc. Nat. Acad. Sci. 57*, 170–177.

Young, E., and Sinsheimer, R. (1967), Vegetative bacteriophage λ DNA II. Physical characterization and replication, *J. Mol. Biol. 30*, 165–200.

Zimm, B. H., and Crothers, D. M. (1962), Simplified rotating cylinder viscometer for DNA, *Proc. Nat. Acad. Sci. 48*, 905–911.

NOTE ADDED IN PROOF

It has recently been discovered (Loveless, 1969; Lawley and Thatcher, 1970) that the exact nature of the products produced by methylating (alkylating) agents depends on the detailed nature of the alkylating agent. Since some alkylation products may not lead to depurination, these new findings point out the fact that physical studies can give data of only limited value in considering the potential mutagenicity of a compound.

Loveless, A. (1969), Possible relevance of O-6 alkylation of deoxyguanosine to the mutagenicity and carcinogenicity of nitrosamines and nitrosamides, *Nature 223*, 206–207.

Lawley, P. and Thatcher, C. (1970), Methylation of deoxyribonucleic acid in cultured mammalian cells by *N*-methyl-*N*'-nitro *N*-nitrosoguanidine, *Biochem. J. 116*, 693–707.

CHAPTER 6

Effects on DNA: Transforming Principle

Roger M. Herriott*

Department of Biochemistry
School of Hygiene and Public Health
The Johns Hopkins University
Baltimore, Maryland

I. INTRODUCTION

Genetic transformation in bacteria involves the transfer of hereditary determinants from ruptured or lysed cells of one strain to a recipient strain which then develops characteristics of the donor cell line. It was this experiment which led to the discovery that DNA is the carrier of heritable properties (Avery *et al.*, 1944). DNA can be purified and stored and will confer the specific properties soon after being incorporated by recipient cells. DNA from transformable bacteria is particularly suited for studies of mutagenesis for it is quite stable and its genetic characteristics can be assayed relatively precisely.

The purpose of this chapter is to describe the procedures necessary to perform quantitative transformation experiments and to review the studies of chemical mutagens with DNA. A glossary of some of the abbreviations and unusual terms is to be found at the end of this chapter (p. 216).

* The writer acknowledges support from Public Health Service research grant AI-01218 from the National Institutes of Allergy and Infectious Diseases; and from Atomic Energy Commission contract AT(30-1)-1371.

TABLE 1. Genetic Markers of Transformable Bacteria

Organism and genetic marker	Symbol	Concentration of selective agent (μg/ml)	References
D. pneumoniae			
Capsular	SI–SIII		Avery *et al.* (1944)
Penicillin (benzyl)	pen	0.05 (units)	Hotchkiss (1951)
Mannitol	man	3000	Hotchkiss and Marmur (1954)
Amylomaltase	mal		Lacks and Hotchkiss (1960)
Streptomycin	str	150	Hotchkiss and Marmur (1954)
Streptomycin dependence	str-d	500–1000	Ravin and Mishra (1965)
Bryamycin	bry	10	Marmur and Lane (1960)
Erythromycin	ery$_2$	1	Ravin and Iyer (1961)
M-protein	M		Feingold and Austrian (1966)
Sulfanilamide	Sul *adb*	800	Hotchkiss and Evans (1958)
Sulfanilamide	Sul *a*	20	Hotchkiss and Evans (1958)
Sulfanilamide	Sul *d*	80	Hotchkiss and Evans (1958)
Sulfanilamide	Sul *b*	15	Hotchkiss and Evans (1958)
Para aminosalicylic	pas	4	Hotchkiss and Evans (1958)
Para nitrobenzoic	nob	10–50	Hotchkiss and Evans (1958)
Micrococcin	mic	0.1	Ottolenghi and Hotchkiss (1962)
Amethopterin	am	0.5	Ottolenghi and Hotchkiss (1962)
Aminopterin	ami-A	2.4	Sicard and Ephrussi-Taylor (1965
Uracil	ura		Morse and Lerman (1969)
Optochin	Q^r	15	Lerman and Tolmach (1959)
H. influenzae			
Streptomycin	str	2000	Alexander and Leidy (1953)
Streptomycin	str	100	Hsu and Herriott (1961)
Streptomycin	str	10	Hsu and Herriott (1961)
Novobiocin[a]	nov	2.5	Goodgal and Herriott (1961)
Novobiocin[a]	nov	25	Voll and Goodgal (1966)
Kanamycin	kan	8	Goodgal and Herriott (1961)
Erythromycin	ery	10	Goodgal and Herriott (1961)
Viomycin	vio	150	Goodgal and Herriott (1961)
B. subtilis			
Tryptophan	try	20	Spizizen (1958)
Indole	ind	20	Spizizen (1958)
Histidine	his	20	Nester and Lederberg (1961)
Tyrosine	tyr	10–20	Freese and Strack (1962)
Arginine	arg	20	Mahler *et al.* (1963)
Sulfanilamide	sul	200	Ephrati-Elizur (1968)
Erythromycin	ery		Takahashi (1966)
Streptomycin	str	1000	Frazer and McDonald (1966)

TABLE 1. (Continued)

Organism and genetic marker	Symbol	Concentration of selective agent (μg/ml)	References
Streptomycin dependence	str-d	1000	Frazer and McDonald (1966)
Methionine	met		Barat *et al.* (1965)
Leucine	leu		Barat *et al.* (1965)
Isoleucine	ileu		Barat *et al.* (1965)
Valine	val		Barat *et al.* (1965)
Amylase	amy		Yuki and Ueda (1968)

[a] Novobiocin is the generic name of Cathomycin, a trade name.

II. BIOLOGICAL SYSTEMS

Most studies of genetic transformation have involved one of three species of bacteria: *Diplococcus pneumoniae, Haemophilus influenzae,* or *Bacillus subtilis.* For this reason, the discussions to follow will concentrate on these three organisms, although recent publications on *Haemophilus streptococci* (Pakula, 1965; Perry, 1968; Dobrzanski and Osowiecki, 1967; Leonard *et al.*, 1967), *Bacillus licheniformis* (Thorne and Stull, 1966; Spizizen and Prestidge, 1969), *Neisseria meningitidis* (Catlin, 1960; Catlin and Cunningham, 1965; Jyssum and Lie, 1965) and *E. coli* (Avadhani *et al.*, 1969) suggest that these organisms might also have been used.

Genetic maps of two of the transformable bacteria are shown in Fig. 1. Segments of the bacterial maps described by several workers are included to show more fine structure and to indicate differences that various workers observe.

Table 1 contains some of the genetic markers for which there are well-described procedures which allow only cells with these markers to survive and form colonies. Together, the maps in Fig. 1 and the data in Table 1 indicate the variety of genetic markers available for transformation and suggest the relative position of each marker.

There are some decided advantages to using transforming DNA to detect mutagenesis. Treatment of purified transforming DNA with mutagen eliminates protection by the cell and subsequent metabolic alteration of the modified DNA. The DNA can also be freed of the mutagen after interaction. Some reactions of mutagen with free DNA are reported to be much faster (25 times) than with comparable genetic units in the cell (Bresler *et al.*, 1965). Purified DNA can be treated in the single- or double-stranded form, thereby covering both phases through which cellular DNA passes. The genetic effects of mutagen on single-stranded (denatured) DNA can

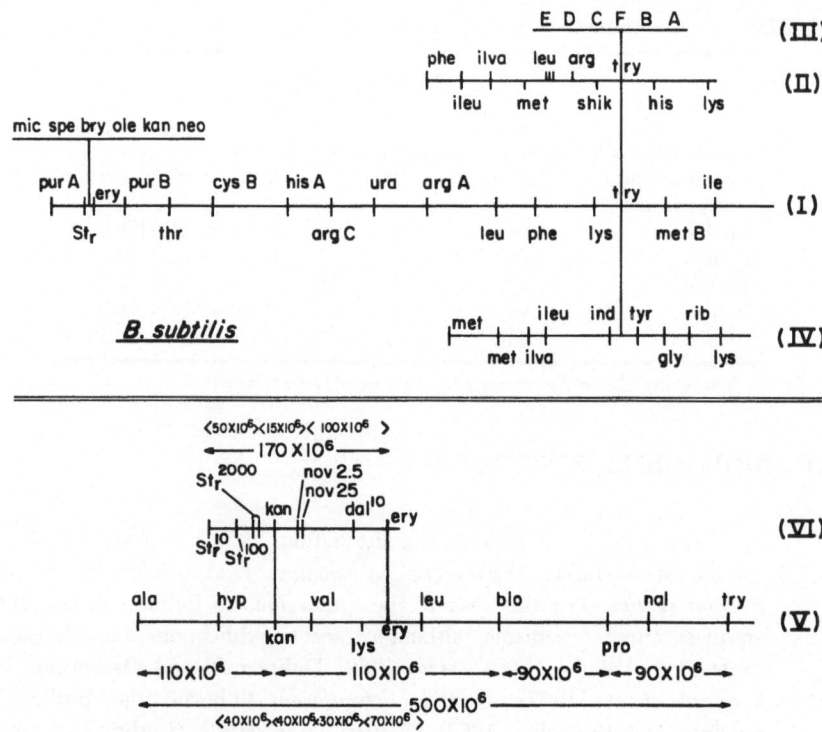

FIGURE 1.   Genetic maps of *B. subtilis* and *H. influenzae*. (I) Smith *et al.* (1968), strain
   BD 1037. (II) Barat *et al.* (1965), strain 168. (III) Carlton and Whitt (1969), strain
   SB 491 (see Table 6). (IV) Kelly and Pritchard (1965), strain 168. (V) Michalka
   and Goodgal (1969), strain Rd. (VI) Bendler (1968), strain Rd. Map distances in *H.*
   influenzae are in molecular weight units (daltons). (Courtesy of Academic Press.)

be evaluated directly (Postel and Goodgal, 1966, 1967; Chilton, 1967) or
after renaturation (Horn and Herriott, 1962). For reasons not fully understood, single-stranded DNA yields mutations faster than native DNA only
in some systems and for some mutagens (Horn and Herriott, 1962; Freese
and Strack, 1962; Strack *et al.*, 1964). In spite of these apparent inconsistencies, the use of purified transforming DNA provides the most direct
approach to correlation of chemical changes with genetic and structural
changes. In contrast to the difficulty of interpreting changes taking place
following exposure of whole cells to mutagen, it is a great advantage to be
able to reduce the components of the system to the two reactants, the
mutagen and DNA.

For a more detailed discussion of genetic transformation in bacteria the reader is referred to reviews by Ravin (1961), Schaeffer (1964), Spizizen *et al.* (1966), and Tomasz (1969). The methodology of transformation is described later in this chapter.

III. MUTAGENESIS

Mutations, as they relate to the present interest, fall into at least four separable classes: transitions, transversions, deletions, and frame shift mutations. The first two were defined by Freese (1959*b*) and the last by Brenner *et al.*, (1961).

Transitions are point mutations (single base change) in which one pyrimidine is replaced by the other, or a purine is replaced by the other. Transitions are readily induced and reverted by base analogs.

Transversions are point mutations in which a pyrimidine replaces a purine or *vice versa*. These occur spontaneously (Freese, 1961) and in conjunction with other types of mutations, by the action of some mutagenic agents. As yet no reagent specifically induces only transversions. Acid (*p*H 4–5) and slightly elevated temperatures or ethylethane sulfonate (EES) yield transversions and also cause their reversion but base analogs do not.

Deletions. As the name implies, these mutations develop when one or more bases in the DNA are modified, as with nitrous acid or UV. They fail to revert to wild type and fail to recombine with certain other mutants.

"Frame Shift" mutations are induced by several agents which cause the cell to insert (or delete) one or more extra bases into the replicated DNA. Since the reading of information from messenger RNA (complementary to DNA) is a succession of 3 base codons (frames) beginning at a fixed point on the RNA, an insertion (or deletion) of an extra base will totally alter the information in the frames following the insertion. Frame shift mutants can be recognized for they are inducible and reverted by acridine dyes (e.g., proflavin) and they are not reverted by base analogs. Streisinger *et al.* (1966) suggested that acridines stabilize the system during replication so the chances of misrepairing are increased.

To identify the class of mutant it is necessary to induce identical mutants with known agents and to test for reversion with these agents. From these properties, tentative classification of the unknown can be suggested.

The mechanism of mutation is understood in very few cases.* Pairing errors, as originally suggested by Watson and Crick (1953), is an attractive hypothesis and it is probably correct, in some instances, but it lacks solid

---

* The amino acid code is degenerate, so some mutations do not lead to a change in the amino acid.

experimental support. It is clear from the work of Freese (1961) that it does not account for spontaneous mutations in T4 phage.

The discovery of excision and repair processes (Setlow and Carrier, 1964; Boyce and Howard-Flanders, 1964) opens the possibility of misrepair as an explanation, but such a frequency of inaccuracy is not customary during replication. Perhaps, the error arises from the action of the excision enzyme rather than the repair system. The error attributable to the polymerase (Speyer, 1965, 1966) is of special interest as is the suggestion noted above of Streisinger *et al.* (1966).

For a more detailed discussion of mutations the reader is referred to the excellent book by Drake (1970), and to other chapters in this volume.

A. Kinetics of Mutagenesis

A gene contains many of each of the four bases, and a chemical mutagen reacting in only one manner will modify whatever bases are free and chemically suited for reaction. A change in a base in one position in the gene may produce a new function, e.g., resistance to an antibiotic, whereas at another it may inactivate or prevent the recognition of the new function. Thus each mutagen simultaneously produces new functions, changes that have little or no effect and changes which block the gene product from functioning. This means that the new functions are under destructive attack as they are being formed. To evaluate and correct for this destructive action, reference or standard markers are included in the system.

In many systems the number of mutants of a given type increases linearly with time of exposure to mutagen (dose). This linearity indicates that a single chemical reaction produces the mutation. If more than one reaction were needed, the curve would rise exponentially.

TABLE 2. Induction of Mutations in Transforming DNA by Nitrous Acid (Litman, 1961)[a]

Time (min)	Antibiotic markers			Mutation frequency (percent)	
	Q^r	Am^r	str^{2000}	Q^r/str	am^r/str
0	12	580	238,000	0.005	0.24
1	148	2280	193,000	0.077	1.2
4	190	3670	61,500	0.31	6
8	245	2570	24,800	1	10

[a] Production of optochin (Q^r) and aminopterin (Am^r) resistances in DNA from cells sensitive to these antibiotics but resistant to a high level of streptomycin (str). The nitrous acid was generated by mixing 1 M $NaNO_2$ with 0.25 M acetate buffer pH 4.2 at 37°C. The final concentrations of antibiotics used for screening were: Q = 7 μg/ml; str = 2 mg/ml; Am = 7×10^{-6} M.

TABLE 3. Mutagens on Transforming DNA

Agent	DNA	Marker formed	Temperature	pH	Concentration (molarity)	Percentage M/hit	References
HA	Sb	flu	70	7.5	1	6	Strack et al. (1964)
HA	Sb	flu	75	6.2	1	4[a]	Freese and Freese (1965)
HA	Sb	flu	80	5.2	1	0.5	Bresler et al. (1968)
MeOHA	Sb	flu	70	7.5	1	3	Freese and Freese (1965)
HZ	Sb	flu	45	5–7.5	1	0.02	Freese and Freese (1966)
HZ	Sb	flu	50	8.7	0.1	0.15	Bresler et al. (1968)
HU	Sb	flu	75	6.2	1	0.15	Freese (1965)
HU	Sb	flu	50	8.7	0.1	0.2	Bresler et al. (1965)
NA	Pn	Am^r	37	4.3	1	1.7	Litman (Bresler et al., 1968)
NA	Pn	Q^r	37	4.3	1	1.7	Litman (Bresler et al., 1968)
NA	Pn	Reversion	25	4.3	0.2	0.5	Lacks (Bresler et al., 1968)
NA	Sb	str^{300}				0.3	Bresler and Perumov (1962)
NA	Sb	flu	27	4.2	0.1	1.7	Strack et al. (1964)
NA	Sb	flu		4.3	0.1	1.7[a]	Strack et al. (1964)
NA	Sb	flu	20	4.4	0.2	0.2	Bresler et al. (1968)
NA	Sb	flu	37	4.5		0.27	Maher et al. (1968)
NA	HI	str^5	37	4.5	0.1–1	>10[a]	Horn and Herriott (1962)
NA	HI	kan^8	37	4.5	0.1–1	>10[a]	Horn and Herriott (1962)
NA	HI	nov^{25}	37	4.5	0.3	0.05[a]	Herriott (1970)
MNNG	Sb	flu	60	6.2	0.2	0.8	Freese and Freese (1966)
MNNG	Sb	flu	60	7	0.05	0.11	Freese et al. (1967)
MNNG	Sb	flu	60	5	0.05	0.3	Freese et al. (1967)
AAF-N-SO₄	Sb	flu	37	7.5		0.1	Maher et al. (1968)
N-AcO-AAF	Sb	flu	37	7.5		0.1	Maher et al. (1968)
N-Bzo-AAF	Sb	flu	37	7.5		0.1	Mahler et al. (1968)
DMS	Sb	flu	25	6.2	1	0.05	Bresler et al. (1968)
H₂O₂	Sb	flu	45	7.5	10^{-3}	0.04–0.1	Freese et al. (1967)
UV	Sb	flu	45	7.5		0.02	Bresler et al. (1968)
pH	Sb	flu	37–55	4.2		0.006	Strack et al. (1964)

[a] Denatured transforming DNA was used in this experiment.

The simultaneous loss of function which is exponential in cells is closer to a hyperbolic course in free DNA and is described by the expression

$$\frac{T}{T_0} = \frac{1}{(1+CD)^2}$$

(Rupert and Goodgal, 1960; Bresler *et al.*, 1968) where T=the number of residual transformants after exposure to a dose D of mutagen. The dose is frequently proportional to the time of exposure. T_0=the number of transformants before exposure. C is a factor dependent on the lengths of the DNA and the recombinational unit carrying the mutated base. When the DNA is large C=0.5. Rearranging the above formula and assuming C=0.5, the number of lethal "hits" or inactivating dose D may be calculated from the fractional transforming activity remaining by the expression

$$D = 2\left(\frac{1}{\sqrt{\dfrac{T_0}{T}}} - 1\right)$$

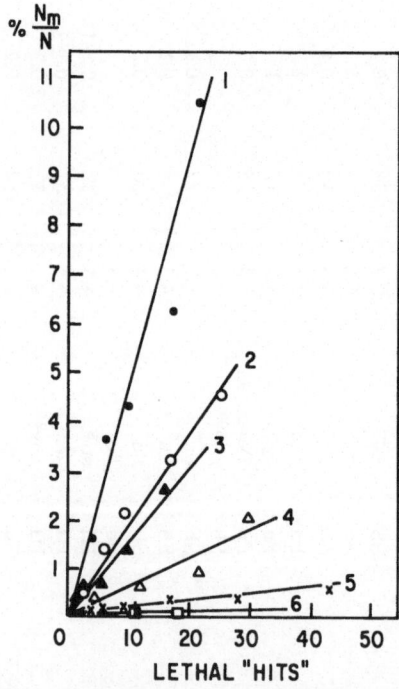

FIGURE 2. Effect of various mutagens on *B. subtilis* DNA. Fluorescent transformants from mutagenized DNA (Bresler *et al.*, 1968; Courtesy of Elsevier Publishing Company). Percent mutants vs lethal hits. Nm = number of mutants; N = number of reference transformants. 1, HA (●); 2, NA (○); 3, HZ (▲); 4, DMS (△); 5, UV irradiation (×); 6, pH 4.2 (□) (data of Strack *et al.*, 1964).

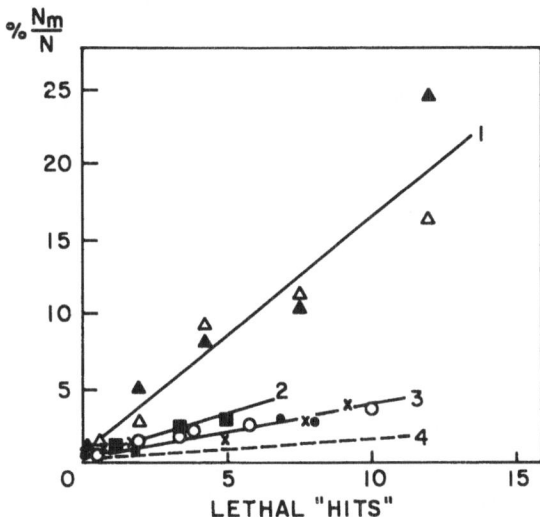

FIGURE 3.   Effect of nitrous acid on pneumococcal and subtilis DNAs (Bresler *et al.*, 1968; courtesy of Elsevier Publishing Company). Percent mutants vs lethal hits. 1, Aminopterin (▲) and optochin resistance (△) (data of Litman, 1961); 2, reversion in amylomaltase locus (□) T, marker (data of Lacks, 1966); 3, str mutants (○, ●, ×); 4, fluorescent mutants. Curves 1 and 2 are for pneumococcal and 3 and 4 for subtilis.

If it is assumed that the reference marker establishes the limit which the mutations might reach under ideal conditions, then the ratio of mutant transformants to reference transformants is a measure of the mutation frequency. The loss of reference markers through action of the mutagen automatically corrects in the ratio for the loss of new mutations by the same process. Hence, the mutants/reference marker is a corrected value of the mutation frequency (see Table 2). Lacks (1966) and Bresler *et al.* (1968) introduce other corrections such as the integration efficiency, which Lacks found to be different in the new and reference markers. A relatively simple measure of the power of a mutagen, which is in general use, is the ratio of mutants to reference markers times 100 divided by the "hits" of inactivation or lethal hits. The values of percentage M/hit for a number of mutagens acting on various transforming DNAs are shown in Table 3 and Figs. 2 and 3.

In practice, the transforming activity of the reference marker does not always decrease uniformly throughout the course of the inactivation, but this may be due to a complication of the assay system and not to the kinetics or the chemistry of mutagenesis. Mechanisms for cellular repair of damage to DNA are known to play a significant role in the deviations (Rupert and Harm, 1966; Setlow, 1966; Howard-Flanders, 1968).

kinetic expression and at two "hits" for Bresler *et al.* (1968). Figure 4 shows data from different laboratories plotted by Bresler *et al.* (1968).

B. Specificity of Mutagens

It seems generally accepted that most induced mutations are transitions (Freese 1959b); i.e., a purine is replaced by the other purine during replication, or a pyrimidine is replaced by the other pyrimidine. Of the remaining mutations, those which are not reversible may be multi-base deletions. Transversions, purine replacement by a pyrimidine or *vice versa*, occur spontaneously and are produced by some mutagens, but no agent produces them exclusively and the mechanism of their formation is unclear.

Studies of the single-stranded bacteriophage S-13 have provided information on the specificity of mutagens (Tessman *et al.*, 1964; Howard and Tessman, 1964; Baker and Tessman, 1968). They suggest a means of deducing the base mutated by the use of a few standard mutagens. This subject is discussed in more detail later.

In the following discussion various mutagens will be considered individually. Significant data are summarized in Table 5.

TABLE 5. Linked Markers

System	1	2	Cotransfer index	Percent linkage[a]	References
D. pneumoniae	str	man		0.4	Hotchkiss and Marmur (1954)
	sul^a	sul d		50	Kent and Hotchkiss (1964)
H. influenzae	str^{2000}	kan^8		70	Herriott (1970)
	str^{2000}	nov$^{2.5}$		40	Goodgal (1961)
	nov$^{2.5}$	nov^{25}		Very high	Voll and Goodgal (1966)
	nal^r	pro$^+$		28	Michalka and Goodgal (1969)
	nal^r	try		2	Michalka and Goodgal (1969)
B. subtilis	aro$_1$	his$_2$	0.57	73	Nester *et al.* (1963)
	aro$_2$	his$_2$	0.3	46	Nester *et al.* (1963)
	his$_2$	try	0.77	87	Nester *et al.* (1963)
	trp$_2$	tyr$_1$	0.41	58	Nester *et al.* (1963)
	his$_2$	flu		50	Nester *et al.* (1963)
	arg	ery	0.035	7	Mahler *et al.* (1963)
	phe$_{96}$	amy$_{12}$	0.21	35	Yuki and Ueda (1968)
B. subtilis	str	ery		57	Takahashi (1966)
	tyr$_1$	try$_2$	0.45	62	Takahashi (1966)
	aro$_2$	tyr$_1$		40	Takahashi (1966)
	met	ile	0.31	47	Takahashi (1966)
	ilva^3	leu^2		80–90	Barat *et al.* (1965)

[a] Linkage is the percent cotransfer of both markers relative to the sum of the less frequent of the individual markers and the doubly marked ones. The values vary wth the care in preparing the DNA (Kelly and Pritchard, 1965). The value is related to the cotransfer index (r) by $\frac{2r}{(1+r)} \times 100$ (Nester *et al.*, 1963).

1. Base Analogs

The base analogs used most have been 5-bromouracil (BU), or, preferably, its deoxynucleoside BUdR, and 2-aminopurine (AP) (Freese, 1959b). BU closely resembles thymine and so substitutes for it during replication (Litman and Pardee, 1956). 2-Aminopurine (AP) replaces adenine (Freese, 1959b) although the mechanism is not clear. These analogs induce both forward and reversion transitions and have been used successfully in phage and bacterial experiments but function only during replication and are not applicable to *in vitro* treatment of transforming DNA.

2. Dyes (Acridines)

Certain dyes—especially the acridines, proflavin, 5-aminoacridine, and acridine orange—have a profound mutagenic effect during replication of bacteriophage. They usually bind strongly and are believed to act by intercalation (fitting between bases in the DNA) (Lerman, 1961, 1964; Brenner *et al.*, 1961). General agreement on the mechanism of acridine mutagenesis has not been reached but it may take place during recombination (Drake, 1970). Both insertion and deletion (frame shift) mutations are produced by the acridines. *In vitro* treatment of DNA with these agents is ineffective.

3. Chemically Reactive Mutagens

a. *Nitrous Acid (NA) (HNO₂)*. NA, made by mixing $NaNO_2$ and buffers at pH 4–5, is a powerful mutagen yielding mutations with all four bases (Tessman *et al.*, 1964). However, its action on thymine is minimal and must be other than deamination, which is considered to be the action on the other three bases. NA produced mutations in some transforming DNAs (Litman and Ephrussi-Taylor, 1959; Nester and Lederberg, 1961; Strack *et al.*, 1964) but failed with DNA from *H. influenzae* (Study, 1961; Horn and Herriott, 1962; Luzzati, 1962). Only when the hemophilus DNA was denatured were mutations produced, but they were numerous (Horn and Herriott, 1962; Postel and Goodgal, 1967). Nitrous acid produces crosslinking of strands in DNA (Geiduschek, 1961),which probably accounts for the failure to find mutations in native hemophilus DNA (Luzzati, 1962), but it is not known why hemophilus DNA crosslinks so much more than pneumococcal and subtilis DNAs. Deamination of guanine should not alter its pairing properties (Freese, 1959a) but studies of polyxanthylic acid (deaminated polyguanylic acid) (Michelson and Monny, 1966) indicate that it does not pair with polycytidylic acid but does interact in non-Watson–Crick fashion with polyadenylic or polyuridylic acids. Such interactions would lead to either transversions or transitions.

b. *Hydroxylamine (HA) (H₂N—OH)*. HA the most specific mutagen. At pH 6–7 it acts only on cytosine, producing C→T transitions (Freese *et al.*, 1961; Schuster, 1961; Tessman *et al.*, 1964). Dilute ($<10^{-2}$ M) solutions of HA inactivate genes without producing mutations due to a peroxy radical reaction which develops in the presence of air (O_2) (Freese and Freese, 1965). Presumably, the inactivation prevents gene duplication somehow. When the concentration of HA is 1 M, the mutagenic action is appreciable. At this high concentration HA is a thousand times more mutagenic on denatured *B. subtilis* transforming DNA than on the native form (Freese and Strack, 1962). The failure of HA to produce mutations in TMV has been difficult to explain (Franklin and Wecker, 1959).

The action of HA on cytosine is most rapid at the 5–6 double bond, but some reports indicate that it is the N-4 amino of this base that is important for mutations (Phillips and Brown, 1967; Janion and Shugar, 1968). Formation of N-4 hydroxyaminocytosine is favored by an acid medium, which has been reported to increase the mutagenic power of HA (Freese and Strack, 1962). More recently, however, the importance of the N-4 hydroxyamino structure has been questioned (Janion and Shugar, 1969). Reaction of HA with adenine to form N-6 hydroxyadenine must also be given some consideration (Budowsky *et al.*, 1969).

   c. Hydrazine (HZ) (H_2N—NH_2). HZ breaks the pyrimidine ring. It is *p*H dependent, acting best near *p*H 8.5. It has not been used extensively. Its action is inhibited by EDTA and resembles HA in being effective at approximately molar concentration (Freese and Freese, 1966).

   d. Hydroxyurethan (HU).

$$\begin{array}{c} H \\ \diagdown \\ HO \diagup \end{array} N-\overset{\overset{\displaystyle O}{\|}}{C}-O\ C_2H_5$$

HU is especially interesting because it was discovered as a consequence of studies of the action of urethans *in vivo* which was not duplicated in *in vitro* tests. Since amines were hydroxylated *in vivo* and cell extracts converted urethan to hydroxyurethan, which was mutagenic *in vitro*, the explanation for the properties of urethan was clear (Boyland and Nery, 1965; Freese, 1965). The mutagenic activity of HU increases with decreasing *p*H, and it produces GC→AT transitions much like HA. This case illustrates the need for both *in vivo* and *in vitro* studies before an understanding may emerge.

   e. Alkylating Agents. The subject of alkylating agents has been reviewed well by others (Krieg, 1963; Lawley, 1966; Loveless, 1966; Van Duuren, 1969). Included among these mutagens are dimethylsulfate (DMS), ethyl methane sulfonate (EMS), ethyl ethanesulfonate (EES), sulfur and nitrogen mustards, ethyleneimine (EI), epoxides (Ep), and β-propiolactone (βp).

Sulfur mustard was the first chemical mutagen (Auerbach and Robson,

1944). It is a powerful inactivator of viruses and nucleic acids (Herriott, 1948), but like the nitrogen mustards it has not shown very strong mutagenic activity. This may be due to the crosslinking action of these agents on the two strands of DNA (Alexander and Stacey, 1958; Geiduschek, 1961). Many of these agents attack the phosphate groups of nucleic acids more rapidly than other groups of the nucleic acid and then are transferred to the bases or hydrolyzed without any appreciable change in genetic properties (Alexander, 1952).

Ethylating agents appear to be superior mutagens to methylating agents (Loveless, 1959). The 7 position of guanine, the 1 and 3 positions of adenine, and the 3 position of cytosine have been alkylated (Reiner and Zamenhof, 1957; Brookes and Lawley, 1961; Bautz and Freese, 1960; Krieg, 1963). Alkylation produces GC→AT transitions if the assay comes soon after reaction, but the alkylated purine is reported to be labile and will depurinate in time (Bautz and Freese, 1960; Krieg, 1963; Tessman et al., 1964). Alkylating agents also produce frame shift mutations and transversions (Lindstrom and Drake, 1970). In double-stranded DNA the depurination may not be mutagenic (Drake, 1969) because repair mechanisms reintroduce the appropriate base.

  f. *Nitrosoguanidine (MNNG).*

$$CH_3 - \underset{\underset{N=O}{|}}{N} - \underset{\underset{N-H}{||}}{C} - N\underset{\diagdown H}{\overset{\diagup NO_2}{\;}}$$

MNNG (*N*-methyl-*N*-nitroso-*N'*-nitroguanidine) is one of the most powerful mutagens when acting on bacteria (Adelberg et al., 1965), but its in vitro mutagenicity is negligible on T4 and S-13 phage (Baker and Tessman, 1968), *H. influenzae* DNA (Patty and Herriott, 1968), and TMV-RNA (Singer and Fraenkel-Conrat, 1967). Nevertheless, it is highly mutagenic for whole TMV virus (Singer and Fraenkel-Conrat, 1967). *B. subtilis* transforming DNA also yielded mutants when the mutagen concentration was raised to 0.05–0.2 M (Freese et al., 1967).

The optimum pH for mutagenicity of MNNG is near pH 6. MNNG produces diazomethane (a methylating agent) in alkaline solution (McKay, 1948) and nitrous acid in acid solution (Zimmerman et al., 1965). Although it is reported to methylate at positions 3 of adenine, 7 of guanine, and 3 of cytosine, the alkylation is negatively correlated with mutagenicity (Singer et al., 1968; Sussmuth and Lingens, 1969). Deamination is still debatable.

4. *Ultraviolet Radiation (UV)*

Ultraviolet radiation is weakly mutagenic but it is known to produce C→T transitions, frame shift mutations, and deletions in phage or bacteria, yet in no case is the mechanism established (Drake, 1970). Purines appear to be unchanged after absorbing 2537 Å radiant energy. Only the pyrimidines in DNA yield photoproducts. They undergo water addition at the 5,6 double bond (Ono *et al.*, 1965), cyclobutyl dimer formation (Beukers and Berends, 1960; Varghese and Wang, 1967), and more recently adduct formation (Wang and Varghese, 1967). Treatment of UV-irradiated DNA with the yeast photoreactivating enzyme (Rupert, 1960; Muhammed, 1966) reduces the quantity of adduct (Wang and Herriott, 1969) and of cyclobutyl dimer (Setlow, 1966) recoverable after acid hydrolysis.

Although there has been a tendency to emphasize the role of thymine in the effects of UV light on DNA, the observations Drake (1963) and Howard and Tessman (1964) indicate that in UV mutagenesis it is cytosine that is affected. Most of the UV mutants of T4 and S-13 arose through a loss of cytosine. In none of 11, S-13 mutants induced by UV light was thymine the modified base. Unfortunately, it is necessary to irradiate the host cells in order to observe the mutations, but it is difficult to see how this might direct the action to cytosine. UV light deaminates cytosine (Dellweg and Wacker, 1962; Daniel and Grimison, 1964) and this would lead to a transition. It is possible that the cytosine (hydrate) is responsible for transition mutations and that other mutations result from dimers, which produce "frameshift" mutations (Drake, 1963). Dimers are reported to produce single-strand gaps in DNA during excision repair (Setlow and Carrier, 1964) and during replication (Rupp and Howard-Flanders, 1968; Miura and Tomizawa, 1968). The mutational error appears to develop during repair (Drake 1969, Witkin, 1969).

Other material on mutagenesis in this and related fields has been covered in earlier reviews (Freese, 1963; Freese and Freese, 1966; Herriott, 1966; Singer and Fraenkel-Conrat, 1969; Drake, 1969; Witkin, 1969).

II. MEASURING MUTAGENESIS

A. Antibiotic-Resistant Mutants

Cellular resistance to antibiotics is frequently a complex phenomenon. In some instances, cells have several independent genes conveying resistance to a given antibiotic (Hotchkiss, 1951; Hotchkiss and Evans, 1958; Hsu and Herriott, 1961; Ravin and Iyer, 1961). Each gene may convey a low level of resistance, and if they act by similar mechanisms the effect may be additive. Occasionally, the combined effect is factorial (Hotchkiss and Evans, 1958; Hsu and Herriott, 1961), which probably means that the mechanisms of resistance are different. Although high-level resistance to an antibiotic can

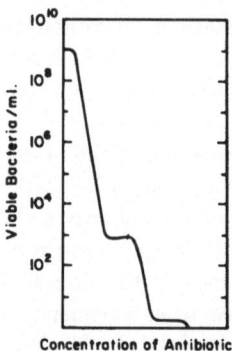

FIGURE 5. Generalized profile of antibiotic resistance in bacteria.

be a single-step mutation, it is probably more frequently the result of a series of single-step mutations (Demerec, 1945; Hotchkiss, 1951; Ravin and Iyer, 1961).

In detecting mutagenic action an easily mutated system is an advantage. For this reason, screening for low levels of resistance to antibiotics is preferred, even though the higher levels, which are less frequent, have the advantage of reducing substantially the level of spontaneous mutants (Horn and Herriott, 1962). The levels of antibiotic used must be determined empirically for each system. Antibiotic resistance profiles of the receptor cells are quickly obtained by plating several concentrations of cells on increasing concentrations of antibiotic (see Fig. 5). The concentration of antibiotic corresponding to the midpoint of that first plateau is presumably that which some natural mutants can tolerate. It is necessary to confirm the genetic and transmissible nature of this resistance. Once this is established, screening for mutations to this level of resistance can proceed. As noted earlier, it is important to include a standard or reference marker in the DNA being exposed to correct for the destructive action of the mutagen (see Table 2).

B. Reversion of Nutritional Auxotrophs

The reversion of nutritional auxotrophs to wild type can be screened directly. DNA from a multiauxotrophic cell line can be analyzed for several revertants with very little additional effort. DNA from such an auxotroph, e.g., his⁻, leu⁻, bio⁻, is treated with mutagen and then used to transform homologous competent multiauxotrophs. Samples of the transformed cells are placed on a series of plates, each containing all necessary nutrients but the one for which reversion is expected. In this way reversion to his⁺, leu⁺, and bio⁺ can be assayed independently and directly. The receptor cells will of course revert spontaneously to independence for each of these markers, and this must be appropriately evaluated. Most auxotroph studies thus far have been in *B. subtilis*, but a number of auxotrophs have been produced recently in *H. influenzae* (Michalka and Goodgal, 1969).

C. Use of Closely Linked Genes

Closely linked genes permit somewhat more sensitive methods of detecting mutagens. DNA from cells carrying a gene (denoted here as S, for standard) that is known to be in or near the gene region concerned is exposed to the mutagen and then used to transform recipient cells. Screening the transformed cells for the S gene immediately eliminates all other cells ($\sim 99\%$) and permits an examination of these transformed S cells, either directly or after replica plating, for mutations in the gene region adjoining the S gene. This double selection ensures a very low level of background cells (often zero) in the controls, thereby increasing the resolving power for new mutations.

The multigenic region (operon) responsible for tryptophan synthesis and the closely linked histidine gene (his$_2$) in *B. subtilis* is the system most studied (Anagnostopoulos and Crawford, 1961; Freese and Strack, 1962; Bresler and Perumov, 1962; Nester *et al.*, 1963; Freese and Freese, 1966; Bresler *et al.*, 1968; Maher *et al.*, 1968). Other linked genes may have comparable use. In *B. subtilis* the tyrosine operon is also close to the histidine and the tryptophan operons (see Fig. 1 and Table 5). The recent mapping of *H. influenzae* for a number of nutritional and new antibiotic genes (Fig. 1) will add breadth to this system (Michalka and Goodgal, 1969). Some of the closely linked markers of the three bacterial systems are noted in Table 5.

Six genes for the enzymes involved in synthesis of tryptophan in *B. subtilis* have been identified and mapped by transformation (Anagnostopoulos and Crawford, 1967; Carlton, 1967). Defective enzymes resulting from mutations can be located by any of three procedures: (a) direct analysis of ruptured mutant cells and comparison of enzyme activity with that from wild-type cells, (b) analysis of the growth response of cells on minimal medium supplemented with one of the known intermediates (cells containing a defective enzyme will grow only if the supplied intermediate is beyond the defect in the synthetic pathway), and (c) analysis of the accumulated intermediates in the colonies or in the supernatant fluid when the cells are supplied ample quantities of minimal nutrients and suboptimal quantities of an end product, e.g., tryptophan, above the nutritional block. This suboptimal quantity of end product keeps the cells growing slowly and accumulating product at the block.

Table 6 lists the enzymes in the tryptophan synthetic pathway of *B. subtilis* and the intermediates on which each mutant grows and indicates which intermediates accumulate when each of the enzymes is defective.

TABLE 6.   Characteristics of Tryptophan Auxotrophs of *B. subtills*[a]

Defective enzyme			Metabolic intermediate	
Number	Name	Map symbol (Fig. 1)	Supports growth	Accumulates
1	Anthranilate synthase	E	10 μg/ml anthranilate	None
2	Phosphoribosyl transferase	D	20 μg/ml indole	Anthranilate
3	Phosphoribosyl anthranilate isomerase	F	20 μg/ml indole	Anthranilate
4	Indole glycerol-phosphate synthase	C	20 μg/ml indole	1-(*o*-Carboxyphenyl-amine)-1-deoxyribulose
5	Tryptophan synthase	A	20 μg/ml indole	Indole glycerol
6	Tryptophan synthase	B	20 μg/ml tryptophan	Indole

[a] See Whitt and Carlton (1968) and Carlton and Whitt (1969) for analytical methods of determining these enzymes and intermediates.

1.   Fluorescent Linked Mutants

The fluorescence of two intermediates in the tryptophan pathway provides a simple means of detecting mutations and hence of detecting mutagens. These intermediates, anthranilate and 1-(*o*-carboxyphenylamino-1-deoxyribulose) (CDR), accumulate as noted in Table 6 when any of the enzymes 2, 3, or 4 of the tryptophan operon are defective. Their accumulation is responsible for the colonies fluorescing when examined under UV light (Anagnostopoulos and Crawford, 1961). Wild-type DNA, after exposure to mutagen, is used to transform competent his⁻ *B. subtilis* which are then plated on minimal medium, lacking histidine but with suboptimal tryptophan added. The latter permits slow growth of the cells, which accumulate intermediates at blocks induced in the tryptophan operon by the mutagen. Only cells transformed to his⁺ will survive, and of those any having a mutation in the closely linked tryptophan operon will accumulate intermediates and fluoresce. The frequency of mutation is determined simply by relating the number of fluorescent colonies to total colonies. Details of the method are described later in this chapter.

2.   Linked Temperature Sensitivity and Other Markers

Temperature-sensitive (ts) markers were important in mapping the T4 bacteriophage genome (Edgar and Lielausis, 1964). Temperature-sensitive markers have been produced in *B. subtilis* transforming DNA (McDonald and Matney, 1963; Yoshikawa, 1966; Nukushina and Ikeda, 1969) and, in theory, could be produced for any marker. This linkage of two properties permits a higher resolution of mutational change for the frequency of spontaneous mutations to two specific markers is very low. The close linkage reduces the probability of damage taking place between them. The reversion of ts-antibiotic resistance to temperature-resistant (tr) antibiotic resistant or the reversion of ts-auxotrophs to tr-auxotrophs should be simple systems to use in the detection of mutagens.

3. Evaluation of Mutagens on Linked Markers

A linear relationship was found between the mutation frequency in linked genes and the dose of mutagen inactivating the reference gene (Freese and Strack, 1962; Freese and Freese, 1966; Bresler et al., (1968). This is readily seen in Figs. 2 and 3. The slopes of the plots (percentage M/hit) reflect the power of the mutagen. The slopes are affected by the pH of the reaction medium but usually not by time, temperature, or concentration of reactants except where the nature of the reaction changes with mutagen concentration, as was found for hydroxylamine (HA) (Freese and Freese, 1966). The mutations per hit for a variety of mutagens acting on linked markers and transforming systems are listed in Table 3.

V. DISCUSSION

Our present understanding of mutagenesis has developed in a large measure from the studies of the rII region of T4 bacteriophage, tobacco mosaic virus nucleic acid (TMV-RNA), and B. subtilis and Escherichia coli. Yet it is clear from the many inconsistencies and gaps in our understanding that the subject is still in an early stage of development.

With only four different bases to be modified by mutagens, dilute solutions of transforming DNA free of other cellular components might be expected to respond in a fairly constant manner to any given mutagenic agent. However, it is apparently more complicated than that. Chemical reactions are influenced by the surrounding structures (neighboring bases), and the assay of genetic properties requires that the DNA reenter a cell where variations such as mutator genes (Yanofsky, 1966), modifying DNA polymerases (Speyer, 1965), repair processes (Howard-Flanders, 1968), and variances in integration and recombination efficiencies (Ephrussi-Taylor et al., 1965; Lacks, 1966; Yoshikawa, 1966) can alter the efficiency of any given treatment and lead to marked differences in results from the various systems and laboratories.

Perhaps one of the conclusions to be reached from studies thus far is that the absence of mutations does not mean the absence of mutagenic activity. When a test fails in one system, try another system! An example or two illustrate the point. Native pneumococcal (Litman, 1961) and subtilis (Strack et al., 1964) DNAs yield mutations readily with NA, but Haemophilus

DNA does not (Study, 1961; Horn and Herriott, 1962; Luzzati, 1962). The base composition of Haemophilus and pneumococcal DNAs is similar. The strands of native Haemophilus DNA appear to crosslink with NA, while strands of pneumococcal DNA do not—or at least much less (Luzzati, 1962). Haemophilus DNA is rapidly mutagenized with NA when the DNA is denatured (Horn and Herriott, 1962). Similar treatment of native and denatured subtilis DNA with NA shows no difference in the two forms of DNA (Strack et al., 1964), but when HA is the mutagen the action on the denatured form is 1000 times that on the native (Freese and Strack, 1962).

MNNG, a powerful mutagen in vivo (Adelberg et al., 1965), has shown disappointingly little mutagenic action in vitro on nucleic acids or phage (Singer and Fraenkel-Conrat, 1967; Patty and Herriott, 1968; Baker and Tessman, 1968). Yet Freese et al. (1967) found it active when the concentration was raised to 0.05–0.2 M and the temperature to 60°C. The in vivo action of MNNG has been attributed to its action at the growing point of DNA (Cerdá-Olmedo et al., 1968). Since single-stranded DNA was not mutagenized in vitro (Patty and Herriott, 1968), this suggests that the polymerase, or some other in vivo structure, is functioning synergistically. Perhaps this agent is being converted to a more active form in vivo much the way urethans are (Boyland and Nery, 1965; Freese, 1965). However, the action of MNNG on subtilis DNA (Freese et al., 1967) was not an in vivo reaction, so other factors must be responsible.

There are occasional chemical differences in natural bases which may be more significant in differences among DNAs than is presently appreciated. Thus, the methylcytosine, methyladenine, hydroxylmethylcytosine, and glucosylated bases may be influencing the rates of mutagenesis.

Reversion frequencies are valuable in distinguishing point mutations from multiple base changes. However, some large changes (additions) revert rather easily and it is frequently not possible to distinguish true revertants from false revertants. The latter may arise from a mutation near but not identical with that of the true revertant, or through a mutation which produces a suppressor substance that gives the unit the phenotypic properties of a revertant. Only a recombinational study can resolve such cases.

The large number of mutations—both forward and revertant—produced for the fine-structure mapping of specific gene regions will undoubtedly yield important information about the action of mutagens as well as the different levels of sensitivity of bases when surrounded by various neighboring bases (Nester et al., 1963; Ravin and Mishra, 1965; Ephrussi-Taylor et al., 1965; Barat et al., 1965; Lacks, 1966; Carlton, 1967; Carlton and Whitt, 1969).

The results found already are of sufficient interest to encourage this type work. Lacks (1966) found that all the pneumococcal mutants fell into four reasonably well-defined groups differing in integration efficiency. The efficiencies ranged from very low (0.05) to moderately low (0.2) to moderate (0.5) and high efficiency (1). UV light was the only mutagen to yield mutants with moderately low efflciency. Since there are only four groups and a given mutagen produces mutation that may have any of at least three different efficiencies, Lacks suggests that the differences may depend on the nature of the bases opposite each other at the mutation site at the time of pairing with the host genome.

Six of 13 NA-induced mutants formed at two "hot spots", indicating that certain G-C are particularly reactive or are available.

In the tryptophan operon of subtilis DNA, seven of ten mutations by NA were on only one (F) of six different enzyme genes. Four of five mutations on the E gene were induced by ICR compounds (Carlton and Whitt, 1969).

A correlation of such information into a logical pattern has not yet been achieved.

VI. SUMMARY STATEMENT

The data presented support the conclusion that mutagens can be detected and quantitative values assigned to their relative strengths with the aid of transforming DNA. Linked genetic markers are advantageous in making such tests but they are not essential. Single-stranded DNA is sometimes more sensitive to mutagens than double-stranded DNA. The weaker the mutagen, the greater the exposure necessary to obtain an effect. DNA is very stable and its genetic properties can be assayed in quite low concentrations, so there are no obvious obstacles to its use as a detector of mutagens.

VII. EXPERIMENTAL METHODS

A. Preparation of Transforming DNA

Many laboratories have special procedures for preparing and purifying transforming DNA, but that described by Marmur (1961) appears to encompass the essential features of many procedures and has yielded active transforming DNA preparations for several bacterial systems. The steps in this procedure and some of the variations are now presented. They are also summarized in Table 7.

1. Harvesting Cells

Cells from an appropriate strain are grown under optimal conditions to

TABLE 7. Variations of Marmur Method of DNA Preparation

Treatment	Marmur	Pn[a]	HI[b]	Sb[c]	B. lich.[a]
Strain		R-36 A	Rd	Sb-25	9945 A MMR-1
Concentration when harvested		$1\text{-}3 \times 10^8$	1×10^{10}	5×10^9	700-1000 µg/ml
Concentration for lysis		$1\text{-}3 \times 10^{11}$	5×10^{10}		
Lytic agent	2% SLS	0.3% deoxycholate	0.3% SLS	Lysozyme + 1% SLS	Lysozyme
Heat at 60°C	10 min		10 min		1-2 hr at 45°C
Pronase	0	0	250 µg/ml		0
Phenol extraction	0	0	0		0
Increased ionic strength	2% perchlorate	0	2 M NaCl		
Ethanol precipitation	+	+	+	+	2°C + 2 vol 95% ethanol
RNAase (µg/ml)	50	100	20	100	
Repeated agitation with chloroform–isoamyl alcohol	+	+	+	+	
Isopropanol precipitation	+	0	+		
Sterilization		80-95% ethanol, 0.45 μ millipore	Heat to 60°C for 10 min or add a trace of I_2	+	
Solvent for storage	0.15 M NaCl		0.2 M NaCl, 50% glycerol, 0.02 M citrate		
Transformations/µg DNA	8-12	8×10^7[g]	$1\text{-}2 \times 10^8$[e]	30-35	4×10^6
Mol. wt. (daltons) $\times 10^6$			(15-30)[f]		8-50
S_{w20}, (Marmur)	24	24	28.5	26	

[a] Gurney and Fox (1968).
[b] Herriott (unpublished).
[c] Kammen et al. (1966).
[d] Thorne and Stull (1966).
[e] Reported also by Postel and Goodgal (1967).
[f] Berns and Thomas (1965) describe a method of obtaining DNA of mol. wt. 400×10^6.
[g] Fox (1957).

the highest concentration in log phase—or to higher concentrations in some instances. The cells are sedimented and 2–3 g of wet paste of cells (5–10 mg DNA) is resuspended in 50 of 0.15 M saline, —1 mM EDTA or 0.15 M saline, —0.02 M Na₃ citrate (SSC). The cells are again sedimented and resuspended in 25 ml of saline EDTA or SSC. The cell concentration in the case of *H. influenzae* is adjusted to 5×10^{10}/ml (~ 80 μg DNA/ml).

2. Lysis

Cells that do not lyse readily with sodium laurylsulfate (SLS) are first incubated with lysozyme (200 μg/ml) at 37°C for 30–60 min. Lysis is induced by addition of concentrated SLS to bring the final concentration to 0.3–2 percent and warming the suspension to 60°C for 10 min. Lysis is rapid and the solution becomes viscous, as indicated by the slow rise of air bubbles. Heating to 60°C destroys some enzymes and denatures proteins in the lysate but does not harm DNA.

3. Protein Removal

Some workers remove the bulk of the protein from the lysate with two consecutive, gentle extractions with equal volumes of redistilled phenol or by digestion at 37–45°C with 250 μg/ml of pronase (CalBiochem). A third method of removing protein depends on a reaction between chloroform and protein. Raising the ionic strength has some advantages when shaking the solution with chloroform–isoamyl alcohol (10:1). High salt concentrations protect DNA and presumably dissociate DNA–protein ionic complexes which might otherwise be carried into the chloroform–protein gel. Vigorous shaking for 1/2 hr followed by centrifugation permits the separation of the upper aqueous layer containing the DNA. Shaking this DNA solution with fresh chloroform–isoamyl alcohol is repeated until no gel or precipitate forms at the interface. The DNA is then precipitated by ethanol and redissolved, or ethanol precipitation and redissolving is interspersed between treatments with chloroform for removal of protein.

4. Ethanol Precipitation of DNA

DNA in the aqueous layer which contains a high salt concentration is readily precipitated by Marmur's method of overlaying the solution, in a beaker or graduate, with an equal volume of cold 95% ethanol. The precipitate of stringy nucleic acid formed at the interface is wound up on a rod used to stir the two phases until the liquids are homogeneous. In some instances, the DNA does not form strings which wind up on a rod. Such precipitates must be centrifuged.

5. Dissolution of DNA and Digestion of the RNA

The stringy nucleic acid wound up on a rod should be transferred quickly to 50–100 ml of SSC or 0.05 M sodium acetate solution and stirred gently until completely dissolved. This may require several days of stirring, but the time is less if the precipitate is pulled into shreds with clean forceps or cut up with clean scissors. A solution of RNAase (Worthington Crystalline), adjusted to pH 5 and heated 10 min at 80°C to destroy contaminating DNAase, is added to the DNA solution to a level of 25–100 μg/ml (pH 7–8) and incubated at 37°C for 0.5–6 hr.

6.   Protein Removal and DNA Precipitation

As in step 3, the salt concentration of the solution from step 5, after RNAase digestion, is raised to 1–2 M and the solution is then shaken with chloroform and isoamyl alcohol until the evidence of protein (a stable gel or film at the interface between the phases) is negligible. Ethanol precipitation is carried out as in step 4. Redissolving in 9 or 18 ml of 0.015 M NaCl–0.0015 M Na₃ citrate is then followed with 1 or 2 ml of 3 M Na acetate + 1 mM EDTA, pH 7.0, and while the solution is rapidly stirred 0.54 vol of isopropyl alcohol is added slowly. The DNA precipitates, leaving much of, but not all, the RNA behind. The DNA is picked up on a rod or sedimented. The precipitate is quickly transferred to 0.15 M saline to yield a concentration of 200–500 μg/ml or to 0.015 M NaCl–0.0015 M EDTA if it is to be reprecipitated. If cloudy, the solution is sedimented at 5000–10,000 g for 1/2 hr and the supernatant solution is drawn off or decanted. Repeated precipitation or digestion with RNAase is necessary to remove the RNA.

7.   Sterilization and Preservation

DNA prepared as described keeps a year or more at 4°C. Effective sterilization is obtained by any of several procedures including (a) filtration through 0.45 μ millipore filters, although this tends to shear the DNA which would unlink some markers, (b) suspension of precipitated DNA in 80–95% ethanol for 15–30 min before dissolving in sterile saline (Hotchkiss, 1966), (c) heating to 60°C for 10 min. Addition of a few drops of chloroform or phenol will keep organisms from growing, as will an equal volume of sterile glycerol or traces of iodine. Iodine is especially useful for destroying DNAase or other enzymes which might have been inadvertently introduced (e.g., from hands). Iodine has no detectable effect on the biological property of DNA (Hsu, 1964). Raising the salt concentration to 0.2–2 M stabilizes the DNA and a chelating agent, 1 mM EDTA or 0.02 M Na₃ citrate, inhibits metal dependent nucleases.

8.   Properties of DNA

DNA is most reliably estimated by the diphenylamine method of Burton (1956). The absorbance at 260 nm of native DNA (E_1 μg/ml) = 0.022. The ratio of absorbancies of a DNA preparation at 260 and 230 nm is 2.3 $\pm$ 0.1

when the protein concentration is negligible. The RNA concentration is determined by the orcinol test (Mejbaum, 1939). Purified DNA gives a color by this test which is 9–10 percent of that of RNA. Correction for this is made in estimating the RNA content of DNA preparations.

B. Biological Assays

1. Transformation

The course of transformation can be put into an equation:

$$B + nD \rightleftharpoons B{\cdot}C^n \longrightarrow BD^n \xrightarrow[\text{of DNAs}]{\text{pairing}} \xrightarrow[\text{linkage}]{\text{chemical}} \downarrow$$

$$\text{transformant} \xleftarrow[\substack{\text{(expression)} \\ \text{(replication)}}]{\text{integration}}$$

Where B = bacteria (competent), D = segments of DNA, n = number of segments. B·Dn represents a temporary reversible stage which quickly changes to an irreversible form, BD^n. A competent cell can take up several (n) pieces of DNA, as evidenced by the fact that it can be simultaneously transformed to more than one independent unlinked marker. The initial steps of the above reaction follow second order kinetics with the rate being dependent on the concentrations of B and of D. This can be seen in the experimental results with *H. influenzae* shown in Figs. 6 and 7. While only the curve labeled "initial rate" is a rate curve, the others indicate that the

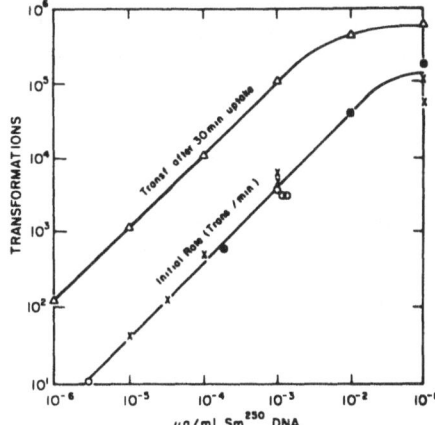

FIGURE 6. Effect of DNA concentration on the number of streptomycin-resistant transformants in 1×10^8/ml *H. influenzae*. Transformation carried out in BHI–saline, 37°C.

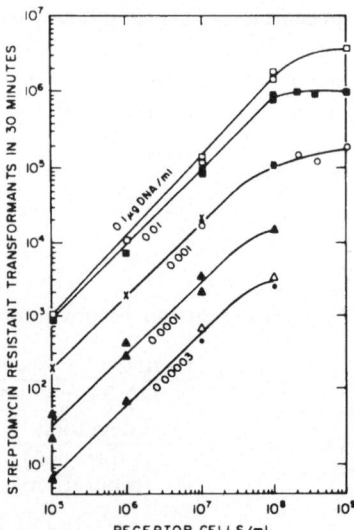

FIGURE 7. Effect of cell concentration and DNA concentration on the number of streptomycin-resistant transformants at 37°C in BHI–saline. Time for uptake of DNA was 30 min. The concentration of DNA in the transformation mixture is noted on the individual curves.

number of transformants per milliliter is directly related to both the DNA and the cell concentrations.

Quantitative determinations of transforming units are most easily understood when the DNA concentration is limiting and the cell concentration, if not in excess, is at least such that most if not all the DNA is taken up in a 30-min period at 37°C, while the cells are at full competence. Such conditions prevail when the range of concentration of DNA is 10^{-6} to 10^{-2} μg/ml in the transformation mixture, and the cells are about 3×10^8/ml. At lower concentrations of cells the number of transformants remains proportional to the DNA concentration but the DNA is not completely taken up in a given period so there is a lower number of transformants per unit DNA added. For many purposes, this is not important. The concentration for *D. pneumoniae* and *B. subtilis* needed for uptake of all the DNA is probably below the 3×10^8/ml noted for *H. influenzae*, because they are larger cells.

2. Competence

Cells, as usually cultivated, show a very low level of transformation. For useful levels of transformation, the cells must be put into a special medium or conditions must be established which promote the development of competence. In several species, development of competence is accompanied by the appearance of a high molecular weight, transmissible factor which promotes competence development in noncompetent cultures (Pakula, 1965; Tomasz, 1969). The action of these factors or the nature of either development of competence or the uptake of DNA by competent cells is far from clear. There is agreement among workers that all systems require protein synthesis during development of competence. Inhibition of protein synthesis blocks competence development.

3. A Measure of Competence

Among the many possible measures of competence (Herriott *et al.*, 1970*b*) the simplest and most practical is the determination of the fraction or percent of the bacterial population that transforms in the presence of an excess (1 μg/ml) of transforming DNA during a 30-min exposure at 37°C. As seen in Table 8, most preparations of competent cells show that 1 percent of the cells transform. Levels of 10 percent and higher have been found. The higher level does not mean, however, that such a culture has ten times as many competent cells, for there is evidence that the competence of individual cells can increase during competence development (Spencer and Herriott, 1965).

The reason such a small fraction of a competent population of cells transforms to a given marker is generally not because only a small fraction of cells is competent but is because of competition of other segments of DNA which produce no transformation but are taken in by cells. A single genetic marker represents no more than 1 percent of the genome of the cell from which the DNA was prepared.

In general, the number of transformants on a plate should be 100–400 per plate, so the standard deviation $= \sqrt{N}$, where N is the number of colonies, is not more than 5–10 percent.

Controls in which the DNA is omitted, and unless the DNA is sterile when the cells are omitted, must be included. Diluents for DNA may be sterile medium or simply buffered saline, but to retain cell viability the diluent must contain some (10 percent) growth medium or protective agents such as Tween-80. DNA solutions greater than 10 μg/ml may be heated to 60°C for 10–30 min without appreciable loss of function if the NaCl concentration is 0.2 M or greater and the *p*H is neutral.

4. Diplococcus pneumoniae

a.   Preparation of Competent Cells (Gurney and Fox, 1968).* Organism: subline R_6 from R_{36A} of Avery *et al.* (1944). Into a half-filled flask of competence medium (described below) at 37°C, wild-type cells to 5 × 10^5 per milliliter are introduced and incubated without agitation at 37°C for 4–5 hr or until the cell count reaches 1–2 × 10^8/ml. Glycerol to a level of 10 percent

---

* Courtesy of Academic Press.

TABLE 8. Competent Cells of Various Systems

Organism	Medium	Stages in development	Time (hr)	Cell concentration Initial	Cell concentration Final	Percent transformed	References
D. pneumococci	Defined	Single	4–5	5×10^5	$1–2 \times 10^8$		Gurney and Fox (1968)
D. pneumococci	Defined	Single		1×10^7	5×10^7	2–5	Tomasz and Hotchkiss (1964)
H. influenzae	BHI$^+$	Two	1 + 0.3	1×10^9	$2–3 \times 10^9$	1–3	Barnhart and Herriott (1963)
H. influenzae	MIc→MIV Stage I	Two	1.7	1×10^9	2×10^9	1–2	Herriott et al. (1970b)
B. subtilis	Penassay Stage II	Two	1.5	$4–5 \times 10^7$		0.5–1.5	Kammen et al. (1966)
B. subtilis	defined	Two	2.5	2×10^7	$1–2 \times 10^8$	3	Wilson and Bott (1968)
Group H. streptococci (Challis)	Complex	Single	3	1×10^6	1×10^7	10	Pakula (1965)
Group H. streptococci	Defined	Single	1	1×10^8	1.3×10^8	10	Leonard et al. (1967)
B. licheniformis	Defined					0.1–1	Spizizen and Prestidge (1969)
B. licheniformis	Partially defined						
Neisseria	Complex	Two-stage	2–5	3×10^8		3–4	Thorne and Stull (1966)
						1	Sparling (1966)

is added and the culture distributed into vials and frozen at $-30°C$. The reader is referred to a recent paper on competence (Tomasz, 1969).

  b. Competence Medium (Gurney and Fox, 1968). To 1 liter distilled water is added 2 g sodium acetate; 5 g vitamin-free casamino acids (salt free); 6 mg L-tryptophan; 11 mg L-cysteine HCl; 5.7 g K_2HPO_4; 10 μg $MnCl_2$; 2 g glucose; 25 ml of vitamin mixture (8.7 g K_2HPO_4; 10 μg $MnCl_2$; 2 g glucose; 25 ml of vitamin mixture (8.0 ml solution x, 2 ml solution y, 10 mg choline, 10 mg $CaCl_2$, 2 H_2O water to 50 ml, sterilized by filtration and stored in refrigerator); 25 mg glutamine; 300 mg sodium pyruvate; 340 mg sucrose; 20 mg adenosine; 20 mg uridine; 850 mg serum albumin fraction V (Armour Pharmaceutical Co., Kankakee, Ill.); and 25 ml fresh yeast extract (1 lb Fleishman's bakers yeast is slowly, during 5 min, added to 1 liter of boiling distilled water; the solution is coarsely filtered and then sterile filtered; it keeps indefinitely under refrigeration).

  Solution x (weights are in milligrams): Biotin 0.015, nicotinic acid 15, pyridosine 15, Ca pantothenate 60, thiamine 15, riboflavin 7, water to 100 ml. Sterilize by filtration. Store in refrigerator.

  Solution y (weights are in milligrams): $FeSO_4 \cdot 7 \cdot H_2O$, 50; $CuSO_4 \cdot 5$ H_2O, 50; $ZnSO_4 \cdot 7$ H_2O, 50; $MnCl_2 \cdot 4$ H_2O, 30; concentrated HCl 1 ml; water to 100 ml. Sterilize by boiling.

  c. Transformation (Pneumococcus). Competent cells, fresh or frozen samples, thawed, at $1 \times 10^8/ml$ in competence medium are incubated with appropriate concentrations of DNA at 30°C for 30 min. Addition of 1 $\mu g/ml$ of Worthington pancreatic DNAase terminates uptake of DNA and incubation is continued for 30 min at 37°C, and, after appropriate dilution, samples are plated in selective media containing 1% agar at 40°C. The colonies are counted after 40 hr at 37°C (see Table 8).

  Defined media for the growth and study of *D. pneumoniae* have been described by Sicard (1964) and Morse and Lerman (1969).

5. Bacillus Subtilis

  a. Preparation of Competent B. subtilis (Wojnar, in Kammen et al., 1966*. A culture of SB 25 (try_2^- his$^-$) is grown with aeration in Penassay broth (Difco) to a population of $2 \times 10^8/ml$ and diluted in Panassay broth containing 5% glycerol to about $10^5/ml$. The cells are distributed in 0.5 ml portions, frozen in acetone dry ice, and stored at $-17°C$. After thawing, 20–40 cells are inoculated into 20 ml of Penassay broth and grown for 16 hr at 37°C. After sedimentation the cells are resuspended in a minimal medium I for subtilis Sb MM-I (described below) at a concentration of $1.2–1.5 \times 10^8/ml$ (110–130 Klett units) (green filter) and agitated at 37°C for 4–5 hr when the culture has reached $4–5 \times 10^8/ml$ (250 Klett units).

---

* Courtesy of Elsevier Publishing Company.

The cells are sedimented and rsuspended in warm Sb MM-II (corresponds to "transformation" medium: Kammen *et al.*, 1966) the volume of which is ten times that of the supernatant decanted from the cells. The resuspended cells are shaken or aerated for 90 min at 37°C, at which time they are at maximum competence, i.e., 0.5–1.5% of cells can be transformed to one marker by an excess of DNA.

  b. Transformation. Competent cells are exposed in Sb MM-II to transforming DNA and incubated for the desired time, after which exposure is terminated by the addition of 1/100 vol of 2 mg/ml pancreatic deoxyribonuclease for 1–2 min. All dilutions are made in minimal salts (MS) + 0.5% glucose. Plating is in MS + 0.5% glucose + nutrient supplements (NS) and 1.5% Difco Noble agar (see Table 8).

  c. Subtilis Media. Sb-Ms (*minimal salts*): Ammonium sulfate, 0.2%, dipotassium phosphate, 1.4%; monopotassium phosphate, 0.6%; sodium citrate·2 H_2O, 0.1%; magnesium sulfate·7 H_2O, 0.002%.

  Sb MM-I (growth): Sb-MS, 0.5% glucose, 50 μg/ml each of tryptophan and histidine, and 0.02% casein hydrolysate.

  Sb MM-II (growth): Sb-MS + 0.5% glucose, 5 μg/ml each of tryptophan and hisitidine, 20 μg/ml of α,α'-bipryridyl, and an additional 5 μmoles/ml of $MgCl_2$.

  Sb NS (nutritional supplement): A concentrated stock solution is prepared with concentrations of each constituent expressed as mg/liter. L *amino acids:* arginine HCl 15, isoleucine 60, leucine 30, lysine HCl 50, methionine 100, phenylalanine 60, threonine 40, valine 70, cystine (DL) 20, tyrosine 60, serine 15, asparagine 10, glycine 75, alanine 100, proline 150, glutamine 250, and *p*-aminobenzoate 50.

  Vitamins: thiamine HCl, nicotinic acid, calcium pantothenate, pyridoxal HCl, and riboflavin 0.05 each. The stock solution is sterile filtered and added to agar media at the level of 40 ml per liter of medium.

6. Haemophilus influenzae

  Organism: Strains derived from the rough (R) type *d* nonencapsulated one islated by Alexander and Leidy (1953) are used in a number of laboratories.

  a. Preparation of Competent Cells. Competent *H. influenzae* can be prepared in a commercial medium (Barnhart and Herriott, 1963) or in chemically defined media (Herriott *et al.* 1970*a,b*). Only the former will be described in order to simplify the work.

  Medium (BHI⁺): Autocalved Difco brain–heart infusion (BHI) supplemented, after cooling, with sterile 10 μg/ml hemin and 4 μg/ml NAD and 3×10^{-4} M Ca^{2+} has been effective. Several other commercial preparations of BHI have not been effective in promoting competence, although they support growth. If medium containing these supplements (BHI$^+$) stands at 37°C for 24 hr, more fresh NAD must be added.

Stock hemin: 0.1 g hemin (EK Co.) and 0.1 g L-histidine are mixed in 100 ml 4% v/v aqueous 2,2',2"-nitriloethanol. This is sterilized by heating to 70°C for 10 min. It keeps for 3 weeks without loss of activity if refrigerated. After that, it is about half as effective.

Stock NAD: 10 mg/ml of NAD (Nutritional Biochemicals Co.) in water is frozen in 1-ml samples and kept at −17°C. A sample is thawed; 9 ml sterile water is added and diluted appropriately into the medium.

Development of competence: 0.5 ml of cells (usually, but not rstricted to, wild type) at 1–3 × 10^9/ml frozen in BHI$^+$ is added to 25–50 ml BHI$^+$ in a 500-ml Erlenmeyer flask and rotated at 175 rpm at 37°C until the cell density reaches 1 × 10^9/ml (∼21/2 hr). The rotation is stopped but the flask is kept at 37°C for 60 min, after which it is rotated for 20 min or until the suspension reaches 3 × 10^9/ml. Glycerol is added to bring the concentration to 15 percent and samples are frozen and stored at −65°C. Frozen cells lose competence very slowly, remaining useful for at least a month. Once thawed, they keep for many hours at 4°C and less at room temperature.

In the presence of an excess (1 μg/ml) of marked (antibiotic resistance) DNA, approximately 0.5–4 percent of the cells prepared, as described above, become transformed in 30 min at 37°C.

b. Transformation. To 1.7 ml BHI$^+$ is added 0.1 ml of buffered saline containing 1 × 10^{-4} μg/ml DNA and 0.2 ml of competent cells. This suspension is incubated without aeration for 30 min, after which 0.5 ml and 1 ml are placed in plates and mixed with 10 ml BHI$^+$–agar at 43°C. These plates are incubated at 37°C for 90 min to permit expression of the marker and are then overlayed with 10 ml BHI$^+$–agar containing the selective agent (antibiotic) at double the final concentration.

The plates are counted after 18–24 hr incubation at 37°C. Longer time is required if screening for multiple antibiotics. Under optimal conditions, approximately 1–2 × 10^8 transformants can be expected from the uptake of 1/μg of DNA (see Table 8).

c. Minimal Medium for Nutritional Mutants (Michalka and Goodgal, 1969). Modification of the synthetic medium of Talmadge and Herriott (1960) led to a less complex medium in which *H. influenzae* grow. To facilitate handling, a series of concentrated stock solutions are mixed in the order given to prepare the medium (Table 9).

Distilled water is added to the mixture of the above components to bring them to final volume and the *p*H is adjusted to 7.2 with 0.05 ml of concentrated HCl per 100 ml of medium. For solid medium, the mixture of the components is warmed to 45°C and 1.25 percent Bacto agar is added

TABLE 9. Concentrated Stock Solutions

Stock solution	Component and concentration in stock	Dilution to make HI
1	Sodium chloride, 120 mg/ml Potassium sulfate, 40 mg/ml Magnesium chloride, 2 mg/ml Calcium chloride, 0.2 mg/ml L-arginine (Nutritional Biochemicals), 6 mg/ml L-cysteine (Calbiochem.), 4 mg/ml Thiamine hydrochloride (Calbiochem), 0.04 mg/ml Solution acidified slightly (pH 2 to 4) with concentrated HCl, sterilized by filtration	1:20
2	NaH_2PO_4–Na_2HPO_4 buffer, pH 7.0, 0.5 M, sterilized by autoclaving	1:20
3	3% glucose (w/v), sterilized by autoclaving	1:100
4	Uracil (Nutritional Biochemicals), 10 mg/ml L-aspartic acid (Nutritional Biochemicals), 50 mg/ml Monosodium glutamate (Nutritional Biochemicals), 150 mg/ml Sodium hydroxide, 58 mg/ml (no further treatment was necessary because of the concentration of sodium hydroxide)	1:100
5	Nicotanamide adenine dinycleotide (Nutritional Biochemicals), 10 mg/ml, sterilized by filtration	1:1000
6	Hemin (Eastman), 10 mg/ml in 4% triethanolamine (by volume), sterilized by heating at 65°C for 15 min	1:500
7	Calcium pantothenate (Nutritional Biochemicals), 10 mg/ml, pH 5.0–7.0, sterilized by filtration	1:1000

to final volume. The pH is adjusted to 7.2 with 0.08 ml of concentrated HCl per 100 ml of medium.

Biotin (Nutritional Biochemicals), hypoxanthine (Calbiochem), leucine (Nutritional Biochemicals), alanine (Calbiochem), valine (Calbiochem), lysine (Calbiochem), proline (Nutritional Biochemicals), and tryptophan (Calbiochem) are used as nutritional supplements in the minimal medium, and stocks are prepared at 200 × the final concentration used in the medium. The amino acids are dissolved in 0.1 N–NaOH at a concentration of 10 mg/ml, with the exception of valine, which is prepared at a concentration of 4 mg/ml. No further sterilization is necessary.

Hypoxanthine is dissolved at 10 mg/ml in 0.3 N–NaOH.

The stock solution of biotin is prepared by saturation in 50% ethanol slightly acidified (pH 2–5) with HCl.

Kanamycin sulfate (Bristol) is prepared in distilled water at a concentration of 10 mg/ml. The final concentration in growth medium is 7 μg/ml. Nalidixic acid (Sterling-Winthrop) at a concentration of 10 mg/ml in 0.1 N NaOH is diluted in growth medium to a final concentration of 3 μg/ml. Novobiocin (Upjohn) (10 mg/ml) is dissolved in distilled water and used in growth medium at a final concentration of 20 μg/ml. Streptomycin sulfate (Nutritional Biochemicals) is dissolved (250 mg/ml) in distilled water and used in growth medium at a final concentration of 200 μg/ml.

A strain of Rd *H. influenzae* was selected which grew in minimal medium (strain 1). To obtain mutants this strain is grown with aeration at 37°C in BHI broth in a density of approximately 10^9 cells/ml. Ten milliliters of cells is centrifuged and resuspended in 0.5 ml of a solution which consists of 5 ml of stock solution No. 1 and 1 ml of a 0.2 M sodium acetate–acetic acid buffer of pH 5.0. The final pH of the solution is 6.0. Two milligrams N-methyl-N-nitroso-N'-nitroguanidine (K & K Laboratories) is dissolved in 0.5 ml of the 0.2 M acetate buffer, pH 5.00, and 0.1 ml of this solution is added to the resuspended cells. The suspension is incubated at 37°C for 30 min and diluted to 5 ml with BHI broth, to which glycerol is added to a final concentration of 15% by volume. The suspension is dispensed into vials and frozen at $-70°$C.

A viable count of the frozen culture is determined. On the next day a duplicate vial is thawed, diluted, and spread on BHI plates to yield approximately 150 clones per plate. After overnight incuvation at 37°C, the clones are replicated onto minimal agar plates, and minimal agar plates supplemented with nutrients noted above. Presumed mutants are picked and tested.

Only those mutants which were not leaky and had extremely low reversion frequencies ($<1/10^{-10}$) were used in the following studies.

A mutant requiring hypoxanthine (hyp$^-$) was consecutively transformed with mixtures of DNA from eight different auxotrophs and one antibiotic-resistant strain. This led to a recipient (competent) cell with resistance to 2000 μg streptomycin and requiring a variety of amino acids in addition to those in MM. Wild-type DNA transformed these cells to independence of one or more amino acids, which can be determined by placing on plates supplemented with all but one of the required amino acids or factors.

7. Biological Assay of Single-Stranded (Denatured) Transforming DNA

a. H. influenzae (Postel and Goodgal, 1967). Usual competent *H. influenzae* are centrifuged and resuspended in 0.15 M NaCl–0.3 percent Difco Bacto peptone + HCl to pH 4.4 + 3 × 10^{-4} M EDTA. Alkali-denatured and neutralized DNA is added at desired level and the mixture incubated 10 min at 37°C after which a predetermined quantity of NaOH to neutrality is added then incubated 15–20 min at 34°C. This mixture is now diluted and plated as in the case of native DNA. The viability of cells decreases to 50 percent of the initial value by this treatment.

Under the above conditions, 0.01 μg str DNA yielded 1.8×10^5 transformants with 5×10^8/ml competent cells. Radio-labeled DNA studies showed elution of some DNA following neutralization to pH 7. Correcting for this, the authors reported 1.3×10^8 transformants per microgram DNA remaining with the cells. This is close to the figure obtained with native DNA.

   b. *B. subtilis (Chilton, 1967).*   Competent *B. subtilis* are mixed with single-stranded DNA at a final level of 0.1 μg/ml in the presence of 10^{-3} EDTA. After incubation for 30 min at 37°C, the reaction is stopped with the addition of 0.05 ml of 10 μg/ml DNAase in 0.1 M $MgSO_4$. After 10 min at 37°C the cells are plated in the usual manner. A control in which EDTA is omitted serves to evaluate the native or residual transforming activity in the preparation of single-stranded DNA.

8. Determining Fluorescent Mutants in B. Subtilis

As noted in the text, mutations in the genes of three of the different enzymes in the biosynthetic pathway of tryptophan in *B. subtilis* will cause the cells to accumulate intermediates which fluoresce (Anagnostopoulos and Crawford, 1961; Carlton, 1967). The details of the detection of fluorescent mutants is presented in the following paragraphs.

   a. *Procedure for Detecting Mutant B. subtilis which Fluoresce.*[*]   Prototrophic DNA, 1–10 μg/ml in 0.1 M NaCl and dilute buffer, is exposed to the test mutagen. Samples are withdrawn and used for transformation. One-tenth milliliter of treated DNA ($<$0.5 μg/ml) and 1 ml competent cells are mixed in transformation medium and shaken for 2 hr and then diluted in 0.1 M phosphate buffer, pH 7, and plated on week-old minimal-medium plates for *B. subtilis* (Sb-MM) supplemented with 1.2 μg/ml of tryptophan and 250 μg/ml of simulated casein hydrolysate minus the histidine when the recipient cells are strain H-25 and plus histidine when they are strain T3, which has a block in the tryptophan synthetase (Anagnostopoulos and Crawford, 1961). T3 does not grow on indole. The plates are incubated 29 ± 2 hr at 37°C and examined in a dark room with the aid of a mineral light R-51 from Ultraviolet Products, Inc. Placement of the plates on a turntable covered with black velvet aids in detecting fluorescent colonies as the plate is turned at different angles.

The effectiveness of the mutagen and the dose of mutagen applied will determine how many plates will be needed to observe a significant number of fluorescent mutants. Since experience has shown that each plate should

---

[*] Portions of this are unpublished directions from Dr. Ernst Freese.

contain 100–200 colonies, a 1% frequency of mutation, which is quite high, means only a few mutant colonies per plate at low doses of mutagen. The total number of mutants observed can be increased by using many plates or by extending the exposure to mutagen so there are more mutations and greater inactivation. The latter calls for an adjustment in dilution so that the number of transformants per plate is kept up near the optimum. If the mutagen is not very effective, i.e., percentage M/hit = 0.01, it will be necessary to use both increased dose of mutagen and many plates. Fortunately, the fluorescent colonies are easily identified so large numbers of plates can be examined in a short time. These experiments demand meticulous care to exclude contaminants and a special need for controls to prove that the small number of observed colonies are not contaminants.

9. Reversion Tests on Transformed Cells

Just as mutagens have some specificity for certain bases in forward mutations, they also show a specificity for reversion to the original base. Since frequently this means that a different mutagen is required for reversion the result is useful in identifying the base initially changed. Revertants also suggest that the initial change involves only a single base and is therefore a point mutation. In both instances, it assumes that reversion is true and not a phenotypic illusion, which is not infrequent.

Assay of reversion is especially easy using a procedure adapted from Allen and Yanofsky (1963). Single-colony isolates of mutant organisms are grown overnight, centrifuged, and washed; 5×10^8 cells are then placed on plates containing selective medium on which only the revertants will grow. This may be minimal medium plus all supplements but the one for which reversion is being screened. In some instances, a small quantity of solid mutagen is dropped directly onto the center of the plate where it dissolves and diffuses laterally to surrounding organisms (Freese and Strack, 1962). Others have placed a sterile paper disk (S & S # 740 E) in the center of the plate and put a drop of concentrated mutagen onto this paper. After incubation, colonies appearing around the site of the mutagen are scored. The concentrated mutagens used by Carlton and Whitt (1969) were 2 AP, 40 mg/ml; 5 BudR, 8 mg/ml; HA, 0.25 M; NA, 0.25 M; MMNG, 1 mg/ml; ICR-170, 1 mg/ml; ICR-191, 1 mg/ml; EMS, undiluted.

C. Denaturation of DNA

DNA at 10–50 μg/ml is denatured by *heating to 100°C for 10 min* followed by quick chilling in ice water. It is also obtained by exposure to alkali (*p*H 12–13) for a few minutes at 25°C, after which the *p*H is brought to neutrality by 1.8 equivalents of KH_2PO_4. There is usually a residual activity after denaturation which represents 0.5–5% of the initial transforming activity. This has been found to be crosslinked by some unknown mechanism (Mulder and Doty, 1968). Denaturation is accompanied by a ~30% increase in absorbancy at 260 mμ. At concentrations above 20 μg/ml, denatured DNA tends to separate slowly from saline solution.

Denaturation is also induced by acid (Bunville and Geiduschek, 1960) below pH 3 and by organic solvents (Duggan, 1961; Herskovits et al., 1961) and inorganic metal ions (Eichorn and Clark, 1965; Hiai, 1965).

Some denaturation procedures introduce mutational changes. Use of glycol at much lower temperatures minimizes mutations (Duggan, 1961).

D. Renaturation of Denatured DNA

Renaturation of DNA, discovered by Marmur and Lane (1960), is readily obtained without special regard to the method of denaturation. The extent of recovery is favored by NaCl concentration up to 0.4 M, 1 hr at 65°C, and by DNA concentration of 10–50 μg/ml, although lower concentrations will renature. Thirty to fifty percent of the initial transforming activity is recoverable following renaturation. Renaturation follows second order kinetics.

E. Determining the Base Changes in Mutants and Mutations in Transforming DNA

As interest in mutagenesis increases, workers will wish to identify the base changes in DNA that are responsible for phenotypic changes. Base changes in mutants of the single-stranded DNA phage S-13 have been identified by the use of mutagens known to react with certain bases (Tessman et al., 1964). Developments in the handling of DNA suggest that the scheme of Tessman might be applicable to transforming DNA.

The great advantage of the S-13 phage lay in its being single-stranded; hence a point mutation could be attributed to a single base and not a base pair as in double-stranded DNA. The separation of complementary strands of transforming DNA by alkali denaturation and density gradient sedimentation (Peterson and Guild, 1968; Goodgal and Notani, 1968) with the aid of poly G (Summers and Szybalski, 1968) or fractional elution from MAK columns (Roger, 1968) was the first development. These separations are quite good but should not be considered as complete. The next development was the direct assay for transformation with single-stranded DNA (Postel and Goodgal, 1967; Chilton, 1967).

A sample of each of the strands of wild type DNA is treated with a general mutagen such as nitrous acid (NA) which is capable of acting on three or four bases. If this does not produce the mutation, then the change being sought may represent a deletion or multipoint mutation. If the mutation is produced, it might be expected to be more frequent in one strand than in the other because the bases are different.

Samples of each of the strands are next treated with hydroxylamine (HA), which is specific for cytosine. If no mutation is seen in either strand, G-C is eliminated and that leaves A-T. If mutations with HA are seen, they should be present in only one strand preparation and that strand will contain the cytosine. If HA failed to produce mutants, it will be necessary to treat samples of the strands with ethyl methanesulfonate (EMS). This agent has a weak mutagenic action toward thymine and none toward adenine, according to Tessman *et al.* (1964). Comparing the results of all three mutagens on the two strands should provide a clear answer regarding the specific base in each strand responsible for the mutation.

If closely linked markers are used, the spontaneous mutations are so sharply reduced that even low frequencies of induced mutations can be evaluated. The following example illustrates the point. If it is desired to determine the base pair responsible or try_2^- in the *B. subtilis* DNA, use can be made of this marker being closely linked to the histidine biosynthesis operon (his^+). Thus, if the DNA from a try^- his^+ cell is denatured and the strands are separated quantitatively and samples of each strand treated with mutagens as noted above, they can then be tested biologically by the single-strand scheme of Chilton (1967) in try^- his^- recipient cells. Only transformants in which mutations have been induced from try^- to try^+ will grow and form colonies. It may be necessary to grow the transformants in the presene of supplementary amino acids for an hour to promote integration before transferring the cells to minimal medium plates with no supplementary amino acids.

To ensure that the mutation is in fact a true revertant and not a new suppressor mutation which procues a phenotypic change like the revertant, the map distance between the new mutation and one or two reference markers should be determined and compared with the known distances between the original mutation and the reference markers. This can be done by the usual recombinational procedures (Lacks, 1966) or by estimating the co-transfer index or linkage with DNAs of different sizes obtained by shearing and fractionating in a sucrose gradient (Bendler, 1968).

VIII. ACKNOWLEDGMENTS

I am indebted to Drs. J. W. Drake, S. E. Bresler, John Scocca, Jack Michalka, Jon Ranhand, M. J. Voll, and E. Cabrera-Juarez for their suggestions.

The writer is grateful to the following publishers: Academic Press, Inc.; Elsevier Publishing Company; and American Elsevier Publishing

Company, Inc., for permission to reproduce material previously published in their journals.

The writer is personally grateful to the following authors: J. Michalka and S. H. Goodgal; H. O. Kammen, R. N. Beloff, and E. S. Canellakis; S. E. Bresler, V. L. Kalinin, and D. A. Perumov; J. W. Bendler, III; and T. Gurney, Jr., and M. S. Fox.

IX. REFERENCES

Adelberg, E. A., Mandel, M., and Chen, G. C. C. (1965), *Biochem. Biophys. Res. Commun. 18*, 788.
Alexander, H. E., and Leidy, G. (1953), *J. Exp. Med. 97*, 17.
Alexander, P. (1952), *Nature 169*, 226.
Alexander, P., and Stacey, K. A. (1958), *Ann. N.Y. Acad. Sci. 68*, 1225.
Allen, M. K., and Yanofsky, C. (1963), *Genetics 48*, 1065.
Anagnostopoulos, C., and Crawford, I. P. (1961), *Proc. Nat. Acad. Sci. 47*, 378.
Anagnostopoulos, C., and Crawford, I. P. (1967), *Compt. Rend. 265*, 93.
Auerbach, C., and Robson, J. M. (1944), *Nature 154*, 81.
Avadhani, N. G., Mehta, B. M., and Rege, D. U. (1969), *J. Mol. Biol. 42*, 413.
Avery, O. T., MacLeod, C. M., and McCarty, M. (1944), *J. Exp. Med. 79*, 137.
Baker, R., and Tessman, I. (1968), *J. Mol. Biol. 35*, 439.
Barat, M., Anagnostopolos, C., and Schneider, A. M. (1965), *J. Bacteriol. 90*, 357.
Barnhart, B. J., and Herriott, R. M. (1963), *Biochim. Biophys. Acta 76*, 25.
Bautz, E., and Freese, E. (1960), *Proc. Nat. Acad. Sci. 46*, 1585.
Bendler, J. (1968), Ph.D. dissertation, Department of Biochemistry, The Johns Hopkins University, Baltimore, Md.
Berns, K., and Thomas, C. A., Jr. (1965), *J. Mol. Biol. 11*, 476.
Beukers, R., and Berends, W. (1960), *Biochim. Biophys. Acta 41*, 550.
Boyce, R. P., and Howard-Flanders, P. (1964), *Proc. Nat. Acad. Sci. 51*, 293.
Boyland, E., and Nery, R. (1965), *Biochem. J. 94*, 198.
Brenner, S., Barnett, L., Crick, F. H. C., and Orgel, A. (1961), *J. Mol. Biol. 3*, 121.
Bresler, S. E., and Perumov, D. A. (1962), *Biokhimia 27*, 927.
Bresler, S. E., Kriviskii, A. S., Perumov, D. A., and Chernick, T. P. (1965), *Genetika 5*, 53.
Bresler, S. E., Kalinin, V. L., and Perumov, D. A. (1968), *Mutation Res. 5*, 1.
Brookes, P., and Lawley, P. D. (1961), *Biochem. J. 80*, 496.
Budowsky, E. I., Sverdlov, E. D., and Monastryskaya, G. S. (1969), *J. Mol. Biol. 44*, 205.
Bunville, L. G., and Geiduschek, E. P. (1960), *Biochem. Biophys. Res. Commun. 2*, 287.
Burton, K. (1956), *Biochem. J. 62*, 315.
Carlton, B. (1967), *J. Bacteriol. 94*, 660.
Carlton, B. C., and Whitt, D. D. (1969), *Genetics 62*, 445.
Catlin, B. W. (1960), *J. Bacteriol. 79*, 579.
Catlin, B. W., and Cunningham, L. S. (1965), *J. Gen. Microbiol. 37*, 341.
Cerdá-Olmedo, E., Hanawalt, P. C., and Guerola, N. (1968,, *J. Mol. Biol. 33*, 705.
Chilton, M. D. (1967), *Science 157*, 817.

Daniel, M., and Grimison, A. (1964), *Biochem. Biophys. Res. Commun. 16*, 428.

Dellweg, H., and Wacker, A. (1962), *Naturforschung 17b*, 827.

Demerec, M. (1945), *Proc. Acad. Sci. 31*, 16.

Dobrzanski, W. T., and Osowiecki, H. (1967), *J. Gen. Microbiol. 48*, 299.

Drake, J. W. (1963), *J. Mol. Biol. 6*, 268.

Drake, J. W. (1969), *Ann. Rev. Genet. 3*, 247.

Drake, J. W. (1970), "The Molecular Basis of Mutation," Holden-Day, San Francisco.

Duggan, E. L. (1961), *Biochem. Biophys. Res. Commun. 6*, 93.

Edgar, R. S., and Lielausis, I. (1964), *Genetics 49*, 649.

Eichorn, G. L., and Clark, P. (1965), *Proc. Nat. Acad. Sci. 53*, 586.

Ephrati-Elizur, E. (1968), *Genet. Res. Camb. 11*, 83.

Ephrussi-Taylor, H., Sicard, A. M., and Kammen, R. (1965), *Genetics 51*, 455.

Feingold, D. S., and Austrian, R. (1966), *J. Bacteriol. 92*, 952.

Fox, M. S., 1957. Biochim. Biophys. Acta *26:* 83.

Franklin, R. M., and Wecker, E. (1959), *Nature 184*, 343.

Frazer, S. J., and McDonald, W. C. (1966), *J. Bacteriol. 92*, 1582.

Freese, E. (1959a), *Brookhaven Symp. Biol. 12*, 63.

Freese, E. (1959b), *J. Mol. Biol. 1*, 87.

Inc., New York.

Freese, E., and Freese, E. B. (1965), *Biochemistry 4*, 2419.

Freese, E., and Freese, E. B. (1966), *Radiation Res. Suppl. 6*, 97.

Freese, E., and Strack, H. B. (1962), *Proc. Nat. Acad. Sci. 48*, 1976.

Freese, E., Bautz, E., and Bautz-Freese, E. (1961), *Proc. Nat. Acad. Sci. 47*, 845.

Freese, E. B. (1961), *Proc. Nat. Acad. Sci. 47*, 540.

Freese, E. B. (1965), *Genetics 51*, 953.

Freese, E. B., Gerson, J., Tuber, H., Rhaese, H., and Freese, E. (1967), *Mutation Res. 4*, 517.

Geiduschek, E. P. (1961), *Proc. Nat. Acad. Sci. 47*, 950.

Goodgal, S. H. (1961), *J. Gen. Physiol. 45*, 205.

Goodgal, S. H., and Herriott, R. M. (1961), *J. Gen. Physiol. 44*, 1201.

Goodgal, S. H., and Notani, N. (1968), *J. Mol. Biol. 35*, 449.

Gurney, T., Jr., and Fox, M. S. (1968), *J. Mol. Biol. 32*, 83.

Herriott, R. M. (1948), *J. Gen. Physiol. 32*, 221.

Herriott, R. M. (1966), *Cancer Res. 26*, 1971.

Herriott, R. M. (1970), Unpublished results.

Herriott, R. M., Meyer, E. Y., Vogt, M., and Modan, M. (1970a), *J. Bacteriol. 101*, 513.

Herriott, R. M., Meyer, E. Y., and Vogt, M. (1970b), *J. Bacteriol. 101*, 517.

Herskovits, T. T., Singer, S. J., and Geiduschek, E. P. (1961), *Arch. Biochem. Biophys. 94*, 99.

Hiai, S. (1965), *J. Mol. Biol. 11*, 672.

Horn, E. E., and Herriott, R. M. (1962), *Proc. Nat. Acad. Sci. 48*, 1409.

Hotchkiss, R. D. (1951), *Cold Spring Harbor Symp. Quant. Biol. 16*, 457.

Hotchkiss, R. D. (1966), in "Procedures in Nucleic Acid Research" (G. L. Cantoni and D. R. Davies, eds.) p. 541, Harper and Row, New York.

Hotchkiss, R. D., and Evans, A. H. (1958), *Cold Spring Harbor Symp. Quant. Biol. 23*, 85.

Hotchkiss, R. D., and Marmur, J. (1954), *Proc. Nat. Acad. Sci. 40*, 55.

Howard, B. D., and Tessman, I. (1964), *J. Mol. Biol. 9*, 372.

Howard-Flanders, P. (1968), *Ann. Rev. Biochem. 37*, 175.

Hsu, Y. C. (1964), *Nature 203*, 152.

Hsu, Y. C., and Herriott, R. M. (1961), *J. Gen. Physiol. 45*, 197.

Janion, C., and Shugar, D. (1968), *Acta Biochim. Polonica 15*, 107.

Janion, C., and Shugar, D. (1969), *Acta Biochim. Polonica 16*, 219.

Jyssum, K., and Lie, S. (1965), *Acta Pathol. Microbiol. Scand. 63*, 306.

Kammen, H. O., Beloff, R. N., and Canellakis, E. S. (1966), *Biochim. Biophys.* Acta *123*, 39.

Kelly, M. S., and Pritchard, R. H. (1965), *J. Bactdriol. 89*, 1314.

Kent, J. L., and Hotchkiss, R. D. (1964), *J. Mol. Biol. 9*, 308.

Krieg, D. R. (1963), *in* "Progress in Nucleic Acid Research" (J. N. Davidson and W. E. Cohn, eds.) p. 125, Academic Press, New York.

Lacks, S. (1966), *Genetics 53*, 207.

Lacks, S., and Hotchkiss, R. D. (1960), *Biochim. Biophys. Acta 39*, 508.

Lawley, P. D. (1966) *in* "Progress in Nucleic Acid Research and Molecular Biology" (J. N. Davidson and W. E. Cohn, eds.) Vol. 5, p. 89, Academic Press, New York.

Leonard, C. G., Corley, D. C., and Cole, R. M. (1967), *Biochem. Biophys. Res. Commun. 26*, 181.

Lerman, L. S. (1961), *J. Mol. Biol. 3*, 18.

Lerman, L. S. (1964), *J. Mol. Biol. 10*, 367.

Lerman, L. S., and Tolmach, L. J. (1959), *Biochim. Biophys. Acta 33*, 371.

Lindstrom, D. M., and Drake, J. W. (1970), *Proc. Natl. Acad. Sci. 65*, 617.

Litman, R. (1961), *J. Chim. Phys. 58*, 997.

Litman, R., and Ephrussi-Taylor, H. (1959), *Compt. Rend. 249*, 838.

Litman, R., and Pardee, A. (1956), *Nature 178*, 529.

Loveless, A. (1959), *Proc. Roy. Soc. London, Series B 150*, 497.

Loveless, A. (1966), "Genetics and Allied Effects of Alkylating Agents," Pennsylvania State University Press, University Park, Pa.

Luzzati, D. (1962), *Biochem. Biophys. Res. Commun. 9*, 508.

Maher, V. M., Miller, E. C., Miller, J. A., and Szybalski, W. (1968), *Mol. Pharmacol. 4*, 411.

Mahler, I., Neumann, J., and Marmur, J. (1963), *Biochim. Biophys. Acta 72*, 69.

Marmur, J. (1961), *J. Mol. Biol. 3*, 208.

Marmur, J., and Grossman, L. (1961), *Proc. Nat. Acad. Sci. 47*, 778.

Marmur, J., and Lane, D. (1960), *Proc. Nat. Acad. Sci. 46*, 453.

McDonald, W., and Matney, J. S. (1963), *J. Bacteriol. 85*, 218.

McKay, A. F. (1948), *J. Am. Chem. Soc. 70*, 1974.

Mejbaum, W. (1939), *Z. Physiol. Chem. 258*, 117.

Michalka, J., and Goodgal, S. H. (1969), *J. Mol. Biol. 45*, 407.

Michelson, A. M. and Monny, C. (1966), *Biochim. Biophys. Acta. 129*, 460.

Miura, A., and Tomizawa, J. (1968), *Mol. Gen. Genet. 103*, 1.

Morse, H. G., and Lerman, L. S. (1969), *Genetics 61*, 41.

Muhammed, A. (1966), *J. Biol. Chem. 241*, 516.

Mulder, C., and Doty, P. (1968), *J. Mol. Biol. 32*, 423.

Nester, E. W., and Lederberg, J. (1961), *Proc. Nat. Acad. Sci. 47*, 52.

Nester, E. W., Schafer, M., and Lederberg, J. (1963), *Genetics 48*, 523.

Nukushina, J. I., and Ikeda, Y. (1969), *Genetics 63*, 63.

Ono, J., Wilson, R. G., and Grossman, L. (1965), *J. Mol. Biol. 11*, 600.

Ottolenghi, E., and Hotchkiss, R. D. (1962), *J. Exp. Med. 116*, 491.

Pakula, R. (1965), *J. Bacteriol. 90*, 1320.

Patty, R., and Herriott, R. M. (1968), Unpublished results.

Perry, D. (1968), *J. Bacteriol. 95*, 132.

Peterson, J. M., and Guild, W. R. (1968), *J. Bacteriol. 96*, 1991.

Phillips, J. H., and Brown, D. M. (1967), *in* "Progress in Nucleic Research and Molecular Biology" (J. N. Davidson and W. E. Cohn, eds.) Vol. 7, p. 349, Academic Press, New York.

Postel, E. H., and Goodgal, S. H. (1966), *J. Mol. Biol. 16*, 317.

Postel, E. H., and Goodgal, S. H. (1967), *J. Mol. Biol. 28*, 247.

Ravin, A. W. (1961), *Adv. Genet. 10*, 62.

Ravin, A. W., and Iyer, V. N. (1961), *J. Gen. Microbiol. 26*, 277.

Ravin, A. W., and Mishra, A. K. (1965), *J. Bacteriol. 90*, 1161.

Reiner, B., and Zamenhof, S. (1957), *J. Biol. Chem. 228*, 475.

Roger, M. (1968), *Proc. Nat. Acad. Sci. 59*, 200.

Rupert, C. S. (1960), *J. Gen. Physiol. 43*, 573.

Rupert, C. S., and Goodgal, S. H. (1960), *Nature 185*, 556.

Rupert, C. S., and Harm, W. (1966), *Adv. Radiat. Biol. 2*, 1.

Rupp, W. D., and Howard-Flanders, P. (1968), *J. Mol. Biol. 31*, 291.

Schaeffer, P. (1964) *in* "The Bacteria" (I. C. Gunsalus and R. Y. Stanier, eds.) Vol. 5, p. 87, Academic Press, New York.

Schuster, H. (1961), *J. Mol. Biol. 3*, 447.

Setlow, R. B. (1966), *Science 153*, 379.

Setlow, R. B., and Carrier, W. L. (1964), *Proc. Nat. Acad. Sci. 51*, 226.

Sicard, A. M. (1964), *Genetics 50*, 31.

Sicard, A. M. and Ephrussi-Taylor, H. (1965), *Genetics 52*, 1207.

Singer, B., and Fraenkel-Conrat, H. (1967), *Proc. Nat. Acad. Sci. 58*, 234.

Singer, B., and Fraenkel-Conrat, H. (1969), *in* "Progress in Nucleic Acid Research Academic Press, New York.

Singer, B., Fraenkel-Conrat, H., Greenberg, J., and Michelson, A. M. (1968), *Science 160*, 1235.

Smith, I., Dubnau, D., Morrell, P., and Marmur, J. (1968), *J. Mol. Biol. 33*, 123.

Sparling, P. F. (1966), *J. Bacteriol. 92*, 1364.

Spencer, H. T., and Herriott, R. M. (1965), *J. Bacteriol. 90*, 911.

Speyer, F. (1965), *Biochem. Biophys. Res. Commun. 21*, 6.

Speyer, J. F., Karam, J. D., and Lenny, A. B. (1966), *Cold Spring Harbor Symp. Quant. Biol. 31*, 693.

Spizizen, J. (1958), *Proc. Nat. Acad. Sci. 44*, 1072.

Spizizen, J., and Prestidge, L. (1969), *J. Bacteriol. 99*, 70.

Spizizen, J., Reilly, B. E., and Evans, A. H. (1966), *Ann. Rev. Microbiol. 20*, 371.

Strack, H. B., Freese, E. B., and Freese, E. (1964), *Mutation Res. 1*, 10.

Streisinger, G., Okada, Y., Emrich, J., Newton, J., Tsugita, A., Terzaghi, E., and Inouye, M. (1966), *Cold Spring Harbor Symp. Quant. Biol. 31*, 77.

Study, J. (1961), *Biochem. Biophys. Res. Commun. 6*, 328.

Summers, W. C., and Szybalski, W. (1968), *Virology 34*, 9.

Sussmuth, Q., and Lingens, F. (1969), *Z. Naturforsch. 24B*, 903.

Takahashi, I. (1966), *J. Bacteriol. 91*, 101.

Talmadge, M. B., and Herriott, R. M. (1960), *Biochem. Biophys. Res. Commun. 2*, 203.

Tessman, I., Poddar, R. K., and Kumar, S. (1964), *J. Mol. Biol. 9*, 352.

Thorne, C. B., and Stull, H. B. (1966), *J. Bacteriol. 91*, 1012.

Tomasz, A. (1969), *Ann. Rev. Genet. 3*, 217.

Tomasz, A., and Hotchkiss, R. D. (1964), *Proc. Nat. Acad. Sci. 51*, 480.

Van Duuren, B. L. (1969), *Ann. N. Y. Acad. Sci. 163*, 593.

Varghese, A., and Wang, S. Y. (1967), *Nature 213*, 909.

Vielmetter, W., and Schuster, H. (1960), *Biochem. Biophys. Res. Commun. 2*, 324.

Voll, M. J., and Goodgal, S. H. (1966), *Biochim. Biophys. Acta 119*, 65.

Wang, S. Y., and Herriott, R. M. (1969), Unpublished results.

Wang, S. Y., and Varghese, A. (1967), *Biochem. Biophys. Res. Commun. 29*, 5439.

Watson, J. D., and Crick, F. H. C. (1953), *Nature 171*, 964.

Whitt, D. D., and Carlton, B. C. (1968), *J. Bacteriol. 96*, 1273.

Wilson, G. A., and Bott, K. F. (1968), *J. Bacteriol. 95*, 1439.

Witkin, E. (1969), *Ann. Rev. Microbiol. 23*, 487.

Yanofsky, C. (1966), *Proc. Nat. Acad. Sci. 55*, 274.

Yoshikawa, H. (1966), *Genetics 54*, 1201.

Yuki, S., and Ueda, Y. (1968), *Japan. J. Genet. 43*, 121.

Zimmerman, F. K., Schwaier, R., and von Laer, U. (1965), *Z. Vererb. 97*, 68.

GLOSSARY OF TERMS

Genetic Markers

Nutritional

bio = biotin
pur A = purine A
pur B = purine B
ura = uracil
hyp = hypoxanthine
man = mannitol
mal = maltose
rib = ribose
aro = aromatic
ala = alanine
arg = arginine A
arg B = arginine B
cys B = cystine B
gly = glycine
flu = fluorescent
his = histidine
ind = indole
leu = isoleucine
ilva = isoleucine-valine
leu = leucine
lys = lysine
met = methionine
phe = phenylalanine
pro = proline
shik = shikimic acid
thr = threonine
try = tryptophan
tyr = tyrosine

Antibiotic

bry = bryamycin
ery = erythromycin
kan = kanamycin
neo = neomycin
ole = oleomycin
spe = spectromycin
str = streptomycin
str-d = streptomycin dependent
vio = viomycin
nov = novobiocin
mic = micrococcin
am = aminopterin, amethopterin
dal = dalicin
pen = penicillin (benzyl)
pas = *p*-aminosalicylic acid
nob = *p*-nitrobenzoic acid
sul = sulfanilamide
Q = optochin
nal = naldixic acid
Superscript figures are antibiotic concentrations

Other

M = specific protein
S-IᴸS-VIII = specific polysaccharide capsules

Reagents

AP = aminopurine
BudR = 5-bromodeoxyuridine
HA = hydroxylamine
MeOHA = o-methylhydroxylamine
NA = nitrous acid
HU = N-hydroxylurethan
HZ = hydrazine
MNNG = N-methyl-N'-nitro-N-nitrosoguanidine
AAF-N-SO$_4$ = 2-acetyl aminofluorene, N-sulfate
N-AcO-AAT = N-acetoxy-2-acetylaminofluorene
N-B$_2$O-AAF = N-benzoyloxy-2-acetylaminofluorene
EMS = ethyl methanesulfonate
EES = ethyl ethanesulfonate
DES = diethyl sulfate
DMS = dimethyl sulfate
SLS = sodium lauryl sulfate
IRC 170, IRC 191 = acridine half-mustards synthesized by Dr. H.
      Creech of The Institute for Cancer Research, Philadelphia, Pa.
SSC = saline sodium citrate

Other Symbols

MM = minimal medium
MS = minimal salts
SbMMI = minimal medium I for *B. subtilis*
HI MM = minimal medium for *H. influenzae*
Pn mm = minimal medium for *Pneumococcus*
BHI = Difco brain–heart infusion
BHI$^+$ = BHI + hemin and NAD needed for growth of *H. influenzae*
MAK = methlated serum albumin–kieselguhr used in column chroma-
      tography

CHAPTER 7

Mutagen Screening with Virulent Bacteriophages*

John W. Drake

Department of Microbiology
University of Illinois
Urbana, Illinois

I. INTRODUCTION

Virus systems, and particularly those employing the T-even bacteriophages, the single-stranded DNA bacteriophages, and the temperate bacteriophage λ, have occupied the center of the molecular biologist's arena for up to three decades. The reasons for the popularity of virus systems are as relevant now as they were in the 1930s. Viruses often excel in those properties which make microorganisms in general highly useful for genetic studies. They are easy to grow. They exhibit extremely short generation times. (The bacteriophage T4 growth cycle, encompassing about 30 min at 37°C and resulting in the release of several hundred progeny particles, corresponds to an average doubling time of about 3 to 4 min.) Genome storage is extremely simple. Virus stocks are stable for many years in the refrigerator and do not accumulate many spontaneous mutations (Drake, 1966). Extremely large populations (for instance, 10^{11} to 10^{14} particles) are easily obtained, and a

* The usual DNA bases are abbreviated A, T, G, and C, while the corresponding base pairs are represented by A:T and G:C. The 5-hydroxymethylcytosine of bacteriophage T4 is also abbreviated as C. Research of the author which is described here was supported by American Cancer Society Grant E59, Public Health Service Grant 04886 from the National Institute of Allergies and Infectious Diseases, and National Science Foundation Grant GB 6998.

variety of techniques have been developed for selecting rare particles representative of events such as mutation. The relative chemical simplicity of bacteriophages, together with certain chemically unique aspects of their reproduction, has made possible a more detailed dissection of their molecular genetics than has been achieved even in the equally well-studied bacterium *Escherichia coli*. The formal genetics of several virus systems has been developed to a highly detailed and sophisticated level, making possible numerous genetic manipulations which render the analysis of processes such as mutation highly profitable. Finally, the pursuit of bacteriophage genetics, including even many aspects of their molecular genetics, is relatively inexpensive, particularly in comparison with metazoan systems.

II. BACTERIOPHAGES AS GENETIC SYSTEMS

The genetic and physiological behavior of bacteriophages has been described in several specialized textbooks, particularly those by Adams (1959), Stent (1963), Luria and Darnell (1967), and Goodheart (1969). Most general textbooks of molecular and microbial genetics also consider virus systems, the most comprehensive treatment being that of Hayes (1968). Molecular mechanisms of mutation have been described in considerable detail, with heavy emphasis on bacteriophage systems, by Drake (1970). The current status of research on mutational mechanisms has also been reviewed by Drake (1969a). These sources clearly reveal that viruses are excellent systems for the study of mutation, complementation, recombination, and many allied phenomena.

Spontaneous rates of mutation are relatively high in viruses when expressed per gene or per base pair. A typical T4 or λ gene mutates in the forward direction at a frequency of about 10^{-5} per DNA duplication, a value about a hundredfold greater than those observed in bacteria, and about a thousandfold greater than those observed in eukaryotes (Drake, 1969b). Spontaneous mutation rates in viruses are therefore easily estimated by direct counting procedures, increases as small as twofold being reliably detected. For instance, the frequency of spontaneous T4r mutants, which are detected by their plaque morphologies, is about 4×10^{-4} in typical stocks. If the mutant frequency is estimated by setting out 100 plates per stock, each plate containing about 1000 plaque-forming units, and screening by eye, then twofold differences in the observed mutant frequencies are very significant. Two such measurements would typically consume less than 2 man-days altogether, including the preparative chores. Selective techniques, of course, may increase the ease of measurement ten- or a hundredfold.

Complementation tests are particularly straightforward in bacteriophage systems because it is generally easy to introduce dissimilar but related genomes into a common cytoplasm by mixed infection. Sexual differentiation is absent, and any related pair of bacteriophages can coinfect. Growth is measured by counting the progeny released after a single cycle of growth, and the average burst size is very sensitive to environmental and to genetic factors, including both intragenic and intergenic complementation.

The manipulative aspects of bacteriophage crosses are also simple. The resolving power of many available screening systems makes fine-scale mapping quite simple. However, the statistical aspects of bacteriophage crosses are complex. Virus particles experience both replication and recombination approximately randomly in time during their intracellular replication, and recombinant frequencies depend upon the duration of the growth cycle, among other factors. Some viruses, particularly those containing single-stranded DNA (such as ϕX174 and S13), recombine at very low rates, making fine-scale mapping difficult. Viruses often exhibit the vagaries of recombination which are collectively known as marker effects: many markers (mutations) used in crosses affect recombination frequencies by their very presence. In addition, bacteriophage recombination frequencies are often poorly reproducible: 50% differences are certainly not unusual. These aspects of bacteriophage genetics need not, however, significantly interfere with the adaptation of virus systems to mutagen screening.

Since viruses exhibit both extracellular and intracellular phases, different aspects of their susceptibility to mutagenic and inactivating agents can conveniently be examined. During its intracellular replication, the virus chromosome (which is typically a single, more or less unclothed DNA duplex) is first subjected to transcription, wherein some of its genes direct the synthesis of messenger RNA, and only later to replication and recombination. Many DNA repair processes begin to function early in the intracellular phase, and certain aspects of the interaction of repair systems with mutagenesis can therefore be separated from the DNA replication process itself. In particular, mutagens which are suspected of acting via interactions with repair systems, or of being replication-dependent for their action, can be applied during appropriately different periods of viral development. Free virus particles, on the other hand, present a mutational target which is not immediately buffered by repair systems and which corresponds approximately to naked DNA. (The physical state of double-stranded bacteriophage DNA is different from that of free DNA. The viral DNA is packed very tightly within the particle, and is partially denatured or dehydrated. Its chemical reactivity is intermediate between those of single-stranded and double-stranded DNAs in solution. The sensitivities of virus particles to chemical and to physical mutagens are therefore sometimes enhanced compared to free DNA (Freese and Strack, 1962), but whether differentially with respect to different mutational pathways is still not yet clear.)

As a result of the biphasic nature of virus replication, the direct and indirect effects of chemical and physical agents upon viruses can often be separated. Either the growth medium or the host cell itself can be pretreated before the virus particles are introduced. For instance, irradiated cells are sometimes mutagenic for unirradiated viruses (Jacob, 1954), and treating cells with hydroxylamine before infection produces different mutational pathways compared to treating free virus particles (Tessman *et al.*, 1965).

The viruses such as bacteriophage S13 which contain single-stranded DNA replicate by means of double-stranded intermediates. These viruses therefore offer an advantage compared to their fatter cousins. When attention is focused upon the chemical nature of the mutational target, it is often easy to distinguish A:T and G:C base pairs in viruses with double-stranded DNA, but it usually difficult to distinguish the individual bases, such as A from T. This difficulty does not arise with viruses such as S13, however, at least when mutagens are applied to free virus particles.

III. MEASURING VIRAL MUTATION RATES

Spontaneous mutation is often measured by growing a population until it contains numerous mutants, measuring the mutant frequency, growing for some additional number of generations, and measuring the new mutant frequency. The mutation rate m per DNA replication is given by

$$m = \frac{0.4343(f'-f)}{\log(N'/N)}$$

where the f values represent the mutant frequencies and the N values the population sizes at the beginning and the end. It is implicit in this formulation that the mutations are replication-dependent, arising either as errors of replication or else as errors (such as errors of repair) which occur side by side with replication. Continuous culture devices such as the chemostat can extend the versatility of this type of measurement for cellular systems, but usually not for viral systems.

When a mutagen is employed, a variation on the above procedure involves splitting the initial population into portions which are either treated or else maintained as a control. The question of *whether* mutagenesis was induced is answered simply by comparing the two final mutant frequencies. The accuracy of this comparison depends upon the sample sizes, and upon the sophistication with which the control is handled. An agent which slows the replication of the treated sample, for instance, or which selects against the mutant phenotype, may induce unrecognized mutations, and demands a carefully designed control. The question of *how much* mutagenesis was induced is answered by comparing the initial and final populations, as described in the equation.

Another widely employed variation of the procedure is suitable for estimating whether mutagenesis is present or absent, but is not particularly suitable for the determination of mutation rates themselves. Several parallel cultures are grown from small inocula, the mutant frequencies of each are measured, and these values are averaged. The result is often called a *mutation* (or *reversion*) *index*. Unless many cultures are prepared, even threefold differences are not very significant. This method is particularly sensitive to the time at which the first mutant clone originates during the growth of a particular culture. An early mutation, compared to the time when the population size is equal to the reciprocal of the mutation rate, produces a "jackpot" culture containing an unusually high mutant frequency. Unusually low mutant frequencies, on the other hand, are very unlikely to be observed. For these reasons, it is often better to use the median rather than the mean mutant frequency when calculating the mutation index, especially when the median is smaller than the mean.

An alternative general method, and one which is particularly useful when growth conditions select for or against the mutant phenotype, is the null fraction method. Each of a number of tubes is inoculated with few enough virue particles so that no tube is likely to receive any spontaneous mutants. The particles are then allowed to replicate, with or without a mutagen. If replication proceeds until some (but not all) of the tubes contain one or more mutants, then the mutation rate, m, can be estimated from the fraction of tubes $F(0)$ without any mutants: $F(0) = e^{-M}$. Here M represents the average number of mutant clones (mutational events) per tube. The average number of generations, G, per tube is determined separately by assaying the total virus content of several tubes, and is simply equal to the average number of particles. The mutation rate is $m = M/G$. If the system is free of complications introduced by selection or by the delayed expression of new mutants, then the accuracy of the measurement can be increased by counting the number of mutants in each tube. An extensive treatment of measurements of this type, and in particular of their potential accuracy, is given by Lea and Coulson (1949).

When mutations are induced in free virus particles, mutation rates are expressed as mutant frequencies per unit of treatment. Since mutagens may be unstable or may be consumed during the course of the reaction, and since the exact conditions of treatment may be difficult to reproduce from laboratory to laboratory, it is both a convenient and a common practice to express mutation rates per lethal hit. Mutagens which affect free virus particles both mutagenize and inactivate (if only because of the induction of lethal mutations). When the inactivation kinetics (single-hit or multiple-hit) have been determined, the observed survival, S, after treatment can be converted into lethal hits. The number of lethal hits, h, is given by $h = -2.3 \log S$ in the case of single-hit kinetics, or by $h = -2.3 \log [1-(1-$

$S)^{1/n}]$ in the case of n-hit kinetics. For values of S beyond the "shoulder" of a multiple-hit curve, the approximation $h = -2.3 \log (S/n)$ will suffice. The number of mutations per lethal hit usually turns out to be easily reproducible.

Accurate measurements of mutation rates, or of the existence of mutagenesis itself, obviously depend in part upon the magnitude of the spontaneous background. It is therefore desirable to use stocks with low backgrounds. One general method to obtain such stocks is to grow several (usually about half a dozen will suffice), and to screen for the best one. As indicated previously, however, the shape of the clone size distribution makes it highly unlikely that a stock will appear exhibiting an unusually low mutant frequency. Another method for obtaining low-background stocks involves antiselection. If the physiology of the mutant phenotype places it at a selective disadvantage under some set of conditions, then stocks grown under these conditions contain a minimum of mutants. The procedure is usually obvious. Since about half of the mutants in a population arise during the final generation, however, antiselection during viral growth is sometimes not very effective.

Selection artifacts are an ever-present danger in mutagen testing, and are particularly likely to occur when a potential mutagen affects growth rates or viabilities. Even the most experienced investigators are misled from time to time. The only reliable principle in avoiding selection artifacts is to consider carefully all of the details of the system. It is obviously difficult, and frequently even impossible, to construct a truly representative control culture when the experimental culture is treated with an agent which produces considerable lethality. Even reconstruction controls, in which artificial mixtures of mutant and wild-type particles are treated together, may fail, because selection may operate not upon the mutant phenotype itself, but upon an intermediate in the production of the mutant (such as a "premutational lesion" or a heteroduplex heterozygote). When free virus particles are treated, the absolute number of mutants (instead of just the mutant frequency) sometimes increases, and selection is then impossible. An examination of the types of mutants which appear may demonstrate that they were induced, instead of being selected from the spontaneous background. This might be achieved, for example, by fine-scale mapping experiments, or by reversion analyses as described in the following. Another useful method to guard against selection artifacts is to examine mutagenesis in both the forward and the reverse directions, also as described in the following.

Plating-density artifacts are common among microorganisms. They consist of variable counting efficiencies according to the number of organisms plated in a single container. Artifacts of this type can appear in bacteriophage systems, and may be present whenever more than about 10^7 particles are plated in selective systems—for instance, when revertants are to be scored. As the number of particles plated increases, the efficiency of plating may decrease because of cell killing, thus underestimating the number of mutants. On the other hand, when the plated suspension contains many inactivated particles, as is frequently the case following mutagenesis, reactivation phenomena can occur which lead to large overestimations of the numbers of mutants. This occurs because coinfection of a cell by a live nonmutant particle plus an inactivated mutant particle can lead to "marker rescue" of the mutation into live progeny particles. The general method for detecting plating density artifacts is to assay mutants at two or more different plating densities and to look for nonproportionalities. This procedure can be tedious if the mutant frequency is low, and is therefore frequently omitted, with occasionally disastrous results.

IV. MUTATIONAL PATHWAYS

The choice of a suitable screening system for detecting mutagenesis requires some understanding of the sensitivity of the assay system, and of the goal of the measurements. These factors become immediately apparent when a choice must be made between screening for forward or for reverse mutations. By *forward mutation* we usually mean the set of all possible mutations which inactivate or seriously modify a gene. It is desirable to know which inactivating mutations occur within the gene itself, compared to mutations which occur entirely outside of the gene. (The latter type can consist, for example, of polar and of regulatory gene mutations.) It is only possible to recognize intragenic mutations with confidence when the system has been studied in considerable detail, and when the properties of many of its mutants have been investigated. As we shall see in the next section, few systems fulfill this requirement. Forward mutations will consist in general of all possible types of point mutations, and of macrolesions. Thus a forward-mutation screening system should be able to detect mutagenesis regardless of the molecular type of lesion produced. This is the main advantage of screening forward mutations. On the other hand, if a mutagen produces a specific type of mutation which is relatively rare among spontaneous mutations, and produces it weakly, its action may be obscured in the spontaneous background of many other molecular types. ·

The direct way out of the dilemma is to study reversion. Many mutants revert along molecular pathways which have been well defined by previous studies. Revertant screening with many such mutants is technically simple and umambiguous, in contrast to forward mutation where leaky mutants are often recovered with variable efficiencies. A set of 20 or so standard tester mutants can often be assembled, at least one of whose members is likely to respond to any given mutagen. Furthermore, the particular subset of mutants reverted by a mutagen offers immediate insights into the molecular pathways which are promoted by that mutagen. In a forward-mutation screening system, similar information about mutational mechanisms can be obtained only by studying the induced mutants in considerable detail. A general drawback to reversion screening systems is the possibility that the mutagen may fail to revert any of the tester mutants, either becaue it produces an exotic type of mutation, or for other entirely unanticipated reasons. A cautious screening program should therefore employ both forward- and reverse-mutation screening tests.

The most useful contemporary taxonomy of mutational lesions divides them on the basis of the chemical configuration of the mutational lesion, rather than on the basis of the gene, organism, or phenotype involved. Mutations generated as if by chromosome breakage, such as deletions, duplications, and rearrangements (including inversions and translocations), are called *macrolesions*. Since the prokaryotes are generally unichromosomal, aberrancies of chromosome segregation are not perpetuated as heritable mutations. A type of macrolesion which could occur, but which has never been observed in nature, consists of an addition of DNA base pairs which are not simply a duplication of a pre-existing sequence, but are instead an entirely new and perhaps nonsense sequence. Except for deletions, which are easily recognized by their inability to revert and by their failure to recombine with two or more point mutations, macrolesions have not been studied very extensively in prokaryotic systems. Deletions are induced in bacteriophage T4 by nitrous acid (Tessman, 1962). They are induced in bacteria by both ultraviolet and X-irradiation (Schwartz and Beckwith, 1969), but ultraviolet irradiation produces very few deletions in bacteriophage T4 (Drake, 1963).

Mutations which involve only one or a few base pairs are called *microlesions* or *point mutations*, and are composed of base-pair substitutions and frameshift mutations. Frameshift mutations consist of deletions and additions of small numbers of base pairs. There is no well-defined boundary between frameshift mutations of large extent and macrolesions of small extent, except that a reverting deletion would usually be considered to be a frameshift mutation, and large additions would perhaps not be reverted by mutagens which readily induce frameshift mutations. Many frameshift mutations in bacteriophage systems are recognized by their susceptibilities to reversion induction by proflavin or other acridines (Orgel and Brenner, 1961; Lerman, 1964) or to reversion induction by specific mutator genes (Greening and Drake, unpublished results). Frameshift mutations are also produced by acridines coupled to alkylating agents (Ames and Whitfield, 1966) and by ultraviolet irradiation (Drake, 1963). Both of these agents, however, also produce base-pair substitutions.

Base-pair substitutions consist of transitions, in which the purine–pyrimidine orientation is preserved (A:T ↔ G:C), and transversions, in which the purine–pyrimidine orientation is reversed (A:T ↔ T:A ↔ G:C ↔ C:G). Transitions are produced with high specificity by base analogs such as 5-bromouracil and 2-aminopurine and (except for deletions) by nitrous acid (Freese, 1959; Howard and Tessman, 1964a; Tessman et al., 1964; Wittmann and Wittman-Liebold, 1966; Drake, 1970). Hydroxylamine specifically induces G:C → A:T transitions, provided that free virus particles are treated (Freese et al., 1961; Champe and Benzer, 1962; Tessman et al., 1964; Drake, 1970). No known mutagen specifically induces transversions, although A:T → C:G transversions are specifically induced by an E. coli mutator gene (Yanofsky et al., 1966a). A number of agents, however, including ultraviolet irradiation, alkylation, low pH, mild heat, and photodynamic action, appear to induce mixtures of transversions and transitions (Drake, 1963, and unpublished results; Howard and Tessman, 1964b; Freese, 1961; Drake and McGuire, 1967a,b). Transversions also constitute a considerable proportion of spontaneous base-pair substitutions (Drake, 1970). It is likely that many chemically and physically induced transversions arise indirectly, as errors of repair.

Mutational pathways can often be relatively easily deduced because of the fact that frameshift mutations usually revert only by frameshift mutagenesis, and base-pair substitutions only by base-pair substitution. Patterns of induced reversion ("reversion analysis") therefore provide important information about mutational mechanisms. This relationship is maintained, however, only when intracistronic reversion is distinguished from extracistronic reversion. The distinction is particularly easy in viral systems, since extracistronic suppression usually proceeds by means of modifications of the specificity of information transfer (transcription and translation), and information transfer is primarily a host-cell function. Frameshift mutations and base-pair substitutions frequently revert, however, by means of intracistronic suppression. Transversions, for instance, often revert by means of transitions, because proteins often tolerate amino acid substitutions: a severely deleterious amino acid substitution may be overcome by inserting yet a third amino acid into the mutant codon, or else by introducing a second compensating amino acid substitution at another position within the same polypeptide (Yanofsky et al., 1966b). It is therefore often difficult to distinguish between base-pair additions and deletions, or between transitions and transversions.

The final appearance of a mutation in any screening system depends not only upon the primary chemical event, such as base damage or enzymatic error, but also upon the probability that the initial lesion will persist to produce a detectable phenotypic alteration. Several enzymatic systems which erase premutational lesions have already been described, and it is likely that a considerably greater number will be discovered in the near future. Paradoxically, such systems are themselves frequently faulty, and may convert lethal lesions into mutations (Witkin, 1969). Mutation rates are therefore observed through a smokescreen of repair systems, which exhibit variable specificities and efficiencies from organism to organism. The fates of premutational events initiated by chemical mutagens also depend upon the properties of the enzymes of DNA replication (Drake and Allen, 1968; Drake et al., 1969; Speyer, 1969). Base-analog mutagenesis, for instance, can be either promoted or abolished by mutations in the DNA polymerase of bacteriophage T4. For reasons which are only partially understood, the frequency with which a particular type of mutational lesion is induced by various chemical mutagens depends strongly upon the functional state of the DNA polymerase, the DNA ligase, and other enzymes of DNA metabolism.

Even if a premutational lesion survives the error-minimizing activities of the enzymatic apparatus of DNA repair and replication, it may not induce an easily detectable phenotypic change. The probability of expression depends strongly upon the molecular type. Frameshift mutations, for example, and probably most macrolesions, produce drastic effects, totally inactivating the genes which they intersect. Base-pair substitutions, on the other hand, are much more variable. One class, the nonsense mutations, inactivates the affected genes by introducing anomalous punctuation signals (particularly "Stop!"). Another class, the missense mutations, produces amino acid substitutions which range from the disastrous to the innocuous. A third class, because of codon degeneracy, produces no changes in amino acid composition; for instance, nearly all transitions within the third positions of codons go unrecognized. Some genes, such as the "purple-adenine" genes of Neurospora (de Serres, personal communication), appear to be extremely sensitive to missense mutations, whereas others, such as the T4rII cistrons (Drake, 1970; Koch and Drake, unpublished results), appear to tolerate many amino acid substitutions.

One aspect of mutagenesis which has become increasingly important during the last decade is our very imperfect ability to predict whether an agent is likely to be mutagenic or to determine how a proven mutagen actually works. Extrapolations based upon the effects of agents upon DNA or its constituents, or upon comparisons with more intensively studied compounds, often prove to be misleading. Major reactions with the DNA bases which strongly suggest specific mutational mechanisms sometimes turn out to be unrelated to the actual mechanism. A number of strongly mutagenic agents, for example, depurinate DNA; these include alkylating agents, acid, and heat (Freese, 1961; Zamenhof, 1960). Depurination itself, however, is at most very weakly mutagenic (Drake, 1970; Drake and Ray, unpublished results). Ultraviolet irradiation preferentially produces G:C → A:T transitions (Drake, 1963; Howard and Tessman, 1964*b*), but does so by producing premutational lesions which often do not even involve G:C base pairs (Witkin, 1969; Meistrich and Drake, unpublished results). In both of these examples, the mutations are produced either by relatively rare reactions or circuitously during repair. Extrapolations from the major chemical reactions of a compound, or from comparisons with apparently related compounds, are therefore useful traffic lights, but do not by themselves prevent accidents. Mutagen screening must continue to be empirical for many years.

V. BACTERIOPHAGE SCREENING SYSTEMS

Optimally useful systems for mutagen screening and analysis employ genes in which mutation is easily scored, preferably in both the forward and reverse directions, and which are representative of the chromosome as a whole. A dispensable gene is desirable for studies of forward mutation, so that all types of mutation which may occur can be recovered. At the same time it is desirable to be able to render the gene indispensable for growth, so that revertants of its mutants can be directly selected. Finally, it is desirable to employ a gene which is relatively sensitive to missense mutation, so that a good proportion of base-pair substitutions can be detected. Discouragingly few systems meet all of these criteria. Some of the best, however, are found in bacteriophages (and in bacteria). We will briefly consider two of the best available systems, and then some of the systems which are only partly developed but which are potentially very useful.

When bacteriophage T4 is plated on various strains of *E. coli*, the wild type produces plaques a few millimeters in diameter with fuzzy edges. A strikingly obvious set of mutants produces much larger plaques with sharp edges. A typical wild-type stock contains roughly 0.05% of these *r* (for rapid lysis) mutants. As indicated in section II, they are easily screened visually. Very little practice is needed for the eye to detect *r* mutants on crowded plates with nearly perfect efficiency, proceeding at the rate of one plate every few seconds. The frequency of *r* mutants can be increased tremendously by mutagens, for instance to 30% or more (Schuster and Vielmetter, 1961). The *r* mutants arise within several regions on the T4 chromosome, the great majority by mutational inactivation of the *r*I or *r*II genes. The *r*I mutants are induced somewhat more efficiently by mutagens which induce base-pair substitutions and the *r*II mutants somewhat more efficiently by mutagens which induce frameshift mutations. The *r*II mutants comprise about 70% of spontaneous *r* mutants and occupy two adjacent cistrons. They are unable to grow in λ lysogens, being restricted by the λ prophage repressor. Stocks of rII mutants can be grown under completely nonselective conditions. The technical aspects of the T4rII system are described in the many references already cited.

The inability of rII mutants to grow on λ lysogens provides the screening system for reversion analysis. Sets of tester mutants are available composed of mutants which revert by A:T → G:C and by G:C → A:T transitions, by various transversions, and also by various types of frameshift mutations. No selective system exists for directly scoring deletions, but these are easily detected in forward-mutation tests. Direct selective tests for duplications are feasible (Weil *et al.*, 1965), but have not yet been fully developed.

A simple reversion spot test is avilable which can be used to screen many compounds using only a few plates, and which can in principle score mutations induced in both free virus particles and replicating chromosomes. Another advantage of spot tests is that a concentration gradient forms on the plate, so that any likely effective concentration is somewhere achieved. We have generally found these tests to be nearly completely reliable, compared to more complicated tests conducted in liquid culture. The test is set up by plating a moderately large number of particles of an rII mutant, together with an approximately equal number of permissive host cells plus a considerable excess of nonselective host cells. When a mutagen is spotted on the plate, it may produce a halo of revertant plaques (compared to the background, which constitutes the spontaneous mutation control). This test is effective with agents which are not highly lethal to the virus or to the cell, but is not well suited for testing agents such as proflavin which must be removed from the system before assaying for mutations.

The T4rII system has two disadvantages. Many base-pair substitutions escape detection because of the tolerance of the rII protein to amino acid substitutions. The efficiency of detection of base-pair substitutions is probably about 5 to 10% when assayed by forward mutation. This difficulty, however, is circumvented by using reversion as well as forward-mutation tests. The rII protein is produced in small and inaccessible amounts (McClain and Champe, 1967), so that amino acid substitutions are assayed only with great difficulty.

Single-stranded DNA bacteriophage S13, a very close relative of φX174, appears not to possess a dispensable gene. In partial compensation, it offers the opportunity of distinguishing between G and C and between A and T as mutational targets. Several host range mutants of S13 have been well characterized. These mutants correspond operationally to T4rII mutants used in reversion tests, in the sense that mutation is scored by differential plating on permissive and nonpermissive host cells. However, only acceptable missense and suppressible nonsense mutations can be studied in this system. Frameshift mutations, or other nonsuppressible types of gene in-activation, produce lethality. Furthermore, since the phenotype of the mutant must be expressed through the coat protein, newly arisen mutants frequently cannot be scored until they have been passaged once to permit mutation expression. The application of this system to the study of muta-tional pathways has been elegantly pursued by Tessman and his collabora-tors. The technical aspects of the system are described in the many refer-ences already cited.

A number of additional viral systems are potentially useful for mutagen screening. The c mutants of temperate bacteriophages such as λ produce clear plaques by virtue of their inability to lysogenize. The λcI mutants can result from any type of mutational inactivation of the cI gene, and are as easily screened visually as are the T4r mutants. This system presents many of the same advantages (and disadvantages) of the rII system, except that a suitable method for screening for revertants has not been fully developed. One potential advantage of the λcI system which is not shared by the T4rII system arises from the fact that the virus sometimes replicates itself, using a set of virus-induced enzymes, whereas at other times it is replicated simply as a portion of the bacterial chromosome, using exclusively host-induced enzymes. Thus two distinct sets of repair and replication enzymes can be compared using the same mutational target. Bacteriophage λ also produces suppressible base-pair substitution mutations whose reversion is easily monitored. The genetics of bacteriophage λ has been reviewed most recently by Dove (1968).

The T4e gene determines the synthesis of the lysozyme which releases particles from infected cells (Streisinger et al., 1961a). The function can be made dispensable by adding egg-white lysozyme. Mutation in the forward direction can be screened by allowing infected cells to lyse normally and then plating the washed, unlysed cells with added lysozyme. Similarly, revertants can be scored by differential plating with and without added lysozyme. Leaky mutants will escape the forward screening system, but can be recognized by a special plaque morphology test (Streisinger et al., 1961b). In addition to intracistronic reversion, e mutants can be induced to revert by mutational inactivation of another nearby gene (Emrich, 1968), and a selective forward-mutation system might be developed with this system. Although manipulation of the e system is technically somewhat more diffi-cult than is manipulation of the rII system, the e system has the compensat-ing advantage that the polypeptide encoded by the gene is easily isolated. The amino acid sequences of the wild-type protein and of many mutants have already been determined (Tsugita and Inouye, 1968; Streisinger et al., 1966).

Animal virus systems may someday become very useful screening systems in mammalian backgrounds. Many animal viruses, particularly the herpes and pox groups, contain double-stranded DNA, and undoubtedly contain dispensable genes. The tissue specificities of animal viruses could perhaps also be adapted to the goals of mutagen testing. The genetics of animal viruses is not very well developed at present, but should be considerably improved in the coming decade.

VI. CONCLUSIONS

Bacteriophage systems, particularly the T4rII system, offer great potential for mutagen screening. They are rapid, reliable, very inexpensive, and readily mastered. An adept novice visiting an appropriate laboratory could reasonably expect to master the basic aspects of one of these systems within a month. The T4rII and S13 systems also provide rapid insights into the molecular classes of mutations being induced.

Scores of compounds have been tested in viral systems, and many mutagens have been detected. When comparative data are available, most compounds which are mutagenic in viral systems are also seen to be mutagenic in bacterial or fungal systems, and *vice versa*. For purposes of extension to mammalian systems, even the host-mediated assay (Legator and Malling, this book) can in principle be adapted to accommodate extant viral systems, thus introducing the pharmacological and organismic complications which ordinarily are not components of microbial systems. A mutagen screening program which purports to be comprehensive should therefore include viral systems, together with other microbial systems.

VII. REFERENCES

Adams, M. H. (1959), "Bacteriophages," Interscience Publishers, Inc., New York.
Ames, B. N., and Whitfield, H. J. (1966), *Cold Spring Harbor Symp. Quant. Biol.* 31, 221.
Champe, S. P., and Benzer, S. (1962), *Proc. Nat. Acad. Sci. 48*, 532.
Dove, W. F. (1968), *Ann. Rev. Genet. 2*, 305.
Drake, J. W. (1963), *J. Mol. Biol. 6*, 268.
Drake, J. W. (1966), *Proc. Nat. Acad. Sci. 55*, 738.
Drake, J. W. (1969a), *Ann. Rev. Genet. 3*, 247.
Drake, J. W. (1969b), *Nature 221*, 1132.
Drake, J. W. (1970), "The Molecular Basis of Mutation," Holden-Day, Inc., San Francisco.
Drake, J. W., and Allen, E. F. (1968), *Cold Spring Harbor Symp. Quant. Biol. 33*, 339.
Drake, J. W., and McGuire, J. (1967a), *Genetics 55*, 387.
Drake, J. W., and McGuire, J. (1967b), *J. Virol. 1*, 260.

Drake, J. W., Allen, E. F., Forsberg, S. A., Preparata, R.-M., and Greening, E. O. (1969), *Nature 221*, 1128.

Emrich, J. (1968), *Virology 35*, 158.

Freese, E. (1959), *Brookhaven Symp. Biol. 12*, 63.

Freese, E., and Strack, H. B. (1962), *Proc. Nat. Acad. Sci. 48*, 1796.

Freese, E., Bautz, E., and Freese, E. B. (1961), *Proc. Nat. Acad. Sci. 47*, 845.

Freese, E. B. (1961), *Proc. Nat. Acad. Sci. 47*, 540.

Goodheart, C. R. (1969), "An Introduction to Virology," W. B. Saunders Co., Phailadelphia.

Hayes, W. (1968), "The Genetics of Bacteria and Their Viruses," 2nd ed., John Wiley & Sons Inc., New York.

Howard, B. D., and Tessman, I. (1964a), *J. Mol. Biol. 9*, 364.

Howard, B. D., and Tessman, I. (1964b), *J. Mol. Biol. 9*, 372.

Jacob, F. (1954), *Compt. Rend. Acad. Sci. 238*, 732.

Lea, D. E., and Coulson, C. A. (1949), *J. Genet. 49*, 264.

Lerman, L. S. (1964), *J. Cell. Comp. Physiol. 64* (Suppl. 1), 1.

Luria, S. E., and Darnell, J. E. (1967), "General Virology," 2nd Ed., John Wiley & Sons, Inc., New York.

McClain, W. H., and Champe, S. P. (1967), *Proc. Nat. Acad. Sci. 58*, 1182.

Orgel, A., and Brenner, S. (1961), *J. Mol. Biol. 3*, 762.

Schuster, H., and Vielmetter, W. (1961), *J. Chim. Phys. 58*, 1005.

Schwartz, D. O., and Beckwith, J. R. (1969), *Genetics 61*, 371.

Speyer, J. F. (1969), *Fed. Proc. 28*, 348.

Streisinger, G., Mukai, F., Dreyer, W. J., Miller, B , and Harrar, G. (1961a), *J. Chim. Phys. 58*, 1064.

Streisinger, G., Mukai, F., Dreyer, W. J., Miller, B., and Horiuchi, S. (1961b), *Cold Spring Harbor Symp. Quant. Biol. 26*, 25.

Streisinger, G., Okada, Y., Emrich, J., Newton, J., Tsugita, A., Terzaghi, E., and Inouye, M. (1966), *Cold Spring Harbor Symp. Quant. Biol. 31*, 77.

Stent, G. S. (1963), "Molecular Biology of Bacterial Viruses," W. H. Freeman and Co., San Francisco.

Tessman, I. (1962), *J. Mol. Biol. 5*, 442.

Tessman, I., Ishiwa, H., and Kumar, S. (1965), *Science 148*, 507.

Tessman, I., Poddar, R. K., and Kumar, S. (1964), *J. Mol. Biol. 9*, 352.

Tsugita, A., and Inouye, M. (1968), *J. Mol. Biol. 37*, 201.

Weil, J., Terzaghi, B., and Crasemann, J. (1965), *Genetic 52*, 683.

Witkin, E. M. (1969), *Ann. Rev. Genet. 3*, 525.

Wittmann, H. G., and Wittmann-Liebold, B. (1966), *Cold Spring Harbor Symp. Quant. Biol. 31*, 163.

Yanofsky, C., Cox, E. C., and Horn, V. (1966a), *Proc. Nat. Acad. Sci. 55*, 274.

Yanofsky, C., Ito, J., and Horn, V. (1966b), *Cold Spring Harbor Symp. Quant. Biol. 31*, 151.

Zamenhof, S. (1960), *Proc. Nat. Acad. Sci. 46*, 101.

CHAPTER 8

Prophage Induction in Lysogenic Bacteria as a Method of Detecting Potential Mutagenic, Carcinogenic, Carcinostatic, and Teratogenic Agents

Bernard Heinemann*

Research Division, Bristol Laboratories
Division of Bristol-Myers Company
Syracuse, New York

I. INTRODUCTION

This chapter deals with the application of the prophage induction phenomenon, found exclusively in certain strains of lysogenic bacteria, to the
detection of a growing list of agents of considerable pharmacological importance. These agents, among other biological properties, may be mutagenic, carcinogenic, carcinostatic, and teratogenic.

Lysogenic bacteria carry the latent form of a bacterial virus, the prophage. A number of physical and chemical agents are now known which are
capable of disrupting this stable, integrated state in some lysogenic strains

* During the writing of this chapter, the author was partially supported by contracts
PH43-64-1159 and NIH 69-35 with Chemotherapy, National Cancer Institute,
National Institutes of Health.

and of eventually transforming the innocuous prophage into infectious phage particles. This is known as prophage induction.

Lwoff (1953), in his classical review of lysogeny, suggested that prophage induction in lysogenic bacteria might serve as a test for detecting compounds with mutagenic, carcinogenic, and carcinostatic properties. Investigators at Bristol Laboratories developed a quantitative assay technique, based on the prophage induction phenomenon, and applied it to systematic screening for potential carcinostatic agents in fermentation broths (Lein et al., 1962; Heinemann and Howard, 1964; Price et al., 1964, 1965). This group clearly established a positive association between prophage induction capability and antineoplastic activity with many different types of chemical agents. This correlation has subsequently been confirmed by a number of investigators (Endo et al., 1963c; Yajima et al., 1963; Marjai and Ivanovics, 1964; Specht, 1965; Gause, 1965; Ikeda and Iijima, 1965; Aoki and Sakai, 1967; Fleck, 1968).

Positive associations of inducing capability with mutagenic activity (Lwoff, 1953; Ikeda and Iijima, 1965), with carcinogenic activity (Lwoff, 1953; Endo et al., 1963a; Price et al., (1965), and with both mutagenic and carcinogenic activities (Epstein and Saporoschetz, 1968) have been noted with a limited number of compounds. It is quite possible that, with proper testing, more mutagenic, carcinogenic, carcinostatic, and also teratogenic activities will eventually be discovered for the known prophage inducing agents. As an example of the usefulness of this test, it may be noted that there was no information available regarding these activities for the antibiotic streptozotocin until a positive induction effect was found (Lein et al., 1962). This result suggested mutagenic (Kolbye and Legator, 1968), carcinogenic (Arison and Feudale, 1967), and carcinostatic (Evans et al., 1965) activities which were subsequently discovered.

Prophage induction is a relatively simple in vitro test system which offers a number of advantages over the use of screens involving laboratory animals for detecting agents capable of mutagenic, carcinogenic, carcinostatic, and teratogenic activities. It is rapid, economical, quantitative, and requires very small quantities of the test agent, and its high degree of sensitivity makes it uniquely capable of detecting weakly active agents as well as those present in natural products in very low concentrations. Once a new agent has been detected by the prophage induction screen, biologists should evaluate the mutagenic, carcinogenic, and teratogenic risks to mankind resulting from exposure to the agent as well as its possible value as a chemotherapeutic agent in cancer. Such studies are usually carried out by various techniques in higher organisms. Determination of the relevance of data obtained from these studies to man requires utmost caution.

This treatise starts (section II) with a brief account of the general properties of "healthy" or "orthodox" lysogenic bacteria and temperate bacteriophages. The conditions controlling phage induction in lysogenic bacteria will be considered and the possible mechanisms will be discussed in section III. Section IV deals with the methodology of prophage induction assay techniques. Some available lysogenic systems for which prophage induction has been demonstrated, based on published observations, are presented. A list of prophage inducing agents obtained from a survey of published observations makes up section V. Those agents which have also been shown to possess mutagenic, carcinogenic, carcinostitic, and teratogenic properties are indicated. In section VI, the presence of prophage inducing agents in the human environment is discussed.

II. GENERAL PROPERTIES OF LYSOGENIC BACTERIA

The past quarter century has witnessed the development of a tremendous volume of imaginative and exciting contributions to microbial genetics from the field of lysogeny. In spite of these numerous contributions to molecular biology, relatively few applications of medical or practical consequence have evolved to date. Lysogenic systems offer numerous mutabilities which might serve as the basis of a screening method for detecting potential mutagenic agents. In this treatise only the prophage induction phenomenon will be considered.

It is beyond the scope of this report to review in detail the mass of data on lysogeny which has accumulated. Instead, in order to make the present discussion reasonably complete, an attempt will be made to present the essential details of "healthy" or "orthodox" lysogeny which are of major relevance to prophage induction assay techniques. For the reader interested in a more comprehensive coverage of the field, there have appeared a numbered of more or less complete treatments, in historical perspective, of the general features of lysogeny (Bertani, 1953, 1958; Jacob and Wollman, 1959a,b; Whitfield, 1962; Stent, 1963; Hayes, 1964). Specialized detailed reviews of the earlier literature on the prophage induction phenomenon (Jacob and Wollman, 1953) and on the genetic aspects of lysogeny are also available (Jacob and Wollman, 1957, 1961). Within the past decade there has been a tremendous increase in the literature on the genetics, biochemistry, and regulatory mechanisms of the lysogenic state, and these have been treated in more rcent reviews (Campbell, 1967; Thomas, 1968; Echols and Joyner, 1968; Signer, 1968; Dove, 1968). Excellent recent reviews of the newer concept which considers temperate phages as episomes rather than as viral elements are also available (Campbell, 1962; Driskell-Zamenhof, 1964; Scaife, 1967).

Bacteriophages were discovered over 50 years ago by Twort (1915) and d'Herelle (1918). They are viruses which infect sensitive bacteria and may be divided into two broad groups according to the type of response elicited in their host. When a *virulent bacteriophage* infects a sensitive host, lytic growth is initiated which almost invariably results in lysis and death of the host accompanied by the liberation of a burst of new infectious phage particles. When a *temperate bacteriophage* infects its sensitive host, two different responses may occur simultaneously in the cell population, resulting in a somewhat more complex life history than occurs with the virulent phages. The working of this coordinated system, first described both by Bordet (1925) and Bail (1925), makes one of the most fascinating stories in the entire development of bacterial and phage genetics. The true nature of lysogency became evident in 1950 from the work of Lwoff and his collaborators at the Pasteur Institute.

One response involves lytic growth which takes place in some host cells, while in the remainder (1–99%) of the sensitive bacterial population the infecting phage causes a second response resulting in the establishment of a relatively permanent union known as *lysogeny*. The choice between lytic and lysogenic growth appears to be the "result of a race," influenced by growth conditions present in the medium, by the phage and by the host including the stage of the cell in its division cycle and may also be genetically controlled.

Bacteria in which lysogeny becomes established are termed *lysogenic* and are said to be in the *lysogenic state*. Lysogenic bacteria retain the hereditary ability to reproduce indefinitely and to produce, under certain conditions, infectious phage particles. This capacity to form phage particles is perpetuated intracellulary in the genetic systems as a noninfectious entity, known as *prophage*, which is transmitted from one generation to the next.

The regulation of lysogeny can be ascribed to the existence of a repressor protein, elicited by the prophage, which represses all phage genetic activity. Some cells in a sensitive population escape repression upon infection with a temperate phage with consequent synthesis of phage messenger RNA and protein culminating in lytic growth. During lysogeny, messenger RNA synthesis is repressed, the lytic cycle is prevented, and the phage chromosome becomes integrated into the host chromosome at a particular site, residing there as prophage. The prophage is passively replicated with the bacterial chromosome and continues repressor synthesis, thus preventing both the prophage itself and the DNA from a homologous superinfecting phage from replicating autonomously and expressing their lytic function. Prevention of the replication of the superinfecting phage confers on the lysogenic cell a remarkable *immunity* against the initiation of lytic growth which would result from the infection either with the corresponding phage or with its homologous mutants. A superinfecting phage, if unrelated to the carried prophage, may be able to direct the synthesis of phage gene products upon its entrance into the lysogenic cell and integrate its chromosome into the host chromosome at yet another site. It can either replace an existing prophage or coexist with it in a condition which is termed *multiple lysogeny*. Lytic growth is prevented upon attainment of multiple lysogeny by the presence of the repressor.

The lysogenic state usually remains a very stable property of a bacterial strain. In most lysogenic populations, only a very small percentage of cells are *cured* of their lysogeny—that is, lose their prophage.

A very interesting feature of the life cycle of some lysogenic bacteria is the induction of the development of prophage into infectious phage particles under certain environmental conditions and by certain physical and chemical agents. The various aspects of this phage production will be considered in greater detail in the following section.

As a conclusion to this brief account of lysogeny, lysogenic bacteria differ from the corresponding nonlysogenic bacteria in their hereditary capacity to produce phage, in their immunity against the phage that they are able to produce, and in the possession of a repressor which regulates the condition.

III. INDUCTION OF PHAGE PRODUCTION IN LYSOGENIC BACTERIA AND BREAKDOWN OF THE LYSOGENIC STATE

Lysogenic bacteria possess the hereditary property of producing temperate bacteriophages. When the relatively stable prophage–bacterium relationship is disturbed, phage development takes place. The phenomenon whereby the noninfectious prophage is transformed into an infectious phage particle is known as *induction*.

The events that take place in the lysogenic cell following induction are identical to those of the vegetative phase of phage development following infection of a sensitive bacterium with an exogenous free phage particle. These events have been reviewed by Jacob and Wollman (1953).

In this section of the report, the phenomenon of prophage induction in inducible strains will be described and the possible mechanisms of induction will be discussed briefly. Particular attention will be given to conditions controlling induction by physical and chemical agents. This account will deal primarily with lambda phage of *Escherichia coli*, since this has been the most thoroughly studied system.

In exponentially growing cultures of lysogenic bacteria, the stable relationship between prophage and host breaks down in some cells of the population every now and then and a relatively constant number of free phage particles (from 10^{-2} to 10^{-7} per cell per generation) are produced, the number depending on the strain of temperate phage. This phenomenon is termed *spontaneous induction*. In some inducible lysogenic strains, it is possible to effect the release of infectious particles in virtually all members of the population by a temperature shift (Sussman and Jacob, 1962), by zygote formation (Jacob and Wollman, 1954), by thymine starvation in auxotrophic cells (Korn and Weissbach, 1962), or by exposure to certain physical and chemical agents.

Available evidence suggests that induction might involve a direct reversal of the establishment of the prophage state.

Induction requires inactivation of the repressor. Thermal induction of a thermo-inducible prophage probably results from a direct inactivation of the repressor, either reversible or irreversibly. In zygotic induction, immunity could be lifted as the prophage is transferred into the cytoplasm of a sensitive cell. In induction by thymine starvation and by exposure to agents, the repressor is inactivated by more complex processes involving the operation of bacterial genes and protein synthesis.

Induction of prophage development, either spontaneous or by chemical agents or radiation, is probably accomplished by destruction or inhibition of the repressor, permitting vegetative phage multiplication to be initiated irreversibly. The inducing shock produces, an alteration of the bacterial nucleic acid balance in which DNA synthesis is selectively blocked (Heinemann and Howard, 1966). Goldthwait and Jacob (1964) proposed that repressor inhibition of lambda phage is caused by the accumulation of a low molecular weight precursor of DNA resulting from a block of DNA synthesis. Inactivation of the repressor initiates excision of the prophage from the host chromosome, resulting in lytic growth similar to that occurring upon infection. The observation that 1-methyl-3-nitro-1-nitrosoguanidine, a prophage inducing agent, inhibits not only the synthesis of DNA but also of RNA and protein is of considerable interest (Terawaki and Greenberg, 1964).

While the cause of spontaneous induction is not well known, it could result from an alteration of environmental conditions or endogenous production of a physiologically significant amount of prophage inducing agents e.g., colicines, H_2O_2, antibiotics, etc.) in the course of bacterial growth, or both, which may profoundly influence the balance of the phage–bacterium relationship so that the lytic cycle supervenes in a small proportion of the lysogenic population. Further experiments to define this situation more thoroughly are needed.

Upon exposure of a growing culture of inducible lysogenic bacteria to appropriate doses of physical or chemical inducing agents, growth proceeds without bacterial division during a latent period of one or two generations. Each induced cell, or *infectious center*, will produce a single plaque if plated at this point on a lawn of a suitable indicator culture. Upon further incubation in a suitable nutrient medium, following the latent period, the bacterial cells lyse and as many as 500 infectious phage particles may be released per bacterium. Each phage particle is capable of producing a single plaque.

The *aptitude*, or efficiency of induction, of a lysogenic strain is controlled by the nutritional conditions of the system and the physiological state of the culture both before and after the inducing shock.

IV. EXPERIMENTAL PROPHAGE INDUCTION TECHNIQUES

A. Inducible Lysogenic Systems

Strains of lysogenic bacteria are very prevalent and appear to be widely distributed among bacterial species. The prophage of some lysogenic bacteria can be induced by treatment with physical and chemical agents with the eventual release of infectious phage particles. To detect prophage inducing agents, a lysogenic strain carrying the prophage and a sensitive indicator culture on which the induced phage particles will form plaques are needed. Much of the early work on prophage induction was carried out with lysogenic systems of *Bacillus megatherium* and of *Pseudomonas pyocyanea*. Following this, phage lambda of *E. coli* became one of the most extensively investigated of the inducible phages. Investigators rcently involved in screening for prophage inducing agents have favored lysogenic systems utilizing *E. coli* K-12 (λ), or strains derived from it, and *Micrococcus lysodeikticus* 53-40 (N-5).

Comparisons of the relative induction response of various lysogenic systems to prophage inducing agents have been limited. Price *et al.* (1965) found that seven antibiotics capable of inducing *E. coli* W1709 (λ) were incapable of acting as inducers of either *Agrobacterium tumefaciens* B6 ((Ω) or *M. lysodeikticus* 53-40 (N-5). Dudnik (1965) observed that *M. lysodeikticus* 53-40 (N-5) was more sensitive than *E. coli* K-12 (λ) for demonstrating inducing effects with two antibiotic preparations.

It would appear that a good deal more work is required before a particular lysogenic system, or systems, can be recommended as optimal for detection of prophage inducing agents. Different types of inducing agents should be evaluated under assay conditions considered to be optimal to induce development of vegetative phage in the lysogenic system. There is a need for quantitation of assay results. Perhaps a "spectrum" of lysogenic systems will be necessary to detect a variety of inducing agents similar to the use of various test organisms in screening for new antibiotic agents.

The intense earlier studies devoted to the lysogenic state of the gram-negative organisms have been matched more recently by similar efforts with gram-positive organisms, and results suggest that the phage–bacteria relationships may be similar in both groups of organisms.

Table 1 lists some inducible lysogenic systems reported in the literature

TABLE 1. Inducible Lysogenic Systems

Culture	Lysogenic strain	Indicator strain
A. tumefaciens	B6 Ω	B6
Bacillus cereus	6464	B569
B. megatherium	899	PR3E
Bacillus stearothermophilus	1503-4R	4S
Corynebacterium diphtheriae	25	C7
Corynebacterium diphtheriae	C4 (ATCC 11952)	C4 (ATCC 11951)
E. coli	K12 (ATCC 10798)	W3001
E. coli	W1709	W3001
M. lysodeikticus	ML53-40	ML53-5
	(ATCC 15800)	(ATCC 15801)
Mycobacterium butyrium	R1	607B
Pseudomonas aeruginosa	MAC-264	PAE-2-1
Rhizobium meliloti	13 (13)	L5
Salmonella thompson	19 (5/19)	19
Salmonella typhimurium	LT 2 gal$^+$	LT 2 gal$^-$
Streptococcus group A	K56	K56
Xanthomonas campestris	P165	P125

which might be adaptable to quantitative prophage induction assays. Numerous derivatives of the original *E. coli* K-12 (λ) strain have been utilized which are not listed. Unfortunately, relatively few inducible lysogenic bacteria with their indicator strains can be obtained from established culture collections.

B. Prophage Induction Assay Techniques

Several methods for detecting prophage induction in lysogenic systems exposed to chemical agents have been described by investigators. These procedures may be classified into two general types, as follows:

1. The two-stage test: During the first stage, the *induction period*, lysogenic cells are exposed to the agent in liquid medium. The second stage consists of enumerating the plaque-forming units produced which may be either infectious centers (induced bacteria) or free phage particles, or both, on a lawn of an indicator culture. The induction period may be carried out in a nutrient broth medium by exposing the growing lysogenic culture to the agent during incubation for 3 to 5 hr and then enumerating the plaque-forming units produced. Alternatively, the lysogenic culture may be exposed to the agent either in a nutrient medium or in a salts solution for a

Phage designalion	Inducing agent	References
Ω	Ultraviolet	Beardsley (1960)
	Ultraviolet	Altenbern (1962)
M1	Several agents	Clarke and Cowles (1952)
TP-1	Mitomycin C	Welker and Campbell (1965)
	N-methyl-N'-nitro-N-nitrosoguanidine	Kozak and Dobrzanski (1967)
β	Ultraviolet	Groman et al. (1958)
λ	Several agents	Heinemann and Howard (1964)
λ	Several agents	Price et al. (1964)
N-5	Several agents	Field and Naylor (1962)
R1	Ultraviolet	Bowman and Redmond (1959)
φ-MC	Mitomycin C	Yamamoto and Chow (1968)
13	Ultraviolet	Kowalski (1966)
5/19	Several agents	Smith (1953)
P22	Mitomycin C and streptonigrin	Levine and Borthwick (1963)
P7738	N-methyl-N'-nitro-N-nitrosoguanidine	Malke (1967)
P165/P125	Mitomycin C	Sutton and Quadling (1963)

period up to 1.5 hr. A condition of static growth is sought and induction is then terminated by suitable dilution. At this point, prophage induction may be detected by enumerating the number of infectious centers on agar plates with a sensitive indicator. Alternatively, if induction is terminated by dilution with nutrient medium and incubation continued, induced bacteria will undergo lysis and release free phage particles, the number depending on the burst size of the lysogenic strain used. Induction is detected by counting the plaques formed by free phage particles on solid medium with a sensitive indicator culture. Two-stage procedures have been used since the early studies on prophage induction.

2. *The single-stage test:* In an attempt to provide a more rapid procedure for mass screening of compounds, single-stage techniques performed entirely on an agar medium have been suggested (Gado et al., 1965/1966; Mayer et al., 1969). A suitable mixture of the lysogenic and indicator strains is placed on an agar surface. Test agent is applied to the agar surface where it diffuses into the agar and, if active, produces a zone of increased plaque formation, upon incubation of the plate, around the test material. Each plaque represents either an induced bacterium (infectious center) or a free phage particle resulting from spontaneous induction.

Test procedures have been described in which lysis of the lysogenic strain is detected either turbidimetrically in broth medium (Yajima *et al.*, 1963) or on an agar plate (Ikeda and Iijima, 1965). Results obtained by such methods must, of necessity, be confirmed by more definitive procedures since lysis of lysogenic bacteria may also be produced by phenomena other than prophage induction.

In most instances, scant attention has been paid to determining the conditions of maximal sensitivity of test methods requisite for developing a quantitative assay. The number of lysogenic bacteria capable of being induced may vary from 1–100% depending on (1) the concentration of the prophage inducing agent and (2) the nutritional conditions under which the assay is conducted. With the two-stage procedure, results obtained may be quantitative and sufficiently sensitive to detect induction when it occurs in only a small percentage of a lysogenic population. This is particularly useful in purification work with natural products and in working with weakly active agents.

A simple, rapid, quantitative assay for prophage induction utilizing *E. coli* K-12 (λ) in a two-stage procedure was first described by Heinemann and Howard (1964). The sensitivity of this procedure was further improved, to facilitate the detection of chemical inducers in fermentation filtrates, by using a streptomycin-dependent strain, *E. coli* W1709 (λ) (Price *et al.*, 1964). It might be of interest to mention several of the factors which they sought in developing a prophage induction assay capable of giving optimal results.

Lysogenic culture: (1) It should not be a potential pathogen for laboratory personnel. (2) Its growth in broth should be evenly dispersed and nonclumping, to facilitate optical density measurements for standardizing the culture population. The growth cycle should not be prolonged. (3) Its growth should not be stimulated by the test agent while induction is taking place. Such growth would give a proportionately greater release of spontaneously produced phage resulting in a high percentage of false-positives. Growth during the induction period may be prevented by using a lysogenic strain with a specific growth requirement which is eliminated from the induction medium. It is particularly desirable to use such a strain in tests with natural products which may contribute growth nutrients to the system. (4) It should possess maximal sensitivity to the inducing agents and be capable of detecting weakly active agents. (5) Its aptitude to induction should not be reduced upon transfer to the induction medium. (6) Enumeration of free phage particles is most easily accomplished if the lysogenic bacteria are sensitive to an antibiotic to which the indicator bacteria are resistant. Upon plating the two together on antibiotic-containing agar, the presence of the lysogenic bacteria can be ignored since their ability to grow as well as to produce phage is prevented by the antibiotic, while the ability of the phage particles to produce plaques is unaffected.

Induction medium: (1) It should not permit growth of the lysogenic bacterium and maturation of induced phage to occur. (2) It should be a defined medium and have no effect on the inducing agent; e.g., the activity of azaserine is reduced by certain aromatic amino acids present in complex media (Gots *et al.*, 1955).

Maturation medium: It should permit maximum development of free phage particles.

Indicator culture: (1) Its growth should produce a smooth lawn on the agar surface. (2) It should be resistant to an antibiotic to which the lysogenic strain is sensitive.

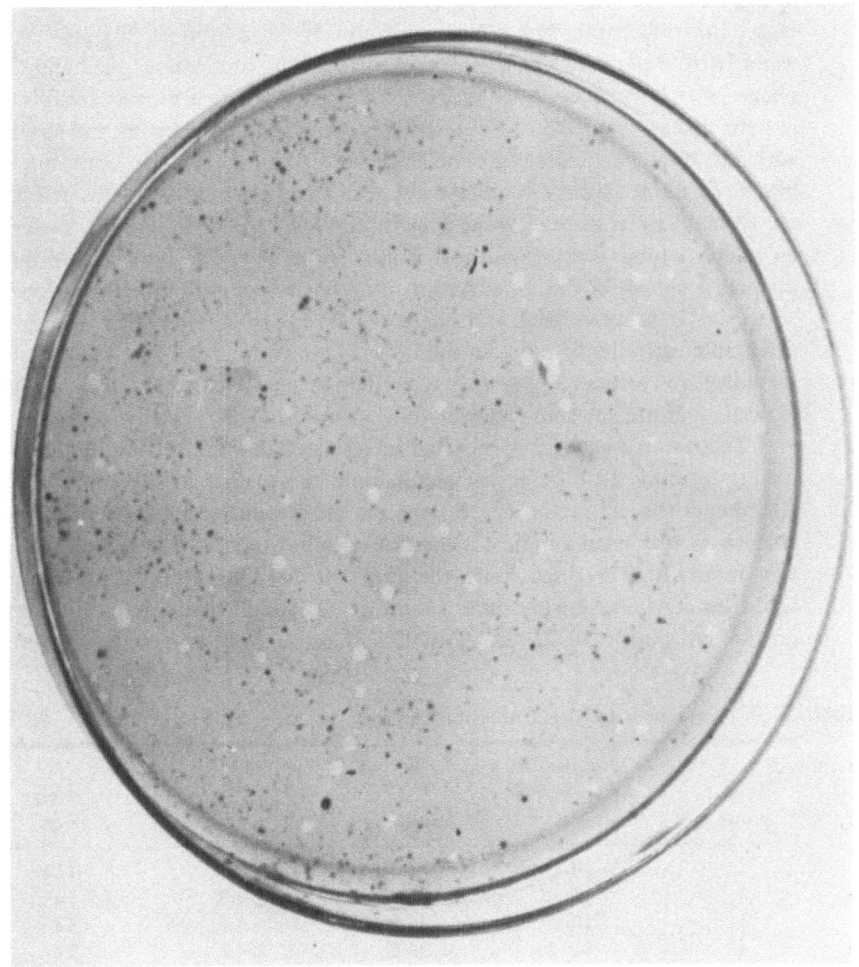

FIGURE 1. A petri plate showing growth of a lawn of *E. coli* W3001 bacteria on which lambda-bacteriophage induced from a lysogenic culture, *E. coli* W1709, has formed plaques.

Other factors which require standardization for optimal induction are the age of the lysogenic bacteria and the duration of the induction and maturation periods.

The assay procedure currently in use in our laboratory utilizes *E. coli* W1709 (λ) and has been described in detail by Price *et al.* (1964). This procedure is described here briefly. The lysogenic strain is streptomycin-dependent, tetracycline-sensitive, and auxotrophic for leucine, threonine, and thiamine. This test consists essentially of inducing a young culture of lysogenic cells in a defined medium devoid of streptomycin for $1\frac{1}{2}$ hr at 37°C in a water bath with the test agent and then growing the induced cells in Heart Infusion Broth with added streptomycin for 2 hr at 37°C in a shaker water bath to permit liberation of free phage. The broth is sampled and the phage particles produced are enumerated by the soft agar overlay technique in petri dishes using Heart Infusion Agar containing tetracycline and seeded with the tetracycline-resistant indicator strain *E. coli* W3001. Counting of plaque-forming λ-phage is carried out after incubation of the plates at 37°C for 18 hr. A petri plate showing growth of a lawn of *E. coli* W3001 bacteria on which lambda-bacteriophage induced from *E. coli* W1709 has formed plaques is shown in Fig. 1. The remaining portion of each sample is returned to the 37°C water bath and incubated for approximately 3 hr to permit detectable turbidity to occur in the tubes. Test samples failing to show such turbidity are considered probably to contain an agent toxic for *E. coli* W1709 (λ) cells and are retested at lower concentrations.

Test sample activity is reported in terms of the ratio of the number of plaque-forming λ-phage in the test sample (T) to that in the control (C). All phage present in the control sample are produced spontaneously. An analysis of test results indicates that samples having a T/C value, or induction index, of 3.0 (three times the spontaneous phage count), or greater, could be considered with some assurance ($P<0.05$) to be active inducers of the lytic cycle in *E. coli* W1709 (λ) cells. A known active material

TABLE 2. Precision of the Induction Assay[a]

Trial number	Number of independent assays	T/C value (av.)	SD[b]	Percentage error
1	10	45.2	5.0	11
2	11	36.7	4.5	12
3	10	38.4	4.8	12
4	10	34.6	4.0	11

[a] Data kindly supplied by Dr. K. E. Price, Bristol Laboratories, Syracuse, N.Y.
[b] Standard deviation was calculated from the formula $S = (X^2/n-1)^{1/2}$ where S = standard deviation, X = the deviation from the mean of the series, and n = the number of assays.

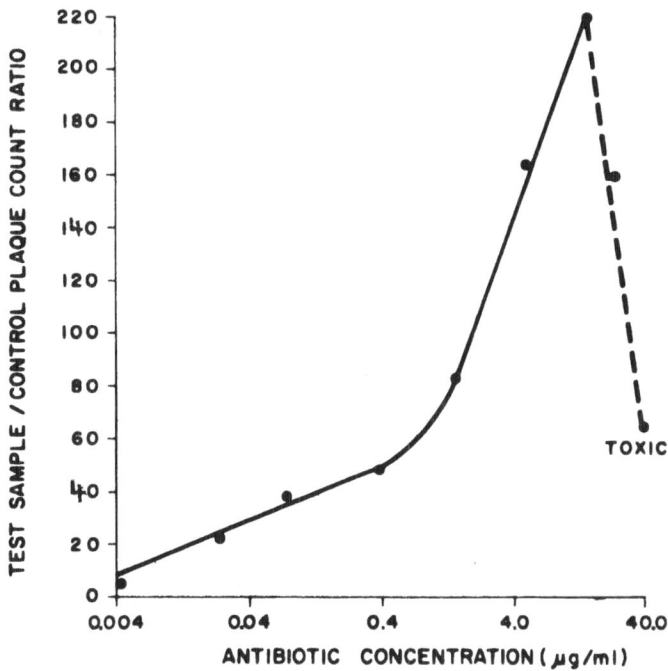

FIGURE 2. Response of *E. coli* W1709 (λ) induction system to various concentrations of mitomycin C.

(mitomycin C, 0.2 ml of an 0.25 μg/ml solution) is included as a positive control in each test.

The reproducibility of the induction assay is better when replicate assays of a given sample are made in a single day than when assays are performed on successive days. As a result, all assays pertaining to a given experiment were performed during a single assay run. The precision of the induction assay is indicated in Table 2. The assays in a trial were performed in the course of a single run on different days. The material assayed consisted of a 0.25 μg/ml (0.05 μg/ml final concentration in the assay) solution of mitomycin C which was kept frozen between trials. In daily assay runs, where comparative data are required, a mitomycin C standard (0.05 μg/ml) is run, and the results accepted if the per cent error is less than 15%.

The dose-response characteristics of prophage inducing agents can vary markedly. Figure 2 illustrates the results obtained with *E. coli* W1709 (λ) using different concentrations of the agent mitomycin C. The "test sample/control" plaque count ratio (*T/C*) is plotted versus antibiotic concentration. It will be seen that the curve has three slopes. At concentrations between

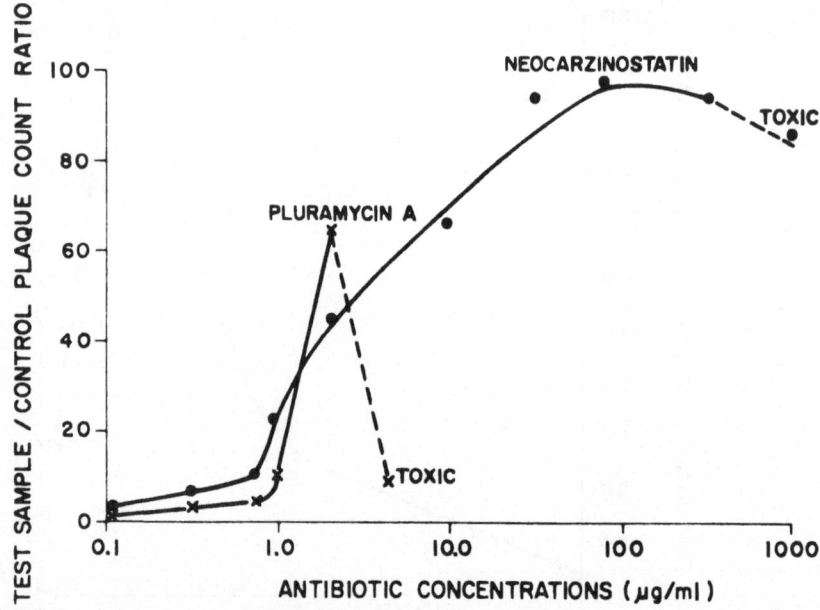

FIGURE 3. Response of *E. coli* W1709 (λ) induction system to various concentrations of neocarzinostatin and pluramycin A (from Price *et al.*, 1965).

0.004 and 0.4 μg/ml, the *T/C* ratio rises slowly. From 0.4 to 6 μg/ml, the slope increases. Above this level of mitomycin C, the curve falls off because of killing of large numbers of organisms due to the antibiotic action of the agent.

In Fig. 3, curves with pluramycin A and neocarzinostatin are shown. Pluramycin A is an extremely toxic antibiotic and inhibited both *E. coli* and lambda phage production in the induction system at a concentration of 4.0 μg/ml. It gave a spikelike dose-response curve with a maximan *T/C* of 65 and was active only over a fourfold range. In sharp contrast, neocarzinostatin was nontoxic for *E. coli* W1709 (λ) cells at concentrations up to 500 μg/ml. It gave a maximal *T/C* that approached 100 and produced significant induction over a greater than thousand fold dosage range.

C. Paper Chromatographic Techniques

A procedure has been described whereby paper chromatograms of agents which induce lambda bacteriophage in *E. coli* can be developed using bioautographs with a lysogenic test system (Heinemann *et al.*, 1967). Such a procedure finds use for resolution of materials with multicomponent inducing activities when found in natural products and in mixtures which are difficult to separate.

V. KNOWN PROPHAGE INDUCING AGENTS

A. Historical

Lwoff and Gutmann (1950) suggested that phage production in lysogenic bacteria might be inducible by external factors. The earliest recorded observation of prophage induction seems to be that of Lwoff *et al.* (1950) with a lysogenic strain of *B. megatherium* exposed to ultraviolet irradiation. Other forms of radiation including X-rays (Latarjet, 1951), γ-rays (Marcovich, 1956), and alpha and deuteron particles (Rappaport, 1958) were subsequently found to possess inducing capability. The group at the Pasteur Institute also observed the phenomenon with chemical agents: nitrogen mustard (Jacob, 1952), hydrogen peroxide, tertiobutyl peroxides, butadiene-1,3-diepoxide, ethyleneimine, and 2,4,6-tris-ethyleneimino-1,3,5-triazine (Lwoff and Jacob, 1952).

Smith (1953) reported induced phage lysis in a lysogenic culture of *Salmonella thompson* by means of nitrogen mustard, mustard gas, sulfathiazole, glutathione, and sodium thiolacetate.

It was subsequently demonstrated that the antibiotics azaserine (Gots *et al.*, 1955), phagolessin (Hall-Asheshov and Asheshov, 1956), and mitomycin C (Otsuji *et al.*, 1959) as well as the antimetabolite aminopterin (Ben-Gurion, 1962) were also effective inducers.

Until 1962, only a very small number of substances had been examined for prophage inducing capability. A report (Lein *et al.*, 1962) observing a correlation between prophage inducing and antineoplastic properties in a group of antibiotics markedly stimulated systematic screening for such agents. As a result, the ranks of chemical inducers of prophage development have been greatly expanded. Many more compounds still remain to be tested.

B. Association of Prophage Inducing Capability with Mutagenic, Carcinogenic, Carcinostatic, and Teratogenic Activities

Table 3 has been prepared to list known prophage inducing agents on the basis of data reported in the literature. The agents are grouped as radiations, alkylating chemicals, products of microbial metabolism, and miscellaneous compounds. Correlation of published data is hindered by the variables occurring within the experimental conditions used by different investigators for assessing prophage induction. As previously suggested, only those agents are included in which induction has been demonstrated by appropriate plating techniques with an indicator strain. The degree of prophage induction is indicated in general terms of "potent" (P), "moderately

TABLE 3. Prophage-Inducing Agents

Agent	Prophage induction	Mutagenic	Carcino-genic
			Effect
A. Radiations			
Ultraviolet rays	+	+	+
X-rays	+	+	+
Gamma rays	+	+	+
Alpha particles	+	+	+
Deuteron particles	+		
B. Alkylating agents			
Aliphatic mustards:			
Sulfur mustard	+	+	+
Nitrogen mustard	P	+	+
Bis (2-chloroethyl) amine HCl	S		+
Tris (2-chloroethyl) amine HCl	P		+
Alanine mustard	P		
Nitromin	S	+	+
Aromatic mustards:			
Chlorambucil	S	+	+
Sarcolysin	S	+	+
Ethyleneimines:			
TEM	M	+	+
TEPA	+	+	
ThioTEPA	S	+	+
Bayer E39	S	+	+
Trenimon	M	+	

studied		
Carcino-static	Terato-genic	References
		Lwoff *et al.* (1950); Altenburg (1930); Findlay (1928)
+	+	Latarjet (1951); Muller (1927); Frieben (1902); Despeignes (1956); Russell (1950)
+	+	Marcovich (1956); Hanson (1928); Shellabarger *et al.* (1957); Gray and Read (1948); Knezevic and Karanovic (1966)
		Rappaport (1958); Ward (1953); Evans (1947)
		Rappaport (1958)
+		Smith (1953); Auerbach *et al.* (1947); Heston (1949); Bass and Freeman (1946)
+	+	Jacob (1952); Auerbach *et al.* (1947); Boyland and Horning (1949); Haddow *et al.* (1948); Haskins (1948)
+		Heinemann and Howard (1964); Boyland and Horning (1949); Goldin *et al.* (1949)
+	+	Specht (1965); Boyland and Horning (1949); Goldin *et al* (1949); Faucounau *et al.* (1963)
+		Endo *et al.* (1963c); Izumi (1954)
+		Specht (1965); Szybalski (1958); Shimpo *et al.* (1960); Ishidate *et al.* (1951)
+	+	Specht (1965); Fahmy and Fahmy (1960); Weisburger (1966); Galton *et al.* (1955); Murphy *et al.* (1958)
+		Specht (1965); Fahmy and Fahmy (1960); Weisburger (1966); Blokhin *et al.* (1958)
+	+	Heinemann and Howard (1964); Epstein and Shafner (1968); Conklin *et al.* (1963); Lewis and Crossley (1950); Didcock *et al.* (1956)
+	+	Price *et al.* (1965); Epstein and Shafner (1968); Sugiura and Stock (1955); Thiersch (1957a)
+	+	Specht (1965); Epstein and Shafner (1968); Weisburger (1966); Buckley *et al.* (1951); Okano *et al.* (1959)
+	+	Specht (1965); Bertram and Hoehne (1959); Weisburger (1966); Domagk (1958); Adams *et al.* (1961)
+		Specht (1965); Lüers and Röhrborn (1965); Blum (1959)

TABLE 3. (Continued)

	Effect		
Agent	Prophage induction	Mutagenic	Carcino-genic
Nitroso compounds:			
1-Methyl-3-nitro-1-nitrosoguanidine	M	+	+
N,N'-dinitroso-N,N'-dimethylterephtal amide	S	+	
N,N'-dinitroso-N,N'-dimethyloxamide	S	+	
Others:			
1,2,3,4-Diepoxybutane	M	+	+
Dimethyl sulfate	S	+	+
1,4-Bis (3-bromopropionyl) piperazine	S		
β-Propiolactone	+	+	+
4-Nitroquinoline-N-oxide	P[a]	+[a]	+[a]
C. Products of microbial metabolism			
Alazopeptin (related to DON)	S	+	
Azaserine	P	+	
Bleomycin (related to phleomycin)	P		
Bruneomycin (related to streptonigrin)	P		
Carcinophilin	P		
Carzinostatin	P		
Chromomycin A3	M		
Colicin E2	P		
Colicin P	+		
Cycloheximide	S		
Daunomycin (synonomous with rubidomycin and rubomycin C)	M		+
DON (6-diazo-5-oxo-L-norleucine, an analog of azaserine)	S		
Gancidin A	P		
Griseolutein A and B (mixture)	M		
Hedamycin (related to pluramycin A)	P		
Illudins S and M	M		

studied		References
Carcino-static	Terato-genic	
+		Allan and McCalla (1966); Mandell and Greenberg (1960); Sugimura *et al.* (1966); Greene and Greenberg (1960)
		Menzel and Geissler (1966); Marquardt *et al.* (1964)
		Menzel and Geissler (1966); Marquardt *et al.* (1964)
+		Heinemann and Howard (1964); Bird and Fahmy (1953); McCammon *et al.* (1957); Rose *et al.* (1950)
		Field and Naylor (1962); Kolmark (1956); Druckrey *et al.* (1966)
+		Heinemann and Howard (1964); Davies *et al.* (1960)
		Field and Naylor (1962); Smith and Srb (1951); Palmes *et al.* (1962)
+[a]		Endo *et al.* (1963a); Okaboyashi *et al.* (1965); Nakahara *et al.* (1957); Moore *et al.* (1960)
+	+	Price *et al.* (1965); Ikeda and Iijima (1965); Suguira (1959/1960); Thiersch (1958)
+	+	Gots *et al.* (1955); Hemmerly and Demerec (1955); Clarke *et al.* (1957); Thiersch (1957b)
+		Aoki and Sakai (1967); Ishizuka *et al.* (1967)
+		Gause (1965); Rossolimo *et al.* (1966)
+	+	Endo *et al.* (1963c); Hata *et al.* (1954); Takaya (1965)
+		Heinemann and Howard (1964); Shoji (1961)
+	+	Yajima *et al.* (1963); Tatsuoka *et al.* (1958); Takaya (1965)
		Nomura (1963)
		Hamon and Peron (1965)
+		Price *et al.* (1965); Reilly *et al.* (1953)
+	+	Price *et al.* (1965); Dubost *et al.* (1963); DiMarco *et al.* (1963)
+	+	Price *et al.* (1965); Clarke *et al.* (1957); Thiersch (1957c)
+		Heinemann and Howard (1964); Aiso *et al.* (1956)
+		Heinemann and Howard (1964); Ogata (1959)
+		Bradner *et al.* (1967)
+		Price *et al.* (1965); Anchel *et al.* (1950)

TABLE 3. (*Continued*)

Agent	Prophage induction	Mutagenic	Carcino-genic
			Effect
Iyomycin B (related to pluramycin A)	M		
Lemonomycin (related to xanthomycin)	P		
Macromomycin (related to carzinostatin)	P		
Megacin C	+		
Mitomalcin (related to carzinostatin)	P		
Mitomycin C	P	+	+
Narangomycin	M		
Peptinogan	S		
Phagolessin A58	+		
Phleomycin	P		
Pluramycin A	M		
Porfiromycin (related to mitomycin)	P	+	
Rubiflavin (related to pluramycin)	P		
Rufochromomycin (related to streptonigrin)	P		
Sarkomycin	S		+
Streptonigrin	P		
Streptovitacin A (related to cycloheximide)	S		
Streptozotocin	P	+	+
Xanthomycin	M		
D. Miscellaneous chemical agents			
Nitrofurans:			
2-[2-(5-Nitro-2-furyl)vinyl] quinoline	P[b]		
3-Amino-6-[2-(5-nitro-2-furyl) vinyl]-1,2,4-triazine HCl (Panfuran)	P		
Nitrofurazone	M[c]	+	
Nihydrazone	M		
Nitrofurantoin	M[c]		
Furazolidone	M[c]	+	
Furaltadone	M[c]		

studied		References
Carcino-static	Terato-genic	
+		Price *et al.* (1965); Nomura *et al.* (1964)
		Heinemann (unpublished data)
+		Heinemann (unpublished data); Chimura *et al.* (1968)
		Holland (1963)
+		Heinemann (unpublished data); McBride *et al.* (1969)
+	+	Otsuji *et al.* (1959); Szybalski (1958); Kawamata *et al.* (1966); Wakaki *et al.* (1958); Tanimura (1961)
+		Heinemann (unpublished data); Rao *et al.* (1964)
+		Price *et al.* (1964); Schmitz *et al.* (1963)
		Hall-Asheshov and Asheshov (1956)
+		Heinemann and Howard (1964); Bradner and Pindell (1962)
+		Heinemann and Howard (1964); Maeda *et al.* (1956)
+		Heinemann and Howard (1964); Ikeda and Iijima (1965); Evans *et al.* (1961)
+		Heinemann (unpublshed data); Aszalos *et al.* (1965)
+	+	Heinemann (unpublished data); Dubost *et al.* (1965); Maraud *et al.* (1963)
+	+	Yajima *et al.* (1963); Dickens and Jones (1965); Umezawa *et al.* (1953); Takaya (1965)
+	+	Heinemann and Howard (1964); Rao and Cullen (1960); Warkany and Takacs (1965)
+		Price *et al.* (1964); Evans *et al.* (1960)
+		Heinemann and Howard (1964); Kolbye and Legator (1968); Arison and Feudale (1967); Evans *et al.* (1965)
		Heinemann and Howard (1964)
+[b]		Miura and Okada (1965)
		Endo *et al.* (1963b)
+		Waterbury and Freedman (1964); Zampieri and Greenberg (1964); Green and Friedgood (1948)
		Waterbury and Freedman (1964)
	+	Waterbury and Freedman (1964); Apgar (1964)
		Waterbury and Freedman (1964); Szybalski (1958)
		Waterbury and Freedman (1964)

TABLE 3. (*Continued*)

Agent	Prophage induction	Mutagenic	Carcinogenic
Vitamin analogs:			
Aminopterin	S	+	
Amethopterin	S	+	
3,5'-Dichloroamethopterin	S		
Thiols and thiol derivatives:			
D,L-Cysteine	M		
D,L-Homocysteine	S		
β-Mercaptoethylamine	S		
D,L-Penicillamine	S		
S-carbamyl-L-cysteine	S		
Dyes (in presence of visible light):			
Acridine orange	M	+	
Thiopyronine	M	+	
Methylene blue	+	+	
Halogenated pyrimidine nucleosides:			
5-Fluorouracil	S	+	
2'-Deoxy-5-fluorouridine	S	+	
2'-Deoxy-5-bromouridine	S	+	
Others:			
Nalidixic acid	M	+	
Oxolinic acid	P		
Leucovorin (UV-irradiated)	S		
Hydroxyurea	+		
Hydrogen peroxide	+	+	+
Urethane	S	+	+
Vincaleukoblastine sulfate	S		

[a] An association between phage induction, mutagenicity, and carcinogenicity has been demonstrated for a series of nitroquinolines and hydroxyaminoquinolines (Endo *et al.*, 1963*a*; Epstein and Saporoschetz, 1968).

studied		References
Carcino-static	Terato-genic	
+	+	Ben-Gurion (1962); Heslot (1960); Bertino (1963); Thiersch (1954)
+	+	Price *et al.* (1965); Heslot (1960); Bertino (1963); LaVelle and LaVelle (1967)
+		Price *et al.* (1965); Goldin *et al.* (1957)
		Heinemann and Howard (1964)
		Heinemann and Howard (1964)
		Heinemann and Howard (1964)
		Heinemann and Howard (1964)
+		Heinemann and Howard (1964); Skinner *et al.* (1958)
+		Smarda *et al.* (1964); Webb and Kubitschek (1963); Korgaonkar and Sukhatankar (1963)
		Geissler and Wacker (1963); Wacker *et al.* (1963)
+	+	Freifelder (1965); Wacker *et al.* (1963); Boyland (1946); Gilman *et al.* (1951)
+	+	Marcovich and Kaplan (1963); Cooper (1964); Heidelberger (1957); Murphy (1960)
+	+	Geissler (1966); Gauze *et al.* (1961); Heidelberger *et al.* (1958); Murphy (1960)
+	+	Price *et al.* (1965); Freese (1959); Kit *et al.* (1958); Murphy (1960)
		Heinemann (unpublished data); Cook *et al.* (1966)
		Heinemann (unpublished data)
		Borek and Rockenbach (1954)
+	+	Price *et al.* (1965); Stearns *et al.* (1963); Murphy and Chaube (1964)
+		Northrup (1958); Dickey *et al.* (1949); Schmidt (1964); Sugiura (1958)
+	+	D'Onofrio and Cavallo (1957); Vogt (1948); Nettleship *et al.* (1943); Elion *et al.* (1960); Sinclair (1950)
+	+	Heinemann and Howard (1964); Johnson *et al.* (1960); Ferm (1963)

[b] Many derivatives are also acitve (Miura and Okada, 1965).
[c] Insertion of a vinyl group between the nitrofuran and the azomethine groups also yields active compounds (Waterbury and Freedman, 1964).

active" (M), "slight" (S), and + when a compound is active, but the concentration is not given on a weight basis. The degree of activity is based on the minimal inducing concentration, as follows:

$$P<1 \ \mu g/ml \qquad M=1-100 \ \mu g/ml \qquad S>100 \ \mu g/ml$$

Representative mutagenic, carcinogenic, carcinostatic, and teratogenic activites were assigned on the basis of data reported in a reasonably thorough, but nonexhaustive, search of the literature. No consideration was given to evaluation of the particular test system employed by the investigator to demonstrate these effects.

Many of the known prophage inducing agents have demonstrated capabilities to cause mutations, to cause and inhibit cancer, and to provoke congenital malformations in various experimental systems. A relationship between the mechanisms responsible for these phenomena has not yet been established. Whether mutational events are involved awaits further investigation. It would be difficult to attempt to find chemical correlations among the great diversity of structures found in these inducing agents.

The known associations existing with the prophage inducing agents are summarized in Table 4. A clear-cut association between inducing capacity and carcinostatic activity was obtained with 70% of the agents tested. This high correlation is the product of intense research efforts in recent years to make new antineoplastic agents available. Associations between mutagenic, carcinogenic, and teratogenic effects are found with 45, 28, and 32%, respectively, of the 89 prophage inducing agents. In all likelihood, more of these effects will eventually be discovered and further increase the percentage of associations.

A number of compounds which could be classified as mutagenic, carcinogenic, carcinostatic, or teratogenic were incapable of prophage induction in *E. coli* test systems (Heinemann and Howard, 1964; Price *et al.*, 1965; Specht, 1965). These results suggest that the associations may be

TABLE 4. Summary of Mutagenic, Carcinogenic, Carcinostatic, and Teratogenic Activities of Agents with Induction Capability

Class of agent	Prophage induction	Number of agents			
		Mutagenic	Carcinogenic	Carcinostatic	Teratogenic
Radiation	5	4	4	2	2
Alkylating	21	17	15	17	7
Microbial metabolite	35	5	4	29	10
Miscellaneous	28	13	2	15	10
Total	89	39	25	63	29

one-way. The relatively few lysogenic systems utilized thus far for screening may be capable of detecting only certain classes of inducing compounds. In addition, special conditions may be required to demonstrate prophage induction with certain classes of agents. For example, the dyes acridine orange, thiopyronine, and methylene blue require the presence of visible light for induction, while the addition of thymidine to the induction system is necessary with certain halogenated nucleosides. It is possible that other classes of prophage inducing agents might eventually be uncovered as the result of (1) testing with a spectrum of lysogenic systems and (2) establishment of necessary conditions for demonstrating the effect.

VI. PROPHAGE INDUCING AGENTS PRESENT IN THE HUMAN ENVIRONMENT

It would be unethical to deliberately use human subjects to test agents which induce mutagenic, carcinogenic, or teratogenic responses. Thus, such knowledge as we have available has usually been obtained indirectly as the result of observations on groups of people accidentally or unwittingly exposed to these agents. No one has yet shown that any agent causes human genetic damage capable of inducing mutations in the germinal cells of man. The prophage inducing agents associated with the induction of birth defects in females include X-rays (Aschenheim, 1920), aminopterin (Meltzer, 1956), nitrofurantoin (Apgar, 1964), and chlorambucil (Shotton and Monie, 1963). Those associated with carcinogenic effects in humans include X-rays (Frieben, 1902), ultraviolet rays (Findlay, 1928), α-rays (Evans, 1947), and mustard gas (Case and Lea, 1955). Varying degrees of antineoplastic effects have been obtained in humans treated with prophage inducing agents.

The concern regarding exposure of the population to an increasing number of synthetic chemicals capable of producing such delayed effects as genetic mutations, cancer, and congenital malformations is understandable. Exposure of man to the majority of synthetics has occurred only in relatively recently times. Of even greater significance may be his exposure to certain natural products which was initiated before the advent of synthetics and continues to present times.

Of the 84 known chemical prophage inducing agents listed in Table 4, 36 (35 products of microbial metabolism and vincaleukoblastine sulfate), or 43%, may be considered to be natural products. Mutagenic, carcinogenic, carcinostatic, and teratogenic effects have been demonstrated with 14, 11, 83, and 31%, respectively of these 36 natural chemical products.

The ever-mounting evidence that natural products, particularly those of microbial origin, may elicit these responses warrants a systematic effort to explore them further. The story of aflatoxin emphasizes this need. While the food supply represents a major route of exposure of the population to natural products, they may also enter the human environment from a variety of other sources, including smoking, cosmetics, and as pollutants. Agents of microbial origin may be introduced into natural products as (1) a consequence of some phase of production, processing, or storage and (2) the result of a contaminating growth.

Few attempts have been made to screen plant or animal natural products for the presence of prophage inducing agents. An unpublished study by the author has indicated the presence of inducing agents in botanical and zoological specimens collected in a marine environment. The implication of these preliminary findings awaits further investigation. Detection of inducing agents when they are prsent in low concentrations in natural products may be extremely difficult and extraction and purification may be tedious. Hopefully, the prophage induction technique may be of value in studying these crude mixtures under complex environmental situations.

At this point I should like to point out briefly a few of the sources through which the population may contact various prophage inducing agents.

Medical sources   Many of the drugs used to control cancer can also induce cancer. Among the prophage inducing agents administered to control tumor growths are radiations, tris(2-chloroethyl)amine HCl, nitromin, chlorambucil, sarcolysin, TEM, TEPA, ThioTEPA, trenimon, 5-fluorouracil, 2'-deoxy-5-fluorouridine, hydroxyurea, urethan, amethopterin, and vincaleukoblastine sulfate. Treatment of cancer patients with drugs which may themselves be a hazard is certainly justifiable until safer and more efficient ones become available. Several antibiotics have been administered as experimental antineoplastic drugs.

A number of prophage inducing agents are used in medical or veterinary practice, or both, for nonmalignant conditions. Among the drugs used are panfuran, nitrofurazone, nitrofurantoin, nihydrazone, furazolidone, furaltadone, penicillamine, methylene blue, nalidixic acid, and hydrogen peroxide. β-Propiolactone is used to sterilize vaccines, grafts, and plasma.

Chemical sources: Among the prophage inducing agents used in the chemical industry are 1,2,3,4-diepoxybutane, TEM, TEPA, dimethyl sulfate, 1-methyl-3-nitro-1-nitrosoguanidine, β-propiolactone, urethan, hydrogen peroxide, and acridine orange. Both TEM and TEPA find use as insect chemosterilants.

Products of microbial metabolism: Many microorganisms found in nature have the capacity to elaborate potent, biologically active substances. Recent attempts to produce potential antineoplastic agents by biosynthesis have made available an ever-increasing number of prophage inducing agents, many with mutagenic, carcinogenic, and teratogenic capabilities. Man may become exposed to these bioagents from "moldy" or contaminated foodstuffs, by microbial action in the digestive tract and on body surfaces, and through the use of fermentation procedures for the preparation of therapeutics, foods, and chemicals. More attention should be focused on this source of potentially hazardous substances.

VII. CONCLUSIONS

Evidence that agents with prophage inducing capability might play a part in the etiology of genetic mutations, cancer, and congenital malformations in humans appears to be increasing. The selectivity shown by the prophage induction system marks it as a useful screen despite its inability, thus far, to detect all active compounds. One may hope that in the years to come this technique will not only make it possible to detect mutagenic, carcinogenic, and teratogenic hazards in the human environment, but also will assist in screening for an antineoplastic agent of therapeutic importance. Systematic screening of chemical agents utilizing this technique is suggested in order to further explore these relationships.

VIII. REFERENCES

Adams, C. E., Hay, M. F., and Lutwak-Mann, C. (1961), *J. Embryol. Exp. Morphol.* *9*, 468.

Aiso, K., Arai, T., Suzuki, M., and Takamizawa, Y. (1956), *J. Antibiot., Series A* *9*, 97.

Allan, R. K., and McCalla, D. R. (1966), *Can. J. Microbiol. 12*, 202.

Altenbern, R. A. (1962), *Biochem. Biophys. Res. Commun. 9*, 109.

Altenburg, E. (1930), *Anat. Rec. 47*, 383.

Anchel, M., Hervey, A., and Robbins, W. J. (1950), *Proc. Nat. Acad. Sci. 36*, 300.

Aoki, H., and Sakai, H. (1967), *J. Antibiot., Series A 20*, 87.

Apgar, V. (1964), *J.A.M.A. 190*, 840.

Arison, R. N., and Feudale, E. L. (1967), *Nature 214*, 1254.

Aschenheim, E. (1920), *Arch. Kinderheilk. 68*, 131.

Aszalos, A., Jelinek, M., and Berk, B. (1965), *Antimicrob. Ag. Chemotherap.—1964*, p. 68.

Auerbach, C., Robson, J., and Carr, J. G. (1947), *Science 105*, 243.

Bail, O. (1925), *Med. Klin. 21*, 1277.

Bass, A. D., and Freeman, M. L. H. (1946), *J. Nat. Cancer Inst. 7*, 171.

Beardsley, R. E. (1960), *J. Bacteriol. 80*, 180.

Ben-Gurion, R. (1962), *Biochem. Biophys. Res. Commun. 8*, 456.

Bertani, G. (1953), *Cold Spring Harbor Symp. Quant. Biol. 16*, 65.

Bertani, G. (1958), *Adv. Virus Res. 5*, 151.

Bertino, J. R. (1963), *Cancer Res. 23*, 1286.

Bertram, C., and Hoehne, G. (1959), *Strahlentherapie Suppl. 43*, 388.

Bird, M. J., and Fahmy, O. G. (1953), *Proc. Roy. Soc., Series B 146*, 556.

Blokhin, N., Larionov, L., Perevodchikova, N., Chebotareva, L., and Merkulova, N (1958), *Ann N Y. Acad. Sci. 68*, 1128.

Blum, K. V. (1959), *Strahlentherapic Suppl. 41*, 396.

Bordet, J. (1925), *Ann. Inst. Pasteur 39*, 717.

Borek, E., and Rockenbach, J. (1954), *Biochim. Biophys. Acta 15*, 140.

Bowman, B. U., and Redmond, W. B. (1959), *Am. Rev. Resp. Dis. 80*, 232.

Boyland, E. (1946), *Biochem. J. 40*, 55.

Boyland, E., and Horning, E. S. (1949), *Brit. J. Cancer 3*, 118.

Bradner, W. T., and Pindell, M. H. (1962), *Nature 196*, 682.

Bradner, W. T., Heinemann, B., and Gourevitch, A. (1967), *Antimicrob. Ag. Chemotherap.—1966*, p. 613.

Buckley, S. M., Stock, C. C., Parker, R. P., Crossley, M. L., Kuh, E., and Seegar, D. R. (1951), *Proc. Soc. Exp. Biol. Med. 78*, 299.

Campbell, A. (1967), *in* "Molecular Genetics" (J. H. Taylor, ed.) Part II, Chapter 8, Academic Press, New York.

Campbell, A. M. (1962), *Adv. Genet. 11*, 101.

Case, R. A. M., and Lea, A. J. (1955), *Brit. J. Preventive Social Med. 8*, 39.

Chimura, H., Ishizuka, M., Hamada, M., Hori, S., Kimura, K., Iwanaga, J., Takeuchi, T., and Umezawa, H. (1968), *J. Antibiot., Series A 21*, 44.

Clarke, D. A., Reilly, H. C., and Stock, C. C. (1957), *Antibiot. Chemotherap. 7*, 653.

Clarke, N. A., and Cowles, P. B. (1952), *J. Bacteriol. 63*, 177.

Conklin, J. W., Upton, A. C., Christenberry, K. W., and MacDonald, T. P. (1963), *Radiation Res. 19*, 156.

Cook, T. M., Goss, W. A., and Dietz, W. H. (1966), *J. Bacteriol. 91*, 780.

Cooper, P. O. (1964), *Virology 22*, 186.

Davies, A. J., Wibin, E. A., Hoppe, E. T., and Depeyster, F. A. (1960), *Surg. Forum 11*, 42.

Despeignes, V. (1956), *Lyon Med. 82*, 428.

d'Herelle, F. (1917), *Compt. Rend. Acad. Sci. 65*, 373.

Dickens, F., and Jones, H. E. H. (1965), *Brit. J. Cancer 19*, 392.

Dickey, F. H., Cleland, G. H., and Lotz, C. (1949), *Proc. Nat. Acad. Sci. 35*, 581.

Didcock, K. A., Jackson, D., and Robson, J. M. (1956), *Brit. J. Pharmacol. 11*, 437.

DiMarco, A., Gaetani, M., Dorigotti, L., Soldati, M., and Bellini, O. (1963), *Tumori 49*, 203.

Domagk, G. (1958), *Ann. N.Y. Acad. Sci. 68*, 1197.

D'Onofrio, F., and Cavallo, G. (1957), *Boll. Ist. Sieroter. Milan 36*, 441.

Dove, W. F. (1968), *Ann. Rev. Genet. 2*, 305.

Driskell-Zamenhof, P. (1964), *in* "The Bacteria" (I. C. Gunsalus and R. Y. Stanier, eds.) Vol. V, Chapter 4, Academic Press, New York.

Druckrey, H., Preussmann, R., Hashed, N., and Ivankovic, S. (1966), *Z. Krebsforsch. 68*, 103.

Dubost, M., Ganter, P., Maral, R., Ninet, L., Pinnert, S., Preud'homme, J., and Werner, G.–H. (1963), *Compt. Rend. Acad. Sci. 257*, 1813.

Dubost, M., Ganter, P., Mancy, D., Maral, R., Ninet, L., and Preud'homme, J. (1965), *Compt. Rend. Acad. Sci. 261*, 4911.

Dudnik, Y. V. (1965), *Antibiotiki 10*, 112.

Echols, H., and Joyner, A. (1968), *in* "The Molecular Basis of Virology" (H. Fraenkel-Conrat, ed.) p. 526, Reinhold, New York.

Elion, G. B., Bieber, S., and Hitchings, G. H. (1960), *Acta Unio Internat. Contra Cancrum 16*, 605.

Endo, H., Ishizawa, M., and Kamiya, T. (1963a), *Nature 198*, 195.

Endo, H., Ishizawa, M., Kamiya, T., and Kuwano, M. (1963b), *Biochim. Biophys. Acta 68*, 502.

Endo, H., Ishizawa, M., Kamiya, T., and Sonoda, S. (1963c), *Nature 198*, 258.

Epstein, S. S., and Saporoschetz, I. B. (1968), *Experientia 24*, 1245.

Epstein, S. S., and Shafner, H. (1968), *Nature 219*, 385.

Evans, J. S., Ceru, J., and Mengel, G. D. (1960), *Antibiot. Ann.*, p. 962.

Evans, J. S., Musser, E. A., and Gray, J. E. (1961), *Antibiot. Chemotherap. 11*, 445.

Evans, J. S., Gerritson, G. C., Mann, K. M., and Owen, S. P. (1965), *Cancer Chemotherap. Rep. 48*, 1.

Evans, R. D. (1947), Presented at the Fourth International Cancer Congress, St. Louis.

Fahmy, O. G., and Fahmy, M. J. (1960), *Genet. Res. 1*, 173.

Faucounau, N., Stoll, R., and Maraud, R. (1963), *Compt. Rend. Soc. Biol. 157*, 1564.

Ferm, V. H. (1963), *Science 141*, 426.

Field, A. K., and Naylor, H. B. (1962), *J. Bacteriol. 84*, 1129.

Findlay, G. M. (1928), *Lancet 215*, 1070.

Fleck, W. (1968), *Z. Allg. Mikrobiol. 8*, 139.

Freese, E. (1959), *J. Mol. Biol. 1*, 87.

Freifelder, D. (1965), *Biochem. Biophys. Res. Commun. 18*, 824.

Frieben (1902), *Fortschr. Geb. Röntgenstr. 6*, 106.

Gado, I., Savtchenko, G., and Horváth, I. (1965/1966), *Acta Microbiol. 12*, 363.

Galton, D. A. G., Israels, L. G., Nabarro, J. D. N., and Till, M. (1955), *Brit. Med. J. 2*, 1172.

Gause, G. F. (1965), *Vestnik Akad. Med. Nauk. SSSR 20*, 46.

Gause, G. F., Kochetkova, G. V., and Vladimirova, G. B. (1961), *Nature 190*, 978.

Geissler, E. (1966), *Biochim. Biophys. Acta 114*, 116.

Geissler, E., and Wacker, A. (1963), *Acta Biol. Med. Ger. 10*, 937.

Gillman, J., Gilbert, C., Spence, I., and Gillman, T. (1951), *S. Afr. Med. Sci. 16*, 125.

Goldin, A., Goldberg, B., Ortega, L. G., Fugmann, R., Faiman, F., and Schoenbach, E. B. (1949), *Cancer 2*, 865.

Goldin, A., Venditti, J. M., Humphreys, S. R., and Mantel, N. (1957), *J. Nat. Cancer Inst. 19*, 1133.

Goldthwait, D., and Jacob, F. (1964), *Compt. Rend. Acad. Sci. 259*, 661.

Gots, J. S., Bird, T. J., and Mudd, S. (1955), *Biochim. Biophys. Acta 17*, 449.

Gray, L. H., and Read, J. (1948), *Brit. J. Radiol. 21*, 5.

Green, M. N., and Friedgood, C. E. (1948), *Proc. Soc. Exp. Biol. Med. 69*, 603.

Greene, M. O., and Greenberg, J. (1960), *Cancer Res. 20*, 1166.

Groman, N. B., Eaton, M., and Booher, Z. K. (1958), *J. Bacteriol. 75*, 320.

Haddow, A., Kon, G. A. R., and Ross, W. C. J. (1948), *Nature 162*, 824.

Hall-Asheshov, E., and Asheshov, I. N. (1956), *J. Gen. Microbiol. 14*, 174.

Hamon, Y., and Peron, Y. (1965), *Compt. Rend. Acad. Sci. 260*, 5948.

Hanson, F. B. (1928), *Anat. Rec. 41*, 99.

Haskins, D. (1948), *Anat. Rec. 102*, 493.

Hata, T., Koga, F., Sano, Y., Kanamori, K., Matsumae, A., Sugawara, R., Hoshi, T., and Shima, T. (1954), *J. Antibiot.*, Series A 7, 107.

Hayes, W. (1964), "The Genetics of Bacteria and Their Viruses," Chapter 17, Blackwell, Oxford.

Heidelberger, C., Chanduri, N. K., Danneberg, P., Mooren, D., Griesbach, L., Dushinsky, R., Schnitzer, R. J., Pleven, E., and Scheiner, J. (1957) *Nature 179* 633.

Heidelberger, C., Griesbach, L., Cruz, O., Schnitzer, R. J., and Grunberg, E. (1958), *Proc. Soc. Exp. Biol. Med. 97*, 470.

Heinemann, B., and Howard, A. J. (1964), *Appl. Microbiol. 12*, 234.

Heinemann, B., and Howard, A. J. (1966), *Antimicrob. Ag. Chemotherap.–1965*, p. 488.

Heinemann, B., Howard, A. J., and Hollister, Z. J. (1967), *Appl. Microbiol. 15*, 723.

Hemmerly, J., and Demerec, M. (1955), *Cancer Res. Suppl. 3*, 65.

Heslot, H. (1960), *Abh. Deut. Akad. Wiss. Berlin, Klin. Med. Wiss.*, p. 109.

Heston, W. E. (1949), *J. Nat. Cancer Inst. 10*, 125.

Holland, I. B. (1963), *Biochem. Biophys. Res. Commun. 13*, 246.

Ikeda, Y., and Iijima, T. (1965), *J. Gen. Appl. Microbiol. 11*, 129.

Ishidate, M., Kobayashi, K., Sakurai, Y., Sato, H., and Yoshida, T. (1951), *Proc. Jap. Acad. 27*, 493.

Ishizuka, M., Takayama, T., Takeuchi, T., and Umezawa, H. (1967), *J. Antibiot., Series A 20*, 15.

Izumi, M. (1954), *Pharm. Bull. 2*, 275.

Jacob, F. (1952), *Compt. Rend. Acad. Sci. 234*, 2238.

Jacob, F., and Wollman, E. L. (1953), *Cold Spring Harbor Symp. Quant. Biol. 16*, 101.

Jacob, F., and Wollman, E. L. (1954), *Compt. Rend. Acad. Sci. 239*, 317.

Jacob, F., and Wollman, E. L. (1957), *in* "The Chemical Basis of Heredity" (W. D. McElroy and B. Glass, eds.) p. 468, The Johns Hopkins Press, Baltimore.

Jacob, F., and Wollman, E. L. (1959a), *in* "The Viruses: Plant and Bacterial Viruses" (F. M. Burnet and W. M. Stanley, eds.) p. 319, Academic Press, New York.

Jacob, F., and Wollman, E. L. (1959b), *in* "Bacteriophages" (M. H. Adams, ed.) p. 365, Interscience, New York.

Jacob, F., and Wollman, E. L. (1961), "Sexuality and the Genetics of Bacteria," Academic Press, New York.

Johnson, L. S., Wright, H. F., Swoboda, G. H., and Vlantis, J. (1960), *Cancer Res. 20*, 1016.

Kawamata, J., Akamatsu, Y., and Ikegami, R. (1966), *Abst. 9th Internat. Cancer Cong.* (October 23), Tokyo, p. 194.

Kit, S., Beck, C., Graham, O. L., and Gross, A. (1958), *Cancer Res. 18*, 598.

Knezevic, Z., and Karanovic, J. (1966), *Bull. Boris Kidric Inst. Nucl. Sci. 17*, 317.

Kolbye, S. M., and Legator, M. S. (1968), *Mutation Res. 6*, 387.

Kolmark, G. (1956), *Compt. Rend. Trav. Lab. Carlsberg, Ser. Physiol. 26*, 205.

Korgaonkar, K. S., and Sukhatankar, J. V. (1963), *Brit. J. Cancer 17*, 471.

Korn, D., and Weissbach, A. (1962), *Biochim. Biophys. Acta 61*, 775.

Kowalski, M. (1966), *Acta Microbiol. Pol. 15*, 119.

Kozak, W., and Dobrzanski, W. T. (1967), *Bull. Acad. Pol. Sci. 15*, 391.

Latarjet, R. (1951), *Ann. Inst. Pasteur 81*, 389.

LaVelle, F. W., and LaVelle, A. (1967), *Exp. Neurol. 17*, 140.

Lein, J., Heinemann, B., and Gourevitch, A. (1962), *Nature 196*, 783.

Levine, M., and Borthwick, M. (1963), *Virology 21*, 568.

Lewis, M. R., and Crossley, M. L. (1950), *Arch. Biochem. 26*, 319.

Lüers, H., and Röhrborn, G. (1965), *Mutation Res. 2*, 29.

Lwoff, A. (1953), *Bacteriol. Rev. 17*, 269.

Lwoff, A., and Gutmann, A. (1950), *Ann. Inst. Pasteur 78*, 711.

Lwoff, A., and Jacob, F. (1952), *Compt. Rend. Acad. Sci. 234*, 2308.

Lwoff, A., Siminovitch, L., and Kjeldgaard, N. (1950), *Compt. Rend. Acad. Sci. 231*, 190.

McBride, T. J., Axelrod, M., Cullen, W. P., Marsh, W. S., Rao, K. V., and Sodano, C. S. (1969), *Proc. Am. Ass. Cancer Res. 10*, 56.

McCammon, C. J., Kotin, P., and Falk, H. L. (1957), *Proc. Am. Ass. Cancer Res. 2*, 229.

Maeda, K., Takeuchi, T., Nitta, K., Yagishita, K., Utahara, R., Osato, T., Ueda, M., Kondo, S., Okami, Y., and Umezawa, H. (1956), *J. Antibiot., Series A 9*, 75.

Malke, H. (1967), *Nature 214*, 811.

Mandell, J., and Greenberg, J. (1960), *Biochem. Biophys. Res. Commun. 3*, 575.

Maraud, R., Coulard, H., and Stoll, R. (1963), *Compt. Rend. Soc. Biol. 157*, 1566.

Marcovich, H. (1956), *Ann. Inst. Pasteur 90*, 458.

Marcovich, H., and Kaplan, H. S. (1963), *Nature 200*, 487.

Marjai, E., and Ivanovics, G. (1964), *Acta Microbiol. 11*, 193.

Marquardt, H., Zimmerman, F. K., and Schwaier, R. (1964), *Z. Vererb. 95*, 82.

Mayer, V. W., Galridge, M. G., and Oswald, E. J. (1969), *Appl. Microbiol. 18*, 697.

Meltzer, H. J. (1956), *J.A.M.A. 161*, 1253.

Menzel, G. R., and Geissler, E. (1966), *Experientia 22*, 800.

Miura, K., and Okada, I. (1965), *Chem. Pharm. Bull. 13*, 525.

Moore, P. R., Mannering, G. J., Teply, L. J., and Kline, B. E. (1960), *Cancer Res. 20*, 628.

Muller, H. J. (1927), *Science 66*, 84.

Murphy, M. L. (1960), *in* "Ciba Foundation Symposium on Congenital Malformation," p. 78, Churchill, Ltd., London.

Murphy, M. L., and Chaube, S. (1964), *Cancer Chemotherap. Rep. 40*, 1.

Murphy, M. L., Del Moro, A., and Lacon, C. (1958), *Ann. N.Y. Acad. Sci. 68*, 762.

Nakahara, W., Fukuoka, F., and Sugimura, T. (1957), *Gann 48*, 129.

Nettleship, A., Henshaw, P. S., and Meyer, H. L. (1943), *J. Nat. Cancer Inst. 4*, 309.

Nomura, M. (1963), *Cold Spring Harbor Symp. Quant. Biol. 28*, 315.

Nomura, S., Yamamoto, H., Matsumae, A., and Hata, T. (1964), *J. Antibiot., Series A 17*, 104.

Northrup, J. H. (1958), *J. Gen. Physiol. 42*, 109.

Ogata, Y. (1959), *J. Antibiot., Series A 12*, 133.

Okaboyashi, T., Ide, A., Yoshimoto, A., and Otsubo, M. (1965), *Chem. Pharm. Bull. 13*, 610.

Okano, K., Fujita, H., Ito, T., Kashiyama, S., Esumi, K., Ito, H., and Toba, T. (1959), *Acta Pathol. Jap. 9*, 644.

Otsuji, N. M., Sekiguchi, M., Iijima, T., and Tekagi, T. (1959), *Nature 184*, 1079.

Palmes, E. D., Orris, L., and Nelson, N. (1962), *Am. Ind. Hyg. Ass. J. 23*, 257.

Price, K. E., Buck, R. E., and Lein, J. (1964), *Appl. Microbiol. 12*, 428.

Price, K. E., Buck, R. E., and Lein, J. (1965), *Antimicrob. Ag. Chemotherap.—1964*, p. 505.

Rao, K. V., and Cullen, W. P. (1960), *Antibiot. Ann.*, p. 950.

Rao, K. V., Marsh, W. S., and Brooks, S. C. (1964), Narangomycin, U.S. Patent 3,155,585, 11/3/64.

Rappaport, H. P. (1958), *Arch. Biochem. Biophys. 76*, 1.

Reilly, H. C., Stock, C. C., Buckley, S. M., and Clarke, D. A. (1953), *Cancer Res. 13*, 684.

Rose, F. L., Hendry, J. A., and Walpole, A. G. (1950), *Nature 165*, 993.

Rossolimo, O. K., Stanislavskaia, M. S., Pevzner, N. S., Shapovalova, S. P., and Lepeshkina, G. N. (1966), *Antibiotiki 11*, 683.

Russell, L. B. (1950), *J. Exp. Zool. 114*, 545.

Scaife, J. (1967), *Ann. Rev. Microbiol. 21*, 601.

Schmidt, F. (1964), *Acta Biol. Med. Ger. 13*, 74.

Schmitz, H., DeVault, R. L., and Hooper, I. R. (1963), *J. Med. Chem. 6*, 613.

Shellabarger, C. J., Cronkite, E. P., Bond, V. P., and Lippincott, S. W. (1957), *Radiation Res. 6*, 501.

Shimpo, K., Narimatsu, E., Higashi, H., and Nishida, K. (1960), *Acta Pathol Jap.* *10*, 303.

Shoji, J. (1961), *J. Antibiot., Series A 14*, 27.

Shotton, D., and Monie, I. W. (1963), *J. Am. Med. Ass. 186*, 74.

Signer, E. (1968), *Ann. Rev. Microbiol. 22*, 451.

Sinclair, J. G. (1950), *Texas Rep. Biol. Med. 8*, 623.

Skinner, C. G., McKenna, G. F., McCord, T. J., and Shive, W. (1958), *Texas Rep. Biol. Med. 16*, 493.

Smarda, J., Koudelka, J., and Kleinwachter, V. (1964), *Experientia 20*, 500.

Smith, H. H., and Srb, A. M. (1951), *Science 114*, 490.

Smith, H. W. (1953), *J. Gen. Microbiol. 8*, 116.

Specht, I. (1965), *Arch. Mikrobiol. 51*, 9.

Stearns, B., Losee, K. A., and Bernstein, J. (1963), *J. Med. Pharm. Chem. 6*, 201.

Stent, G. S. (1963), "Molecular Biology of Bacterial Viruses," Chapter 12, W. H. Freeman & Co., San Francisco.

Sugimura, T., Nagao, M., and Okada, Y. (1966), *Nature 210*, 962.

Sugiura, K. (1958), *Nature 182*, 1310.

Sugiura, K. (1959/1960), *Antibiot. Ann.*, p. 924.

Sugiura, K., and Stock, C. C. (1955), *Cancer Res. 15*, 38.

Sussman, R., and Jacob, F. (1962), *Compt. Rend. Acad. Sci. 254*, 1517.

Sutton, M. D., and Quadling, C. (1963), *Can. J. Microbiol. 9*, 821.

Szybalski, W. (1958), *Ann. N.Y. Acad. Sci. 76*, 475.

Takaya, M. (1965), *J. Osaka City Med. Cent. 14*, 107.

Tanimura, T. (1961), *Acta Anat. Nippon 36*, 354.

Tatsuoka, S., Nakazawa, K., Miyake, A., Kaziwara, K., Aramaki, Y., Shibata, M., Tanabe, K., Hamada, Y., Hitomi, H., Miyamoto, M., Mizuno, K., Watanabe, J., Ishidate, M., Yokotani, H., and Ushikawa, I. (1958), *Gann 49* (Suppl), 23.

Terawaki, A., and Greenberg, J. (1965), *Biochim. Biophys. Acta 95*, 170.

Thiersch, J. B. (1954), *Proc. Soc. Exp. Biol. Med. 87*, 571.

Thiersch, J. B. (1957*a*), *Proc. Soc. Exp. Biol. Med. 94*, 36.

Thiersch, J. B. (1957*b*), *Proc. Soc. Exp. Biol. Med. 94*, 27.

Thiersch, J. B. (1957*c*), *Proc. Soc. Exp. Biol. Med. 94*, 33.

Thiersch, J. B. (1958), *Proc. Soc. Exp. Biol. Med. 97*, 888.

Thomas, R. (1968), *Symp. Soc. Gen. Microbiol. 18*, 315.

Twort, F. W. (1915), *Lancet 189*, 1241.

Umezawa, H., Takeuchi, T., Nitta, K., Yamamoto, T., and Yamaoka, S. (1953), *J. Antibiot., Series A 6*, 101.

Vogt, M. (1948), *Experientia 4*, 68.

Wacker, A., Tuerck, G., and Gerstenberger, A. (1963), *Naturwissenschaften 50*, 377.

Wakaki, S., Harumo, H., Tomioka, T., Shimizu, G., Kato, K., Kamada, H., Kudo, S., and Fujimoto, Y. (1958), *Antibiot. Chemotherap. 8*, 228.

Ward, F. D. (1935), *Genetics 20*, 230.

Warkany, J., and Takacs, E. (1965), *Arch. Pathol. 79*, 65.

Waterbury, W. E., and Freedman, R. (1964), *Can. J. Microbiol. 10*, 932.

Webb, R. B., and Kubitschek, H. E. (1963), *Biochem. Biophys. Res. Commun. 13*, 90.

Weisburger, E. K. (1966), *Pub. Health Rep. 81*, 772.

Welker, N. E., and Campbell, L. L. (1965), *J. Bacteriol. 89*, 175.

Whitfield, J. F. (1962), *Brit. Med. Bull. 18*, 56.

Yajima, K., Katsuhiko, T., and Hata, T. (1963), *Kitasato Arch. Exp. Med. 36*, 57.

Yamamoto, T., and Chow, C. T. (1968), *Can. J. Microbiol. 14*, 667.

Zampieri, A., and Greenberg, J. (1964), *Biochem. Biophys. Res. Commun. 14*, 172.

CHAPTER 9

The Detection of
Chemical Mutagens
with Enteric Bacteria*

Bruce N. Ames

Biochemistry Department
University of California
Berkeley, California

with an addendum by

Charles Yanofsky

Department of Biological Sciences
Stanford University
Stanford, California

I. INTRODUCTION

Mutagens alter DNA. As the DNA of all organisms has the same double helical structure and the same four nucleotides, any organism may be used as an indicator system for mutagens. Bacteria have numerous advantages for the detection of mutagens, and we discuss these and describe various methods and strains. We also discuss the validity of the bacterial tests and their pertinence to human mutagenesis and carcinogenesis.

* This research was supported by AEC grant AT(04-3)-34, Agreement No. 156.

II. DISCUSSION

A. Advantages of Using Bacterial Test Systems for the Detection of Mutagens

Any test system for mutagens should be calibrated against the known mutagens to determine the ease and sensitivity of the test in detecting these compounds before trying new substances. We believe bacteria are the system of choice for mass screening of new compounds on the basis of simplicity, sensitivity, economy, and range of compounds detected.

1. The Simplicity of the Plate Test

The advantages of using bacteria for testing mutagens have been discussed by a number of authors and various methods have been examined and used for testing a variety of compounds (Demerec *et al.*, 1951; Szybalski, 1958). We have found that the method of adding the mutagen directly to a lawn of bacteria on a petri plate (Iyer and Szybalski, 1958) to be the most convenient. In this paper we discuss the use of special bacterial strains that make this method capable of detecting many more types of mutagens and much more sensitive and diagnostic of the different classes of mutagens.

Most of the tests have been designed to be done on petri plates. A small sample (about 1 mg) of the suspected mutagen is placed in the center of a petri plate that has been seeded with a lawn of bacteria that cannot grow because of a mutation. If the mutagen can cause the particular mutation to revert in an occasional bacterium, it will enable that bacterium to grow and form a colony. Therefore, a circle of colonies will appear around the spot of mutagen after about a day and a half of incubation. The mutagen forms a concentration gradient as it diffuses from the point of application, and one often sees a clear central area, where the high concentration of the mutagen has inhibited all the bacteria, surrounded by a zone of mutant colonies. The diffusion of the mutagen in the agar allows one, in effect, to test a wide range of concentrations on a single plate (Fig. 1). Because all the revertants caused by the mutagen appear as a ring around the spot of application, occasional contaminants or some spontaneous mutants randomly distributed around the plate do not interfere with the test, and the interpretation is usually unambiguous

Using this test system a person with minimal training can test several hundred compounds every day. No elaborate equipment is necessary and the cost of supplies is small.

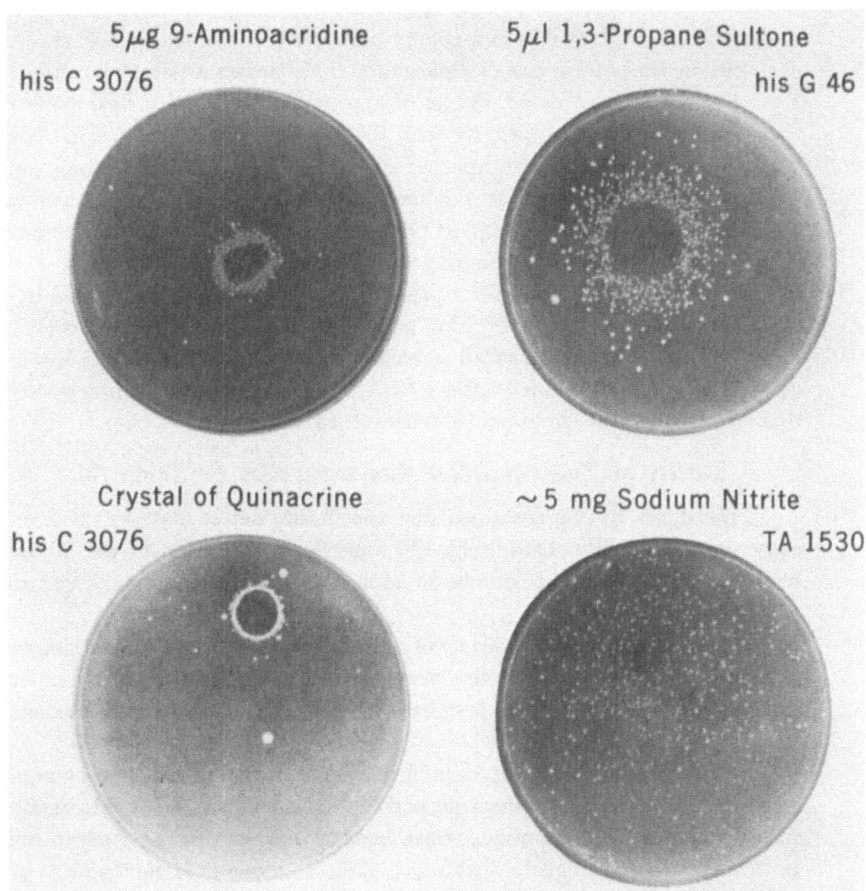

FIGURE 1. Photographs of 90-mm petri plates showing the reversion of tester strains by various mutagens added at a spot on the plate. Each bacterial colony appears as a white spot. Where the mutagen inhibits, the ring of mutants appears outside of the zone of inhibition.

2. Sensitivity: Strains Lacking Repair

A compound that alters only one nucleotide base at random out of a million is a powerful mutagen that can cause a high proportion of lethal events or defective children in a population. This is because every nucleotide is important in the DNA code. Thus one must use large populations in testing for mutagens, as mutation is a rare event for any particular nucleotide or gene. In the bacterial test system even rare mutational events may be detected readily, since about 5×10^8 bacteria are exposed to the mutagen on a petri plate. Even if only a few bacteria are mutated at the appropriate nucleotide, each gives rise to a colony that can be observed in 2 days. For example, in *Escherichia coli* or *Salmonella typhimurium* a point mutation in a gene is caused by a defect in one of approximately 4×10^6 base pairs in the genome. If 5×10^8 bacteria with this point mutation are treated with a mutagen that induces the appropriate nucleotide alteration in only one nucleotide per bacterium at random, then approximately 125 revertant colonies would appear on the petri plate. Depending on the starting mutant, the spontaneous rate might be only a few percent of this.

In bacteria almost all of the primary damage to the DNA caused by a mutagen is repaired by the excision and recombination repair systems (Witkin, 1969) so that only a small percentage of the potential mutations are induced. We have developed tester strains, lacking the excision repair system, that are 10 to 100 times more sensitive to mutagens.

3. A Variety of Types of DNA Alterations Can Be Tested for

Procedures will be described that specifically detect mutagens that will cause (a) base-pair substitutions, (b) insertion or deletion of one or two base pairs, and (c) large deletions. In addition, a general test for mutagenesis will be described.

Protein sequences of a variety of mutant enzymes have been determined in enteric bacteria. Within a few years, a set of mutants each of which responds only to a particular known base-pair change should become available (see addendum).

Simple tests for a variety of nucleic acid alterations are becoming available in *E. coli* and *S. typhimurium* and their phages as so much information and so many genetic techniques have been worked out in these organisms. It is easy to detect agents that cause virus induction (Heinemann *et al.*, 1967, and this volume), phenotypic curing (that is, agents such as streptomycin that cause mistakes in the genetic coding system during protein synthesis; Gorini and Kataja, 1965) (see section G), mistakes in transcribing DNA into RNA, episome curing (Hirota, 1960), insertions of genetic material, DNA replication inhibition, and agents such as caffeine that inhibit repair systems (Sideropoulos and Shankel, 1968).

B. Validity of the Bacterial System as a Test for Mutagens and Carcinogens for Humans

In normal drug testing it is absurd to extrapolate from bacteria to humans. In testing mutagens it is sensible because in both organisms the mutagen is reacting with DNA. This is borne out by the finding that, in general, mutagens for higher organisms are mutagens for microorganisms and *vice versa*, where extensive tests have been done. In some cases in humans a mutagen effective on bacteria may be detoxified, or never reach the DNA of the germ line, or mutations may be repaired effectively; and thus a bacterial mutagen may not appear to be a mutagen for humans. It is worth emphasizing, however, that one molecule of a mutagen is enough to cause a mutation and that if a large population is exposed to a "weak" mutagen it may still be a hazard to the human germ line. Since no repair system is completely effective, there may be no such thing as a completely safe dose of a mutagen.

In addition, a high proportion of the known mutagens are carcinogenic and a high proportion of the known proximal carcinogens are mutagenic. Although the correlation is not absolute and although the theoretical basis is not clear (perhaps only agents causing one type of mutation are carcinogenic), what is clear is that if a substance mutates bacteria there is a high probability that it will turn out to be carcinogenic for humans (more than half of the mutagenic agents listed in section F are carcinogenic). The simple bacterial assay for mutagens should be useful for identifying possible new carcinogens for humans (Szybalski, 1958). It also may be useful in purifying carcinogenic substances from complex natural products such as cigarette smoke, if there is both mutagenic and carcinogenic activity.

My general feeling is that if a compound is a mutagen in any organism, then it should not be used on humans unless there is definitive evidence that it is neither mutagenic nor carcinogenic in animals, or unless the benefit outweighs the possible risk.

C. Validity of a Negative Result in the Bacterial Test System

A class of potential mutagens that is not detected with bacteria is composed of those compounds that are not mutagenic themselves but are metabolized by animals to form a mutagen. The natural compound cycasin is not a mutagen in the bacterial test (Smith, 1966). It is split by the intestinal flora of humans to glucose and the toxic methylazoxymethanol, which is a powerful mutagen, teratogen, and carcinogen (Spatz et al., 1967). The methylazoxymethanol is a mutagen in the bacterial test (Smith, 1966). The host-mediated assay (see chapter in this book) might be useful for detecting compounds that are missed in a direct test if the sensitivity can be increased so that it can detect the known types of mutagens.

If a mutagen is most active at a pH other than pH 7, the pH of the bacterial growth medium in the agar plates, it is possible that it would not be detected on the petri plates. For example, a known mutagen such as nitrite is more active at low pH, where it is in the form of nitrous acid. Geneticists usually use nitrite at a pH about 4.5 when they are using it as a mutagen (e.g., Schwartz and Beckwith, 1969). Even at pH 7, however, strain TA1530 is reverted by nitrite (Fig. 1), though the less sensitive strain hisG46 is not. (The nitrite added as a color fixative to bacon, ham, frankfurters,

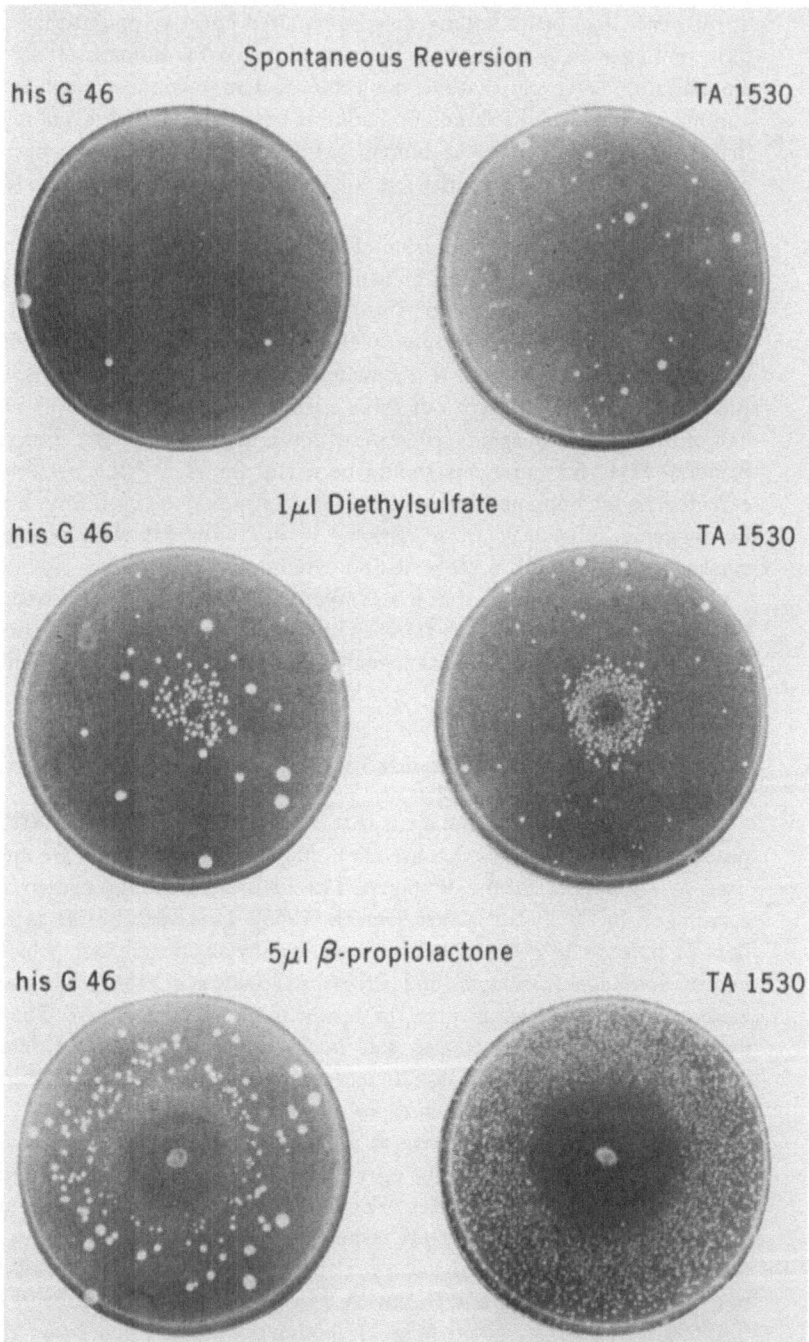

FIGURE 2. Photographs showing the reversion of mutation *hisG46* by various mutagens.

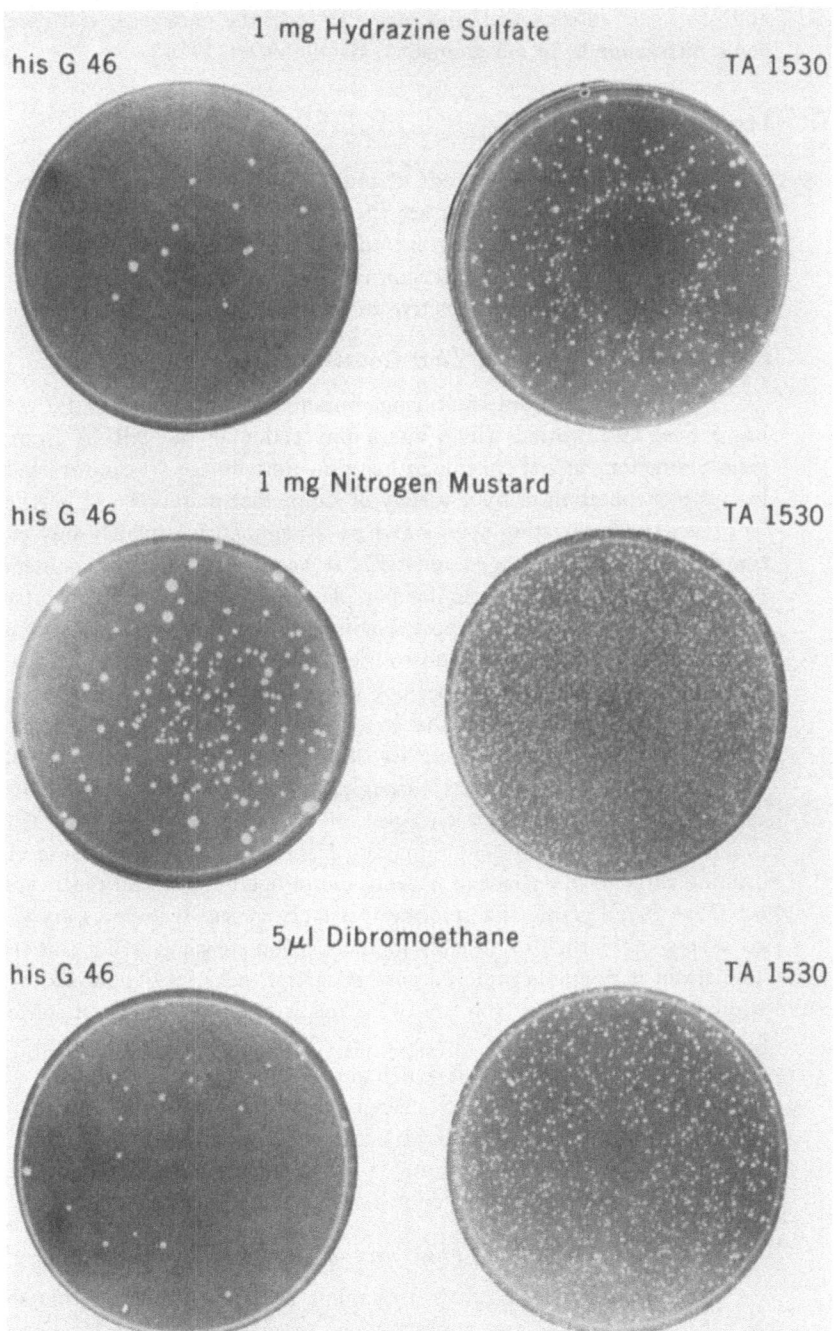

The double mutant *hisG46–uvrB* (TA1530) is compared to the *hisG46* strain.

and fish forms nitrous acid, and may also form the mutagenic and carcinogenic nitrosamines, in our stomachs; Sander *et al.*, 1968.)

D. Tester Strains

We have screened hundreds of mutants against a variety of mutagens
and chosen three mutants that are the most sensitive to reversion by each
class of mutagen. We have then introduced into these three mutants a deletion through the uvrB gene (excision repair); these strains then lack excision
repair and are much more sensitive to reversion by mutagens.

1. Detection of Mutagens That Cause Base-Pair Substitutions

HisG46 is a histidine-requiring mutant of *S. typhimurium* LT-2 that
has a base substitution which alters one codon in the *m*RNA from the
gene coding for the first enzyme of histidine biosynthesis. It can revert either
by a direct mutation or by a variety of suppressor mutations. It is reverted
by a variety of alkylating agents and by 2-aminopurine (which only causes
transitions) and by nitrogen mustard. It has a relatively low spontaneous
reversion rate (about ten colonies per plate), is extremely well reverted by
all the mutagens, and is the most sensitive to reversion of all the strains we
have examined for base-pair substitutions. We have found no mutagen that
reverts any other strain that does not revert G46, with the natural exception
of the acridine-type mutagens that insert and delete bases. It is clear that the
strain will detect agents causing transitions. Furthermore, since the strain
can revert directly and revert through suppressors, I suspect that it will
detect all possible base-pair changes.

TA1530 is a strain containing the hisG46 mutation and also a single
deletion through the galactose operon, biotin operon, excision repair system
for DNA (uvrB gene), and chlorate-resistance genes. It is necessary to add
an excess of biotin (0.1 μmole) to the petri plate when using this strain.
The strain is nonpathogenic because it cannot make its lipopolysaccharide
without the galactose operon, and it especially sensitive to a variety of mutagens because it is lacking the excision repair system. It cannot revert for
gal, uvrB, or *bio* because it is a deletion. Compared to *hisG46*, it is about
seven times more sensitive to reversion by diethylsulfate, about 30 times
more senstive to reversion by hydrazine, about 100 times more sensitive to
reversion by nitrogen mustard and β-propiolactone, and has about a three
times higher spontaneous reversion rate (Fig. 2, left and right).

2. Detection of Mutagens That Add or Delete One or Two Base Pairs

a. HisC207. HisC207 is a histidine-requiring frameshift mutant of
S. typhimurium LT-2 that appears to be missing one or two base pairs in
the aminotransferase gene of the histidine operon. The mutant is reverted

by a variety of acridine (ICR compounds) (Ames and Whitfield, 1966) that can add a base to the DNA in the region of the original lost base pair. Of a large number of different *E. coli* and *S. typhimurium* frameshift-type mutants that have been tried, it is the most sensitive to these ICR compounds. It is not reverted by any of the alkylating agents or agents that cause base-pair substitutions.

TA1531 is a strain containing the *hisC207* mutation and also a single deletion through the galactose operon, biotin operon, uvrB gene, and chlorate-resistance gene (see discussion of TA1530). One microgram of the acridine half-mustard ICR191 (Ames and Whitfield, 1966) gave about 17 times as many revertants with this strain as with *hisC207*. The spontaneous

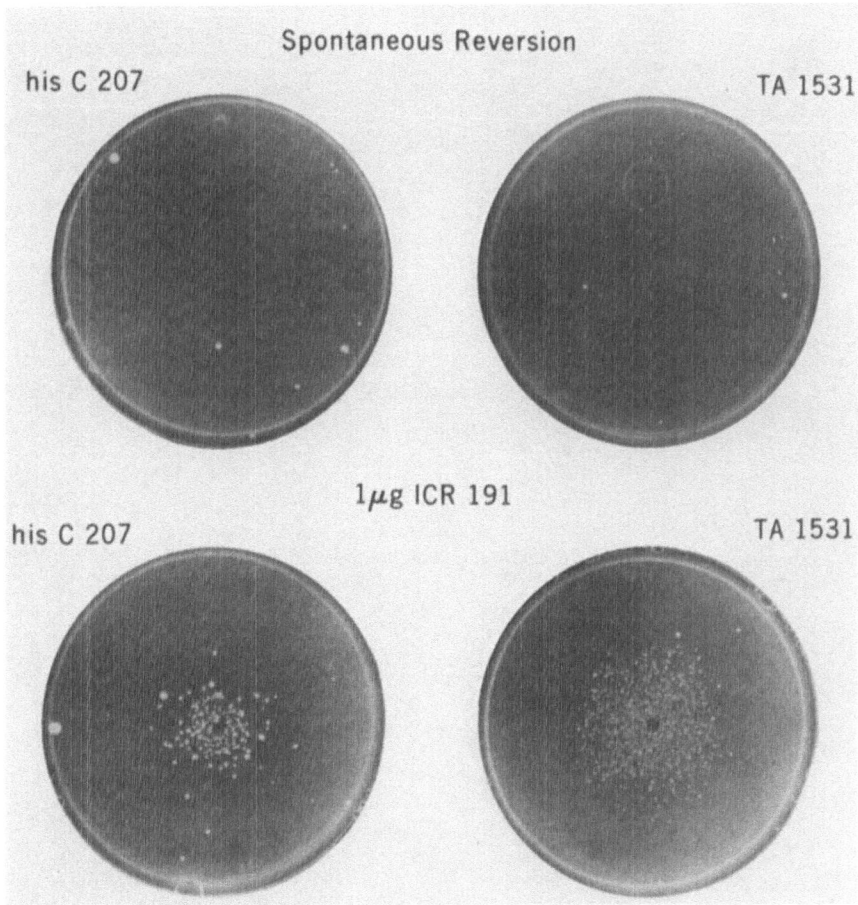

FIGURE 3. Photographs showing the reversion of mutation *hisC207* by various mutagens. The double mutant *hisC207–uvrB* (TA1531) is compared to the *hisC207* strain.

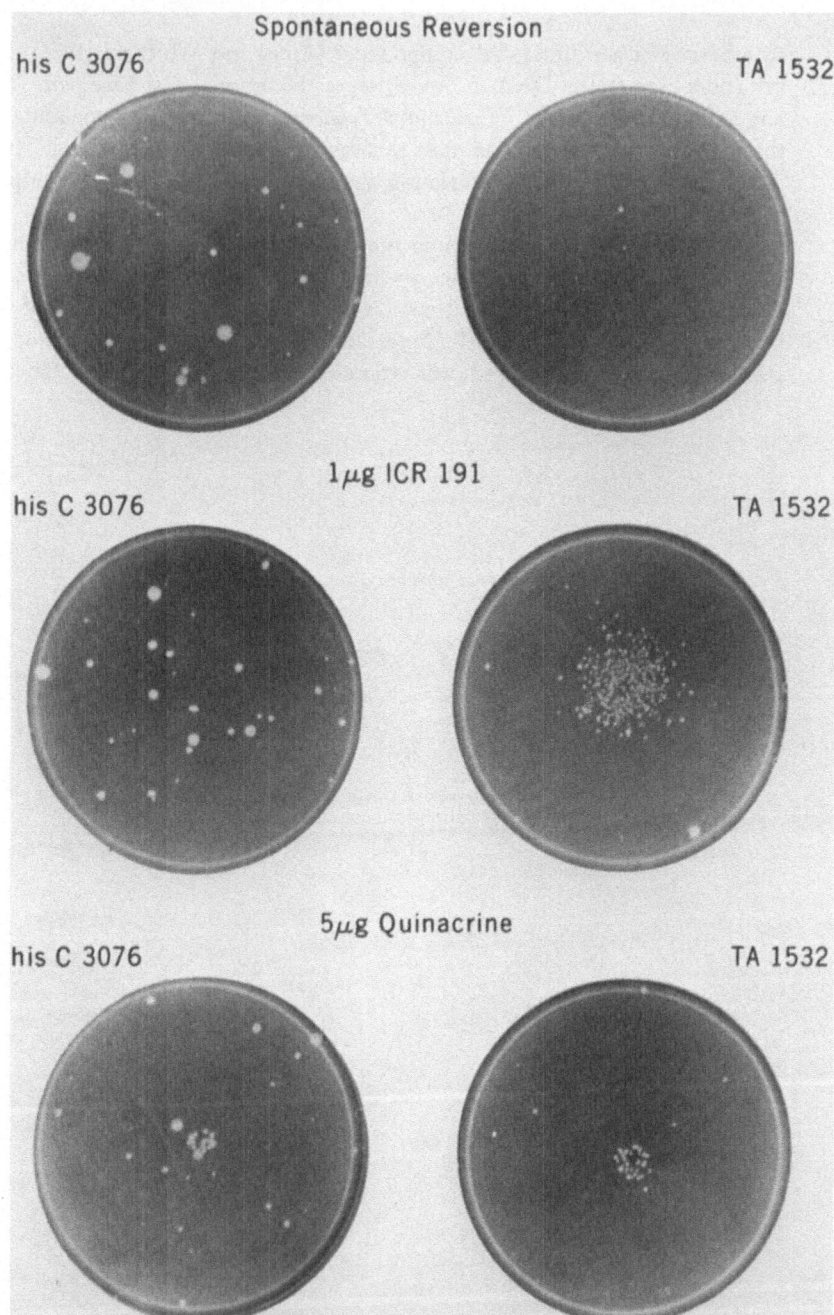

FIGURE 4. Photographs showing the reversion of mutation *hisC3076* by various mutagens. The double mutant *hisC3076–uvrB* (TA1532) is compared to the *hisC3076* strain.

reversion rate of the *hisC207* locus was lower in the TA1531 strain. Figure 3 illustrates some tests with these strains.

   b.   HisC3076.   HisC3076 is a histidine-requiring mutant that appears to have an added base pair in the aminotransferase gene in *Salmonella* (Oeschger and Hartman, 1970). It reverts spontaneously at a low rate, it is reverted by ICR191 and the other ICR acridines, and it is also reverted by quinacrine (atabrine) and 9-aminoacridine. It also is reverted weakly by diethylsulfate, methyl-nitro-nitrosoguanidine and 1,3-propane sultone; there is evidence that these agents can also occasionally delete a base pair (Oeschger and Hartman, 1970; Yourno and Heath, 1969).

   TA1532 is a strain containing the *hisC3076* mutation and a galactose–biotin–uvrB deletion. It is more sensitive to ICR191, about equally sensitive to quinacrine, and has a lower spontaneous reversion rate than *hisC3076* (Fig. 4). It is also more sensitive to reversion by alkylating agents than the parent *hisC3076* strain.

3.   Detection of Mutagens That Cause Large Deletions of the DNA

   The ordinary tester strains will not detect an agent that causes deletions. We have developed a test system that retains the advantages of the positive selection on petri plates and that is specific for agents that cause deletions (Alper and Ames, unpublished results). We have chosen conditions so that two closely linked genes, nitrate reductase and galactokinase, must be deleted in order to have the bacteria grow. The principle of the assay is as follows: Chlorate inhibits *E. coli* and *S. typhimurium* under anaerobic conditions, as it is converted by nitrate reductase to chlorite, which kills the bacteria. Mutations that destroy the nitrate reductase gene enable the bacteria to grow in the presence of chlorate. Galactose kills a galactose epimerase mutant, and mutations that destroy the galactokinase gene eliminate this inhibition by galactose. Single deletions are known that go through both the galactokinase and nitrate reductase genes. Such deletions are the only way for a galactose epimerase mutant to grow on a galactose–glycerol plate with chlorate incubated anaerobically, as both genes have to be mutated for the organism to grow and double point mutations are extremely rare. Thus a crystal of some chemical can be put on a petri plate with a lawn of 10^8 bacteria, and, if the chemical causes deletions, then a ring of mutants should grow up around the crystal after incubation. It would be possible to test the mutants to determine that they are truly deletions and not double point mutants by checking their reversion to growth on nitrate as a nitrogen source under anaerobic conditions.

   An epimerase mutant of *S. typhimurium*, *galE503* (formerly *gal-m*1; Fukasawa and Nikaido, 1961), is grown on nutrient broth. One-tenth milliliter of this culture is added to top agar (section III) and poured on a minimal–2% glycerol plate containing 0.1% galactose, 0.3% potassium chlorate, and 0.1 μmole of biotin. The plates are incubated aerobically for 8 hr at 37°C, overnight at 37°C anaerobically, and then aerobically for another 48 hr. The mutagens are added to the surface of the plate before incubation as in the other test systems. There are usually about ten spontaneous mutants per plate.

E. A General Test for Mutagenesis

Streptomycin resistance is generally used as a test for mutagenesis in that the streptomycin-resistant mutants can grow on a streptomycin plate while the wild-type bacteria can not. This is not a general test of mutagenesis, however, as the streptomycin-resistance mutation is a mutation in a gene for a ribosomal protein and only certain mutations will give the appropriate altered protein—a complete nonfunctional protein would be lethal for the cell (Silengo *et al.*, 1967). Thus nonsense mutations (UAA, UGA, UAG), frameshift mutations, and deletions are not detected in the streptomycin test.

A more general test is that of mutation to resistance to L-azetidine carboxylic acid (Calbiochem, Los Angeles, Calif.). Azetidine carboxylic acid (AC) is a proline analog that kills bacteria because it is incorporated into protein instead of proline. Mutants resistant to AC appear to be missing a functional proline permease, as they will not grow on proline as a nitrogen source. Any mutation that destroys the functioning of the permease gene, then, will result in a colony being formed by the AC-resistant mutant.

The test has not proved satisfactory for adding crystals of mutagens to plates as in the other tests. It can be used for testing a culture that has been mutagenized in liquid culture to see the increase in AC-resistant mutants over the unmutagenized culture. We ordinarily use wild-type *S. typhimurium* LT-2, but strain TA1530 can be used if an excess of histidine and biotin is added to the plate and the culture. A several-hundredfold increase over the spontaneous rate is obtained either by treating the bacteria with an alkylating agent such as methyl-nitro-nitrosoguanidine or by using an agent such as ICR191, which specifically causes frameshift mutations.

The mutagenized and untreated cultures are compared by making pour plates (see section III) on minimal medium and then putting a 6-mm disk of filter paper containing 3 μmoles of AC on the surface of each plate. The plates are incubated for 2 days and the number of colonies from resistant mutants counted within the zone of inhibition. The numbers are corrected for any difference in the number of bacteria plated. With a well-mutagenized culture it is convenient to plate only 10 μl instead of 0.1 ml. Alternatively, 15 μmoles of AC may be incorporated into the minimal agar, instead of using a disk of AC.

F. Agents That Have Been Shown to Be Mutagenic Using These Strains

Radiation: ultraviolet,* fast neutrons, X-rays.*

Methylating agents: N-methyl-N'-nitro-N-nitrosoguanidine,* strep-tozotocin,* methyl methanesulfonate,* N-nitroso-N-methylurethane,* methylazoxymethanol.*

Ethylating agents: diethylsulfate, ethyl methanesulfonate.

Other alkylating agents: β-propiolactone,* β-butyrolactone,* 1,3-propane sultone,* nitrogen mustard,* chloroethylamine,* dibromoethane (ethylenedibromide), ethyleneimine.*

Nonalkylating agents: hydrazine,* hydroxylamine, nitrous acid.

Acridines and acridine-like compounds: quinacrine (atabrine), 9-amino-acridine, and a variety of acridines, aza-acridines, and benzacridines:

Base analogs: 2-aminopurine, 5-bromouracil.

G. Phenotypic Curing

Streptomycin, kanamycin, and neomycin cause phenotypic curing by causing occasional misreading of the messenger RNA (Davies *et al.*, 1965). If a few milligrams of streptomycin is put down on one of the pour plates of the tester strains, a zone of inhibition is observed after incubation. In *hisG46*, but not in *hisC207*, a ring of dense growth peripheral to the zone of inhibition indicates phenotypic curing (Whitfield *et al.*, 1966). The mis-coding results in the occasional misreading of the UAA codon (at the point of mutation) and thus every bacterium can make a minmal amount of func-tional enzyme and grow at some concentration of streptomycin just below the inhibitory concentration. We have also observed this phenomenon with hydrazine. It may be of interest to know which of the compounds being tested as mutagens can also cause miscoding in protein synthesis. This information is apparent on the tester plates with no extra work. The frame-shift mutant *hisC207* serves as a control, as phenotypic curing is not ob-served with mutants having an insertion or deletion of a base pair.

H. Testing of Compounds, Availability of Strains, and Improvements of Procedures

I will be glad to mail the strains to people desiring them and to serve as a clearinghouse for new and improved bacterial tester strains. I will also

---

* The asterisk indicates agents known to be carcinogenic.

keep a list of the response of compounds if investigators will send me their
results.

III. GENERAL METHODS

A. Growth of Bacterial Cultures

The various mutants are grown up at 37°C with shaking in nutrient
broth. We usually grow up the cultures in 5 ml in a 18-mm test tube with a
Bellco plastic closure on top in a New Brunswick rotary shaker. The cultures will grow up overnight and then can be stored in the refrigerator for
several weeks. Permanent cultures of the strains are kept as stabs in small
screw-cap vials of nutrient broth containing 0.6% agar. The threads of the
vials are dipped in molten paraffin before the cap is screwed on, which seals
the tube and prevents drying. These stabs can be kept for many years at room
temperature.

The number of spontaneous revertants in the cultures affects the sensitivity of the tests to some extent, especially with strain TA1530. It is possible to lower the number of spontaneous mutants in a simple way. Due
to the random time of appearance of spontaneous mutants, there is a considerable variation in the number of revertants if a number of parallel cultures are grown up from small inocula. We streak out the culture of the
histidine mutant on a nutrient broth plate, and make nutrient broth cultures
from five to ten single colonies when they are small. Each culture is then
tested to see how many revertants are present, and only the culture with the
fewest revertants is saved. Fresh cultures can be made every few weeks by
inoculating several parallel cultures with a small aliquot containing no revertants and repeating the process. A fully grown nutrient broth culture
should have about 3×10^9 bacteria per milliliter.

We routinely use plugged pipettes and autoclave all materials after use,
even though the *S. typhimurium* LT-2 wild-type strain is not highly pathogenic. This strain has been used by many laboratories for many years and
also in many class experiments. There has been only an occasional case of
diarrhea when someone swallows a mouthful. Those tester strains with a
deletion through the galactose operon are not pathogenic, because of their
inability to make the lipopolysaccharide endotoxin which contains galactose.

B. Pour Plates for Testing Mutagens

In order to obtain a uniform lawn of the histidine-requiring bacteria
(or other tester strains) on a minimal plate, we add 0.1 ml of the nutrient
broth culture to a tube (13 by 100 mm) with 2 ml of molten 0.6% agar

containing 0.5% NaCl (top agar) kept at 45°C. This temperature will keep the top agar molten and not kill bacteria. The tube is then mixed by a brief rotation between the palms and poured onto the center of a petri plate of minimal agar medium (we use the Vogel–Bonner, 1956, minimal medium with 1.5% agar and 2.0% glucose). The plate is then jiggled quickly to cover the surface with the liquid and put on the bench to harden—the mixing and jiggling operation should take no more than 10 sec; otherwise, the soft agar starts to harden and the layer is not smooth. The plates can be prewarmed if the air temperature is low.

A trace (0.1 μmole) of histidine, and an excess of biotin (0.1 μmole), is added to the plate or to the top agar, so that all the bacteria can grow slightly to make a background lawn, so that any inhibition caused by the compounds can be seen. This addition is essential, as growth of the cells is required for mutagenesis in the case of many of the mutagens.

The mutagen may be added to the plate any time after the top agar has hardened (usually 5 min is sufficient). The addition can be delayed for several hours. It is usually convenient to put some crystals (1 to 5 mg) of the mutagen as a point near the edge of the plate with a broken applicator stick that is used as a spatula. If the compound is a liquid, about 5 μl is usually used; this will soak into the agar in a short time. A control of the spontaneous mutation rate should also be run for each tester strain on a separate plate. Plates are incubated upside down for 2 days at 37°C.

If a large number of compounds are being tested, then four or more compounds can be put on the same plate. If any of them inhibits the whole plate, or gives a large number of mutants, then the compounds must be retested individually.

IV. ACKNOWLEDGMENT

I wish to thank Miss Anne Liggett for expert technical assistance and Drs. P. E. Hartman, John R. Roth, George Chang, and Mark Alper for helpful discussions.

V. REFERENCES

Ames, B. N., and Whitfield, H. J. (1966), *Cold Spring Harbor Symp. Quant. Biol.* *31*, 221.

Demerec, M., Bertani, G., and Flint, J. (1951), *Am. Naturalist 85*, 119.

Fukusawa, T., and Nikaido, H. (1961), *Genetics 46*, 1295.

Gorini, L., and Kataja, E., *Biochem. Biophys. Res. Commun.* (1965) *18*, 656.

Heinemann, B., Howard, A. J., and Hollister, Z. J. (1967), *Appl. Microbiol. 15*, 723.

Hirota, Y. (1960), *Proc. Nat. Acad. Sci. 46*, 57.

Iyer, V. N., and Szybalski, W. (1958), *Appl. Microbiol. 6*, 23.

Oeschger, N. S., and Hartman, P. E. (1970), *J. Bacteriol. 101*, 490.

Sander, J., Schweinsberg, F., and Menz, H.-P. (1968), *Hoppe-Seylers Z. Physiol. Chem. 349*, 1691.

Schwartz, D. O.,and Beckwith, J. R. (1969), *Genetics 61*, 371.

Sideropoulos, A. S., and Shankel, D. M. (1968), *J. Bacteriol. 96*, 198.

Silengo, L., Schlessinger, D., Mangiarotti, G., and Apirion, D. (1967), *Mutation Res. 4*, 701.

Smith, D. W. E. (1966), *Science 152*, 1273.

Spatz, M., Dougherty, W. J., and Smith, D. W. E. (1967), *Proc. Soc. Exp. Biol. Med. 124*, 476.

Szybalski, W. (1958), *N. Y. Acad. Sci. Proc. 76*, 475.

Vogel, H. J., and Bonner, D. M. (1956), *J. Biol. Chem. 218*, 97.

Whitfield, H. J., Martin, R. G., and Ames, B. N. (1966), *J. Mol. Biol. 21*, 335.

Witkin, E. (1969), *Ann. Rev. Microbiol. 23*, 437.

Yourno, J., and Heath, S. (1969), *J. Bacteriol. 100*, 460.

NOTE ADDED IN PROOF

1. In Fig. 1, 2 mg of sodium nitrite was used.
2. In the assay for deletions the ingredients in the agar plate should be autoclaved separately. The top agar should have 0.1 ml of 4% glucose added to it. This increases sensitivity by allowing time for gene expression before the galactose kills the cells. The plate should contain no glucose other than this. When this is done about fifty spontaneous mutants are observed per plate.
3. Hycanthone and nitroquinoline-N-oxide* should be added to section F. They specifically cause frame shifts (Hartman *et al.*, *Science*, in press) and can be detected with TA 1532 and *hisD3052* respectively.
4. The carcinogen aminofluorene and a variety of its derivatives, especially nitrosofluorene, cause frame shift mutations and can be detected with strain TA 1533 (*hisD3052* lacking *uvrB*).

Mutagenesis Studies with *Escherichia coli* Mutants with Known Amino Acid (and Base-Pair) Changes

C. Yanofsky

Department of Biological Sciences
Stanford University
Stanford, Califonia

I. INTRODUCTION

Detection of a mutagen-induced change in a gene specifying the amino acid sequence of a protein depends upon recognition of an effect on the functional capacity of the mutationally altered enzyme. If the protein is inactivated by a single amino acid substitution, a limited number of specific amino acid changes will be capable of restoring catalytic activity. We expect that in some instances, because of strict functional requirements, activity, and prototrophy, will be restored by only one amino acid change, the substitution which is the reverse of the one responsible for enzyme inactivity. If a set of strains of the latter type could be obtained, each specifically reverting as a result of a different base-pair change, it should be possible to determine rapidly and unamibiguously the base-change specifices of new mutagens. Unfortunately, very few strains of the type described are available. There are several reasons for this, the most important being that reversion spectra of only a few mutants have been analyzed in terms of the amino acid substitutions which restore function. In addition, strains of the type sought are rare because most mutants are reverted by several different base changes. A further complication is that missense suppressors frequently mutate and restore a small amount of functional protein.

II. MUTANTS WITH AMINO ACID CHANGES IN THE TRYPTOPHAN SYNTHETASE α SUBUNIT (A PROTEIN)

A. Mutants with Changes at Position 48

The amino acid changes at position 48 in the α subunit and presumed corresponding nucleotide changes are summarized in Fig. 1 and Table 1. Since the sole revertant type recovered from these mutants is indistinguishable from the wild type, it is likely that position 48 has a strict requirement and only changes which substitute Glu will be detected. Unfortunately, however, both A3 and A11 are subject to missense suppression. Therefore, it is necessary to examine the types of prototroph that are recovered from these mutants before it can be concluded that a specific base change has occurred (Yanofsky *et al.*, 1966; Drapeau *et al.*, 1968). Mutant A88 presumably will respond only to mutagens that cause T→G changes and apparently neither EMS nor NNG can do this. It should be pointed out, however, that the amino acid(s) at position 48 in revertants of the three strains has not been determined.

B. Mutants With Changes at Position 210

The reversion patterns of mutants A23 and A46 are quite complex, as can be seen in Fig. 2 and Table 1. Furthermore, mutant A23 is subject to missense suppression and mutant A46 gives prototrophs by second-site reversion; i.e., a change of Tyr→Cys at position 174 in the protein. In addition, the α subunits with Ala or Ser at position 210 are functionally indistinguishable from the wild-type protein which has Gly at 210. Thus protein chemistry must be performed to determine which change has occurred

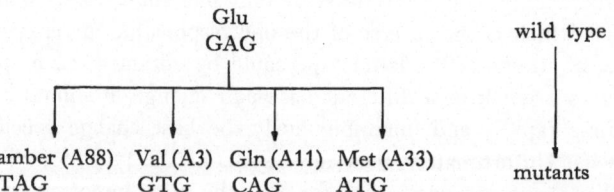

FIGURE 1. Changes at position 48 and probable corresponding base changes. Based on the studies of Yanofsky *et al.* (1966) and Drapeau *et al.* (1968).

TABLE 1. Reversion Analyses[a]

Mutant	Mutagen used to produce	Amino acid change	Position in α subunit	Base change[b] (wild→ mutant)	Reversion to prototrophy by					Other revertant changes[a]	Subject to missense suppression	References
					SP[c]	ICR	NNG	EMS	AP			
A3	UV	Glu→Val	48	A→T	+	0	+	+	0	None	+	Yanofsky et al. (1966); Drapeau et al. (1968)
A11	UV	Glu→Gln	48	G→C	+	0	+	+	0	None	+	Yanofsky et al. (1966); Drapeau et al. (1968)
A33	UV	Glu→Met	48	GA→AT	0	0	0	0	0	None	0	Yanofsky et al. (1966); Drapeau et al. (1968)
A88	UV	Glu→am[e]	48	G→T	+	0	0	0	Very weak	None	0	Yanofsky et al. (1966)
A23	UV	Gly→Arg	210	G→A	+	Weak[f]	+	+	+	Arg→Ser, Thr, Ile	+	Yanofsky et al. (1966); Berger et al. (1968b)
A46	UV	Gly→Glu	210	G→A	+	Weak[f]	+	+	+	Glu→Ala, Val	0	Yanofsky et al. (1966); Berger et al. (1968b)
A58	UV	Gly→Asp	233	G→A	+	0	+	+	+	Asp→Ala	+	Yanofsky et al. (1966)
A78	UV	Gly→Cys	233	G→T	+	0	+	Very weak	0	None	+	Yanofsky et al. (1966)
A9813	UV			−1	+	Weak	0	0	0			Brammar et al. (1967)
A21	UV			−1 or −2	+	+	0	0	0			Berger et al. (1968a)
A540	ICR			−1 or −2	+	+	0	0	0			Berger et al. (1968a)

[a] Reversion tests were performed by plating 0.05 ml (2×10^8 cells) of an overnight L-broth (Lennox, 1955) culture on minimal agar (Vogel and Bonner, 1956) supplemented with 0.1 µg/ml L-tryptophan. In nitrosoguanidine (NNG) and ICR191 (ICR) tests, 100 and 10 µg, respectively, in 0.05 ml were spotted on one section of the agar surface. In the 2-aminopurine (AP) test, 2 mg in 0.05 ml was spread over one third of the agar surface prior to the addition of the bacteria. In the ethyl methanesulfonate (EMS) test, the EMS was diluted with two parts of 0.2 M potassium phosphate buffer at pH 7.0, and 0.05 ml was spread on one third of the agar surface prior to the addition of the bacteria. Plates were scored for revertants after 3 days of incubation at 37°C. [c] Sp = reverts spontaneously. [b] For simplicity only one member of a base pair is mentioned. [d] Known changes at the position of the original amino acid change, other than exact reversal, which restore prototrophy. [e] am = amber. [f] Detected only with high concentrations of ICR191.

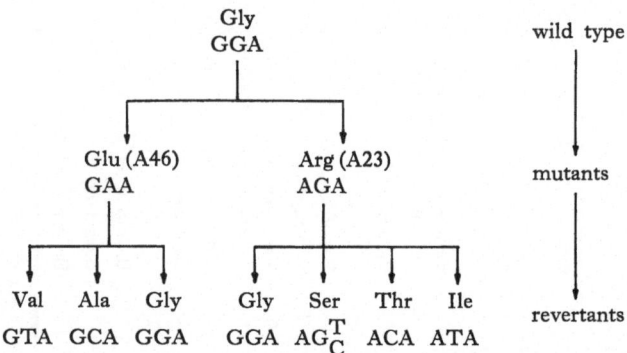

FIGURE 2. Amino acid changes at position 210 and probable corresponding base changes. Based on the studies of Yanofsky *et al.* (1966) and Berger *et al.* (1968*b*).

when a revertant cannot be distinguished from the wild type. On the other hand, the Val revertant of A46 is easily recognized, and therefore changes from A → T (Glu → Val) can be scored, if one is willing to pick revertant colonies and put them through the appropriate tests (Yanofsky *et al.*, 1966). Mutant A23 cannot be used to identify specific base changes; however, it gives prototrophs by G → C, G → T, A → G, and A → T or C changes. Thus if a mutagen does not revert this strain, it can be concluded that the agent is incapable of causing any of the indicated base changes.

C. Mutants With Changes at Position 233

Mutants A58 and A78 appear to have relatively simple reversion patterns; however, both strains revert by second-site reversion and both are subject to missense suppression. Unique base changes are associated with reversion to the wild-type phenotype, A → G (A58) and T → G (A78). These changes can be scored if revertant colonies are picked and examined in appropriate tests (Yanofsky *et al.*, 1966).

D. Other Information

Several additional mutants are listed in Table 1. These strains are partially characterized and have mutagen specificities in reversion tests which may be useful.

There are also somewhat more sophisticated test systems that can be employed in mutagenesis studies. For example, a test can be based on the fact that amber suppressors suppress amber mutants but not ochre mutants. A stock can be prepared which has an ochre alteration and an amber suppressor; the amber suppressor would be one which is known to insert an

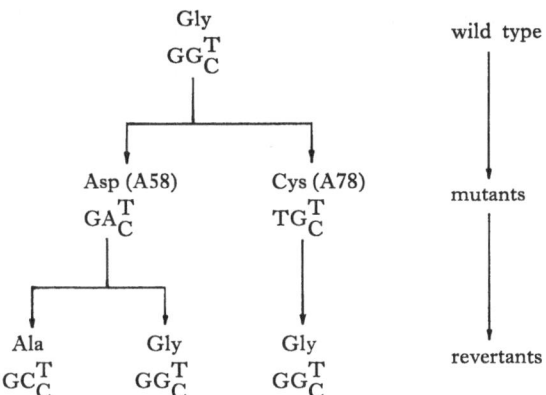

FIGURE 3.　Amino acid changes at position 233 and probable corresponding base changes. Based on the studies of Yanofsky *et al.* (1966) and Guest and Yanofsky (1965).

amino acid which restores function. Mutations from ochre to amber (A → G) can then be scored if a convenient procedure is available for verification that the change is to amber. Similarly, suppressor-containing stocks may be prepared in which the changes UGA (nonsense) ⇌ UAA (ochre) may be scored.

In certain cases, mutator genes may be helpful in establishing the reversion base-change possibilities in a particular mutant. The Treffers mutator gene preferentially causes AT → CG changes (Treffers *et al.*, 1954; Yanofsky *et al.*, 1966). If a mutant is not reverted by the Treffers mutator, obviously this one base-pair change is excluded as a possibility. When additional specific mutator genes are isolated and characterized, this approach should be extremely useful because it would enable the investigator to establish the reversion possibilities of a particular mutant without performing protein structure analyses.

III.　REFERENCES

Berger, H., Brammar, W. J., and Yanofsky, C. (1968a), *J. Bacteriol.* 96, 1672.

Berger, H., Brammar, W. J., and Yanofsky, C. (1968b), *J. Mol. Biol.* 34, 219.

Brammar, W. J., Berger, H., and Yanofsky, C. (1967), *Proc. Nat. Acad. Sci.* 58, 1499.

Drapeau, G. R., Brammar, W. J., and Yanofsky, C. (1968), *J. Mol. Biol.* 35, 357.

Guest, J. R., and Yanofsky, C. (1965), *J. Mol. Biol.* 12, 793.

Lennox, E. S. (1955), *J. Bacteriol.* 100, 390.

Treffers, H. P., Spinelli, V., and Belser, N. O. (1954), *Proc. Nat. Acad. Sci.* 40, 1064.

Vogel, H. J., and Bonner, D. M. (1956), *J. Biol. Chem.* 218, 97.

Yanofsky, C., Ito, J., and Horn, V. (1966), *Cold Spring Harbor Symp. Quant. Biol.* 31, 151.

CHAPTER 10

Mutation Induction in Yeast

R. K. Mortimer*

Donner Laboratory
University of California
Berkeley, California

and

T. R. Manney††

Department of Microbiology
Case Western Reserve University
Cleveland, Ohio

I. INTRODUCTION

As single-cell eukaryotes that can exist either as stable haploids or diploids, the yeasts afford many advantages as test organisms for examination of the genetic effects of chemicals and radiations. The suitable characteristics of these higher fungi include clonability, a short generation time, adaptability to replica plating, and a well-developed genetics. Two species, *Saccharomyces cerevisiae* and *Schizosaccharomyces pombe*, have been used for most of the mutation studies with yeast.

* Financial support from the U.S. Atomic Energy Commission is acknowledged.
† Financial support from the Public Health Service Research Career Development Award 1-K03-GM33672. and Public Health Service Research Grant 1-R01-GM-14399 is acknowledged.
‡ Present address: Donner Laboratory, University of California, Berkeley, California.

II. GENERAL DESCRIPTION OF THE ORGANISMS

Sacch. cerevisiae, the common baking and brewing yeast, normally exists in the diploid phase. The cells, which are ellipsoidal (approximately 5 by 7 μ), divide mitotically by budding. The generation time at 30°C for most strains is in the range of 70 to 90 min. On certain media the diploid cells will divide meiotically to yield an ascus containing four haploid spores. The ascus wall can be removed enzymatically and the spores separated by micromanipulation. In heterothallic strains these spores will divide mitotically as stable haploids. Two of the spores from each ascus are of one mating type, the other two are of the opposite mating. Haploid vegetative cells, if placed in contact with haploid cells of the opposite mating type,

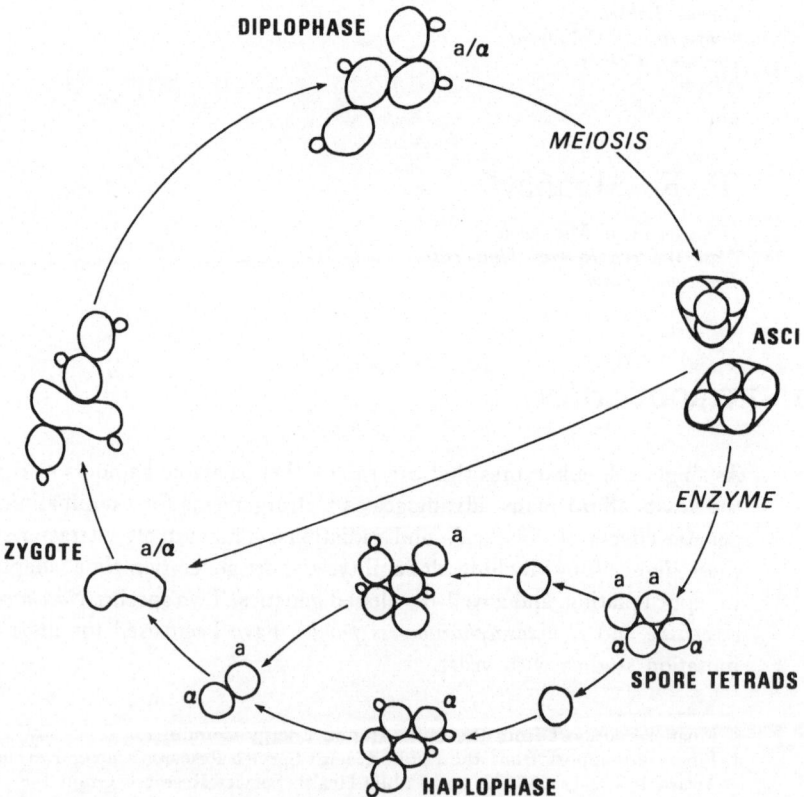

FIGURE 1. Life cycle of heterothallic *Sacch. cerevisiae.*

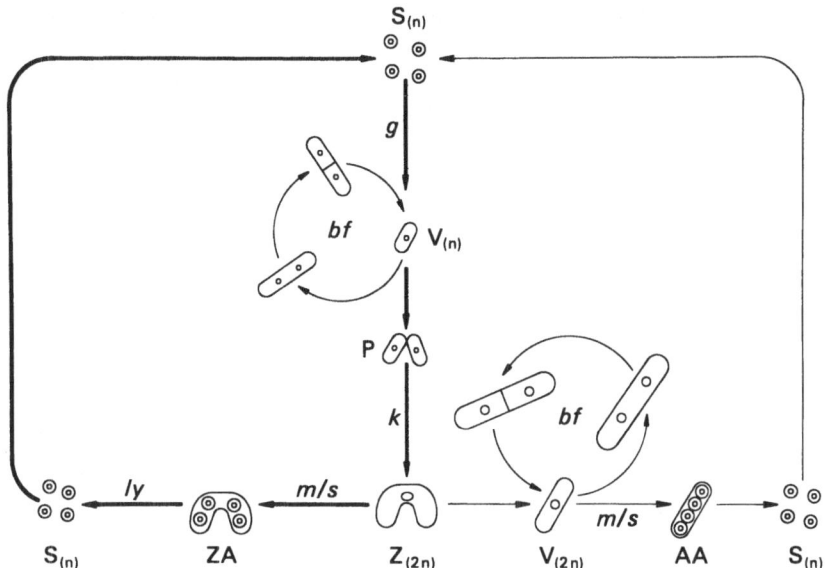

FIGURE 2. Life cycle of *Schiz. pombe.* S(n)—haploid spores, V(n)—haploid vegetative cells, V(2n)—diploid vegetative cells, Z(2n)—diploid zygote, ZA—zygotic ascus, AA—azygotic ascus, g—germination, bf—binary fission, p—plasmogamy, k—karyogamy, m—meiosis, s—sporulation, ly—lysis of the ascus wall. (da Cunha, 1969) Reproduced by permission of the author.

will conjugate to form a diploid zygote which then divides mitotically to reestablish the diploid phase. Mutations induced in haploid cells can be examined genetically by crossing the mutant cells to suitable haploids and examining the segregation of the mutant phenotype in meiotic tetrads formed by the diploid cross. Mutations that involve a single genetic locus are expected to segregate 2:2 in these tetrads. The life cycle of heterothallic strains of *Sacch. cerevisiae* is summarized in Fig. 1.

Cells of *Schiz. pombe* are cylindrical (approximately 9 by 4 μ) and divide by binary fission. The vegetative phase is normally haploid. The haploid cells have a generation time of about 120 min at 30°C. Conjugation will occur between haploid cells of opposite mating type or between two cells of a homothallic strain to yield a diploid zygote. Normally, this zygote then enters meiosis and spore formation with a resultant zygotic ascus containing four haploid spores. These spores can then be isolated and grown vegetatively as haploids. However, by selective procedures a stable diploid phase can be established. These vegetative diploids can be induced to sporulate. Thus, as in *Sacch. cerevisiae*, either haploid or diploid vegetative cells can be used for mutation studies, and genetic changes can be characterized by tetrad analysis. The life cycle of *Schiz. pombe* is presented in Fig. 2.

The minimum haploid chromosome number in *Sacch. cerevisiae*, determined by genetic means, is 17 (Hawthorne and Mortimer, 1968; Resnick, 1969). Six chromosomes have been identified in *Schiz. pombe* (Leupold, 1958; da Cunha, 1969). Because of the small size of the chromosomes, cytogenetic analyses with yeast have so far been relatively unproductive. The genetics of yeasts has been reviewed by Winge and Roberts (1958) and by Mortimer and Hawthorne (1966a, 1969). The most recent of these reviews includes a detailed discussion of the methodology of yeast genetics.

For quantitative evaluation of the mutagenic activity of different physical and chemical agents, a variety of genetic events can be studied in yeast. These include both forward and reverse mutations in haploids and reversional and recombinational events (both intergenic and intragenic) in diploids. A limited selection of publications that illustrate the use of these genetic changes for the study of mutagens will be considered in this article.

III. MUTATION

A. Induction and Isolation of Forward Mutants

Forward mutation in genes controlling the biosynthesis of adenine has been studied extensively in both *Sacch. cerevisiae* and *Schiz. pombe*. In *Sacch. cerevisiae*, mutation at either of two loci, *ade1* or *ade2*, results in red-pigmented mutant colonies. Wild-type cells normally form white to cream-colored colonies. In addition to the red pigmentation, the mutant cells are adenine requiring. Similarly, mutations at either *ade6* or *ade7* in *Schiz. pombe* result in pigmented colonies. In both yeasts, these red-adenine loci are unlinked, in contrast to the analogous purple-adenine mutants, *ad-3A* and *ad-3B*, in *Neurospora*, which are closely linked. To study mutation at the red-adenine genes, mutagen-treated haploid cells are plated onto a complete medium to yield, after incubation, approximately 500 colonies per plate. Mutant colonies or colony sectors can be readily detected by their distinct coloration. Frequencies of chemically induced mutation at these two loci in the range 10^{-4} to 10^{-2} have been reported (Loprieno *et al.*, 1969a; Table 1). The corresponding spontaneous mutation rate has been estimated to be less than 10^{-7} (Loprieno *et al.*, 1969b). Because of the relative ease of plating large numbers of cells and of detecting the red-adenine mutants, this system in yeast should prove to be as valuable as the corresponding one developed for *Neurospora* (de Serres, 1964).

Strains carrying one of the red-adenine mutations provide a genetic background for detection of additional mutations that block formation of the red pigment. Such mutants appear as white colonies or colony sectors,

TABLE 1. Purple Mutation Frequency Induced in *Schiz. pombe* by HA, NA, EMS, and MMS at Comparable Survival Levels[a]

Mutagen[b]	Treatment conditions	Number of colonies analyzed	Percentage survival	Purple mutation frequency $\times 10^{-4}$ survivors
HA (1 M)	pH 6.6, 30°C, 10 hr	6.30×10^4	0.35	14.76
NA (0.01 M)	pH 3.0, 25°C, 14–16 min	3.80×10^4	0.49	26.25
EMS (1 M)	distilled water, 25°C, 2 hr	53.30×10^4	0.44	7.88
MMS (0.1 M)	distilled water, 25°C, 1 hr	80.04×10^4	0.34	2.61

[a] Table 1 from Loprieno *et al.* (1969a). Reproduced by permission of the senior author. (Courtesy of Elsevier Publihing Company.)

[b] The following abbreviations are used in this and subsequent tables: BA, 1,2-benzanthracene; BP, 3,4-benzpyrene; BU, 5-bromouracil; DAP, 2,6-diaminopurine; DBA, 1,2,5,6-dibenzanthracene; DES, diethyl sulfate; EI, ethyleneimine; EMS, ethyl methanesulfonate; 5FU, 5-fluorouracil; ICR, acridine mustard; MMS, methyl methanesulfonate; NA, nitrous acid; NEU, *N*-nitroso-*N*-ethylurethan; NIL, 1-nitrosoimidazolidone-2; NMA, *N*-nitroso-*N*-methylacetamide; NMG, 1-methyl-3-nitro-1-nitrosoguanidine; NMT, *N,N'*-dinitroso-*N,N'*-dimethylterephthalamide; NMU, *N*-nitroso-*N*-methylurethan; NMY, *N*-nitroso-*N*-methylcaprylamide; UV, ultraviolet; X, X-radiation.

while the parental cells form red colonies. Most such white mutants are still adenine dependent and have been shown to be a consequence of an additional mutation at one of several genes that control other steps in adenine biosynthesis (Leupold, 1955; Roman, 1956a). In addition, back-mutation at the red-adenine locus, mutation at a suppressor locus, or mutation to a respiratory-deficient state (genic or cytoplasmic) can block the formation of the red color.

Mutations in genes controlling the biosynthesis of other small molecules normally are not expressed as changes in colony morphology. However, such mutants can be readily detected by replica-plating the colonies that develop from mutagen-treated haploid cells to a minimal medium. Replicas that fail to grow signal a nutritional mutation. Frequencies of such mutations are in the order of 10^{-2} to 10^{-1} after treatment of yeast cells with chemical mutagens (Lingens and Oltmanns, 1964; Lindegren *et al.*, 1965). Because of difficulties in detection of some mutant colonies and sectors following replica-plating, there can be considerable error in estimation of mutation frequencies by this method. The dye Magdala Red (Phloxin B), if included in a medium that contains limiting concentrations of nutrients, is taken up selectively by mutant cells. This permits detection of nutritional mutants on the basis of colony morphology (Horn and Wilkie, 1966; Strömnaes, 1968).

Mutation of *Sacch. cerevisiae* cells to a respiratory-deficient state occurs at a relatively high frequency. These mutant cells form small white colonies (petites) and are unable to grow on nonfermentable substrates. The mutant colonies can be readily detected if the mutagen-treated cells are plated on a medium containing a dye that is taken up selectively by petite colonies (Nagai, 1963). The petite phenotype may be the result of mutation at one of several nuclear genes or due to a cytoplasmic change. The cytoplasmic petites occur spontaneously, and their frequency can be increased greatly by treatment of the cells with certain chemicals, for example, acriflavine, 5-fluorouracil, or ethidium bromide (Tavlitski, 1949; Lacroute, 1963; Slonimski et al., 1968). Genic petites can also be induced by chemical mutagens or ultraviolet light (Sherman, 1963; Mackler et al., 1965). In one recent study, a series of chemically induced petites were examined genetically to determine if they were nuclear or cytoplasmic. Nitrous acid was found to induce almost exclusively genic petites, whereas nitrosoimidazolidone (NIL) and nitrosomethylurethan (NMU) produced both genic and cytoplasmic respiratory-deficient mutants (Schwaier et al., 1968). Thus there are mutagens that appear to act exclusively on nuclear genes, others that induce only cytoplasmic mutations, and some that cause both classes of mutations.

B. Characterization of Forward Mutants

Forward mutants in yeast can be characterized by additional tests such as ability to engage in intragenic complementation, phenotypic reversibility by changes in pH, temperature, or osmotic pressure, senstivity to nonsense suppressors, or revertibility with specific mutagens. From the response of the mutants to these tests, it is possible to infer the type of molecular change that led to the mutation. Complementing mutants or mutants subject to phenotypic reversal are most probably the result of a base-pair substitution event. Similarly, mutants suppressed by nonsense suppressors should be almost entirely due to base-pair substitutions. In contrast, mutants that are the consequence of the addition or deletion of one or more base pairs in a gene generally would fail to respond to these tests. Yeast mutants that show a *meiotic effect*, a large increase in reverse mutation rate during meiotic division compared to mitotic division, are proposed to be of the addition–deletion category (Magni and von Borstel, 1962; Magni and Puglisi, 1966).

C. Reverse Mutations, Suppressors, and Resistance Mutations

Reverse mutation in yeast can be examined either in mutant haploid cells or in diploid cells that are homozygous for a nutritional mutation. Diploid cells are more resistant than haploid cells to inactivation by many mutagens and so present an advantage for experiments in which only a limited number of cells can be used or in which selection of preexisting mutants could interfere (Mortimer *et al.*, 1965). In addition, further genetic examination is facilitated because the revertants can be sporulated directly. Haploid revertants normally must be crossed to another haploid in order to determine if the reversion was due to back-mutation at the original locus or to mutation at a suppressor locus. However, if the mutation under study is of the nonsense type and if other suppressible mutations are carried in the stock, suppressor revertants can be identified as those that lead to simultaneous reversion of two or more nonsense mutations (Hawthorne and Mortimer, 1963; Magni and Puglisi, 1966; Gilmore, 1967; Hawthorne, 1969).

Mutations of the cytochrome *c* gene are ideal for reversion studies because the cytochrome *c* formed by the revertants can be analyzed for amino acid substitutions. For example, one nonsense mutation occurred at a site that normally coded for glutamate. A site revertant of this mutant was found to produce cytochrome *c* that contained tyrosine in place of glutamate at this position (Sherman *et al.*, 1966). Both the original and the reverse mutations can be explained by base-pair substitutions of the transversion type. A series of nonsense suppressors also was found to cause the insertion of tyrosine into this position (Gilmore *et al.*, 1968).

A convenient method for screening chemical mutagens for their ability to induce reversion in yeast has been described by Fink and Lowenstein (1969). A small amount of the mutagen is placed at a point on a growing lawn of cells contained in a petri dish. After a period of incubation, the cells are replica-plated to minimal medium. If the mutagen is effective, a circle of revertants develops on the omission plate, in the region corresponding to the point of application of the mutagen.

Mutation to canavanine resistance can be readily studied in yeast. The resistance mutations are recessive and appear to occur in a gene that encodes an arginine-specific permease (Grenson *et al.*, 1966). Other resistance mutations that can be selected are those that permit growth in the presence of normally toxic concentrations of ethioninine, *p*-fluorophenylalanine, or 5-fluorotryptophan (Surdin *et al.*, 1965; Gits and Grenson, 1966; von Borstel, personal communication).

In Table 2 is summarized information from a selection of mutation studies with yeast. Protocols for the use of different mutagens are presented in many of these papers.

IV. MITOTIC SEGREGATION

A. Relevance to Mutagen Studies

The processes that result in the segregation of heterozygous markers

TABLE 2. Studies of Mutation Induction

Organism	Locus	Mutagen[a]	Method of detection	Characterization of mutants	References
White (wild type) → red (adenine requiring)					
Sacch. cerevisiae	ade1, ade2	UV	Colony morphology	Complementation	Woods and Bevan (1966)
Sacch. cerevisiae	ade1, ade2	NA, NIL	Colony morphology	Complementation; osmotic remediability	Nashed and Jabbur (1966)
Sacch. cerevisiae	ade1, ade2	NA, X, UV	Colony morphology	Leakiness	Raypulis and Kozhin (1966)
Schiz. pombe	ade6, ade7	NMU, NEU	Colony morphology	Specific revertibility	Abbondandolo and Loprieno (1967)
Schiz. pombe	ade6, ade7	HA, NA	Colony morphology	Mutant sectoring	Guglielminetti et al. (1967)
Schiz. pombe	ade6, ade7	NA	Colony morphology	Mitotic stability	Loprieno et al. (1968)
Schiz. pombe	ade6, ade7	Spontaneous	Colony morphology	Leakiness; pH sensitivity; temperature sensitivity	Loprieno et al. (1969b)
Schiz. pombe	ade6, ade7	MMS, EMS, HA, NA	Colony morphology	Leakiness; pH sensitivity; temperature sensitivity; suppressibility; complementation	Loprieno et al. (1969a)
Red (adenine requiring) → white (adenine requiring)					
Sacch. cerevisiae	ade3 to ade8	Spontaneous	Colony morphology	Genetic	Roman (1956a)
Sacch. cerevisiae	ade3 to ade8	Spontaneous	Colony morphology	Fine structure	Roman (1956b)
Schiz. pombe	ade1, 3, 4, 5, 9	NA	Colony morphology	Mutant sectoring	Nasim and Clarke (1965)
Schiz. pombe	ade1, 3, 4, 5, 9	MMS, EMS	Colony morphology	Mutant⁺ sectoring; leakiness	Loprieno (1966)
Prototroph (wild type) → auxotroph					
Sacch. cerevisiae	Nutritional	EMS, NA, UV	Failure to grow on minimal	Requirement	Lingens and Oltmanns (1964)

Organism	Locus	Mutagen	Detection	Notes	Reference
Sacch. cerevisiae	Nutritional	EMS	Failure to grow on minimal	Requirement	Lindegren *et al.* (1965)
Sacch. cerevisiae	Nutritional	NMG	Failure to grow on minimal	Requirement	Lingens and Oltmanns (1966)
Wild type → cytochrome mutants					
Sacch. cereviae	Rho factor	5FU	Indicator agar		Moustacchi and Marcovich (1963)
Sacch. cerevisiae	Rho factor, petite loci	NA, NIL, NMU	Indicator agar	Genic or cytoplasmic	Schwaier *et al.* (1968)
Sacch. cerevisiae	cyt1	UV, NA, NIL, ICR, HA	Indicator agar	Fine structure; specific revertibility	Parker and Sherman (1969)
Wild type → resistant					
Sacch. cerevisiae	can1	5-Aminoacridine	Selective growth		Magni *et al.* (1964)
Sacch. cerevisiae	can1	Spontaneous	Selective growth		Puglisi (1968)
Mutant → revertant					
Sacch. cerevisiae	ade6	NMA, NMU, NMG, NMT, NIL	Selective growth	Locus or suppressor	Schwaier (1965)
Sacch. cerevisiae	ade2, ade6	NA, NMG, NMA, NIL	Selective growth	Locus or suppressor	Zimmermann *et al.* (1966a)
Sacch. cerevisiae	Nonsense suppressors	HA, NA, ICR	Selective growth	Suppressor class	Magni and Puglisi (1966)
Sacch. cerevisiae	Nonsense suppressors	EMS, NA, BU, DAP, UV	Selective growth	Suppressor class	Hawthorne (1969)
Schiz. pombe	met4	NA, NMG, NMU, UV	Selective growth	Locus or suppressor	Loprieno and Clarke (1965)

[a] For abbreviations, see footnote [b], Table 1.

TABLE 3. Studies of Mitotic Segregation in Heterozygotes

Conditions studied[a]	Signal markers	Method of detection	Nature of event studied	References
UV induction	gal1, gal3	Visual selection on galactose–EMB medium	Reciprocal and nonreciprocal	James (1955); James and Lee-Whiting (1955)
Spontaneous occurrence	ade3, ade6	Enrichment for white sectors	Not tested	Roman (1956a, b)
X-ray induction	ade1, ade2, ade6	Visual selection for color	Not tested	Mortimer (1959)
UV induction; coincidence with heteroallelic reversion	Nutritional	Indirect selection by replica-plating	Reciprocal	Fogel and Hurst (1963)
UV induction; coincidence	Nutritional	Indirect selection by replica-plating	Not tested	Wilkie and Lewis (1963)
Mitomycin C; DNA synthesis inhibition	cyh2	Direct selection on actidione	Not tested	Holliday (1964)
UV induction; dark repair; photoreactivation; cell cycle synchrony	Nutritional	Indirect selection by replica-plating	Not tested	Parry and Cox (1965, 1968a, b)
Induction by mutagenic and carcinogenic agents: NA, NMG, DES- NIL, NMT, NMU, NMY, EI	ade2–his8	Visual selection by color	Reciprocal	Zimmermann et al. (1966b); Zimmermann and von Laer (1967)
UV and EMS induction; comparison	ade6	Visual selection by color	Reciprocal and nonreciprocal	Roman (1967, and personal communication)
UV induction in uvs strains	cyh2	Direct selection on actidione	Not tested	Snow (1968)
UV and X-ray induction; cell cycle synchrony	Nutritional	Visual selection for color on Magdala Red	Reciprocal and nonreciprocal	Esposito (1968)
UV and X-ray induction; relationship to meiotic linkage	Nutritional	Indirect selection by replica-plating	Reciprocal and nonreciprocal	Nakai and Mortimer (1969)

a For abbreviations, see footnote b, Table 1.

TABLE 4. Studies of Heteroallelic Reversion at Mitosis

Locus	Inducing agent[a]	Other conditions studied	References
ade1, ade2, ade3, ade4, ade5,7, ade6	Spontaneous	Tetrad analysis of prototrophs	Roman (1956[b])·
ade3	UV	Dose-frequency relationship	Roman and Jacob (1957)
ilv1	UV and	Dose-frequency relationship; segregation of outside markers	Roman and Jacob (1958)
arg4	X-rays	Dose-frequency relationship	Mortimer (1959)
ilv1	Spontaneous	Tetrad analysis of prototrophs; segregation of outside markers	Kakar (1963)
ade6	UV	Full-recovery technique to test reciprocity	Roman (1963)
trp5, arg4	X-rays	Allele mapping	Manney and Mortimer (1964) Manney (1964)
his1	UV and spontaneous	Tetrad analysis of prototrophs; segregation of outside markers	Hurst and Fogel (1964)
ade6	UV	Coincidence with segregation of homozygosity	Fogel and Hurst (1963)
ade2	UV	Photoreactivation; dark repair; cell cycle synchrony; comparison with mutation and segregation of homozygosity	Parry and Cox (1965, 1968*a*, *b*)
his4	X-rays	Allele mapping	Fink (1966)
leu1	UV, X-rays, heavy charged particles	Comparison with mutation and segregation of homozygosity; allele mapping	Nakai and Mortimer (1967)
ade2, trp5, ilv1	NA, NMY, NIL, NMG, BP, BA, DBA, NMU	Liquid holding, coincidence with segregation of homozygosity; tetrad analysis of prototrophs	Zimmermann and Schwaier (1967) Zimmermann (1968*a*, 1969)
ilv1	NMU	Properties of threonine dehydrase from prototrophs	Zimmermann (1968*b*)
his1	UV	Effect of uvs mutations	Snow (1968)
ade6	UV	Effect of homozygosity for mating type	Friis and Roman (1968)

TABLE 4. (Continued)

Locus	Inducing agent[a]	Other conditions studied	References
lys2, tyr1, his7, leu1, met13, tyr3, lys5, ade5	UV, X-rays	Cell cycle synchrony; tetrad analysis of proto- trophs; segregation of outside markers	Esposito (1968)
cyt1, cyt2, cyt3	X-rays	Allele mapping; compari- son with amino acid substitutions in cyto- chrome c	Parker and Sherman (1969)
trp5	Co^{60} gamma rays	Allele mapping	Manney et al. (1969)

[a] For abbreviations, see footnote b, Table 1.

at mitosis are not, strictly speaking, mutations. They are important to a discussion of chemical mutagenesis, however, for several reasons: (1) most mutagenic agents stimulate mitotic segregation to frequencies several thousand times the usual induced mutation frequencies; (2) mitotic segregation facilitates the expression of recessive mutations in diploid cell lines; and (3) these processes may have consequences, during development of multicellular organisms, that are more important than those resulting from mutations.

The relatively stable heterothallic diploid strains of *Sacch. cerevisiae* have provided useful marker systems for studying the genetic mechanisms of mitotic segregation (Tables 3 and 4). These systems now provide sensitive methods for evaluating the chemical and physical agents that stimulate mitotic segregation.

B. Detection of Mitotic Segregation

Two approaches have been used to study mitotic segregation in yeast: segregation of homozygotes from heterozygous diploids, and segregation of prototrophs from noncomplementing (auxotrophic) heteroallelic diploids.

1. Segregation of Homozygotes

In heterozygous diploids the segregation of homozygosity, revealed by the expression of recessive phenotypes, occurs spontaneously at a rate of approximately 10^{-5} to 10^{-4} per cell per generation for a marker not linked to its centromere (MacKay, unpublished). A variety of techniques have been developed for detecting homozygous recessive phenotypes. A direct selective method involves the use of recessive drug-resistance markers.

Segregation of homozygosity for one allele may be either reciprocal or

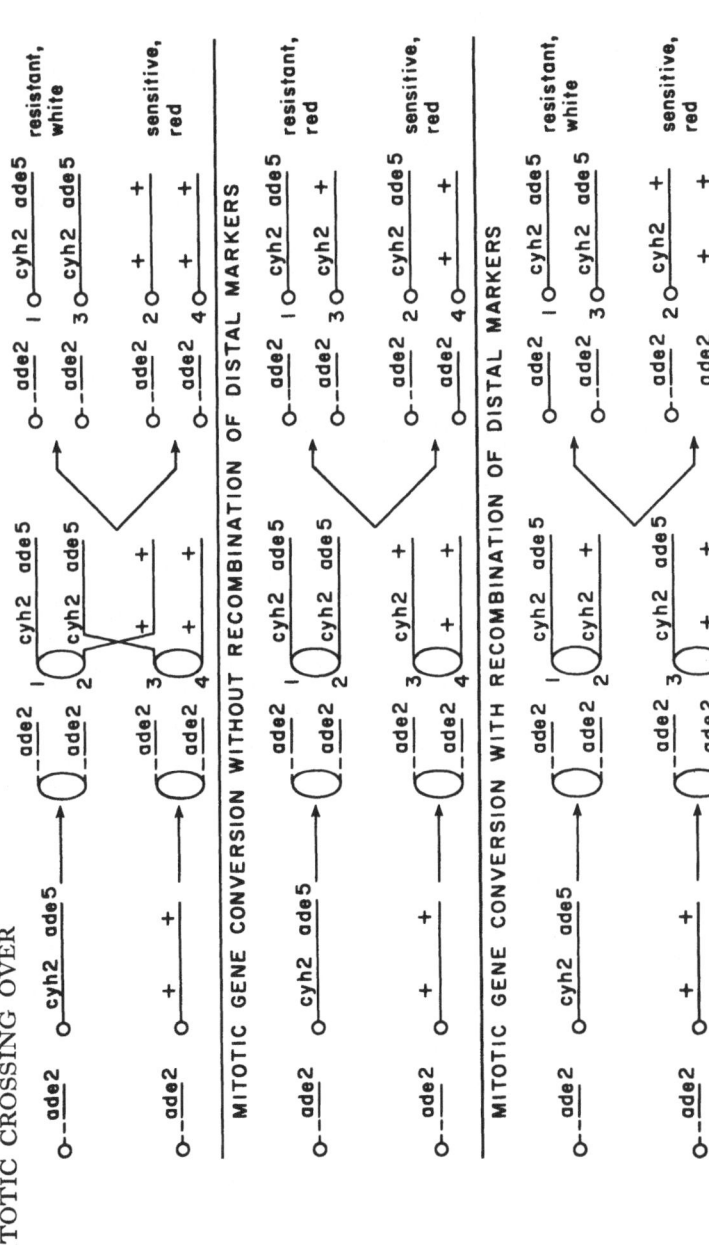

FIGURE 3. Progeny genotypes and phenotypes expected from reciprocal crossing-over and from gene conversion at mitosis. Definitions of genetic symbols: ade, adenine requirement; cyh, resistance to cycloheximide. Numbers designate particular genetic loci; the × designates the wild-type allele at a locus. The phenotypes designated at the right-hand side of the figure refer to resistance or sensitivity to cycloheximide (1 mg per liter) and to the color of the colony on yeast extract–peptone medium.

nonreciprocal with respect to the other allele at the locus. When the segregation is reciprocal, both sides of the resulting sectored colony are homozygous (mitotic crossing-over); when nonreciprocal (mitotic gene conversion), one sector is homozygous and the other heterozygous. Mitotic crossing-over nearly always results in homozygosity for all distal markers on the same chromosome arm, whereas mitotic gene conversion results in distal homozygosity in only a minor fraction of the segregants. In Fig. 3 the general types of segregations observed are illustrated with the selective recessive marker *cyh2* (resistance to cycloheximide) and the distal marker *ade5*. Homozygosity for *ade2* causes accumulation of red pigment in the colonies if *ade5* is heterozygous, but not if it is homozygous. Methods for detection of heterozygosity in the sector having the dominant phenotype will be discussed in a later section.

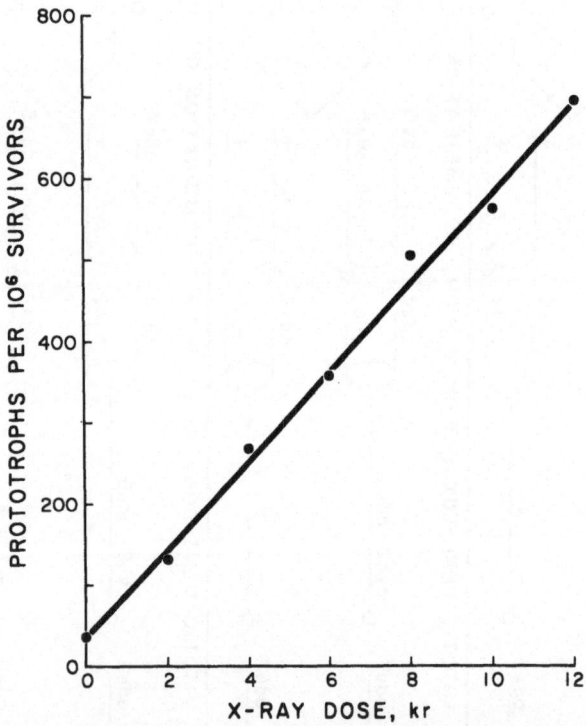

FIGURE 4. Dose-effect relationships for X-ray induction of prototrophs in a hetero-allelic diploid of the genotype *trp5–18/trp5–21*. The sensitivity (slope) corresponds to 5.5 prototrophs per 10^8 survivors per roentgen. The sensitivities for corresponding homoallelic diploids *trp5–18/trp5–18* and *trp5–21/trp5–21* are 0.002 and 0.009, respectively.

2. Segregation of Prototrophy

Diploids carrying noncomplementing allelic mutations (heteroalleles) segregate wild-type progeny at rates that are 10 to 1000 times haploid or homoallelic diploid reversion rates. When auxotrophic markers are used, these wild-type recombinants can be selected as prototrophs. This provides an extremely sensitive method for measuring the frequencies of mitotic events. Figure 4 illustrates the linear dose-effect relationship for X-ray induction of prototrophs in an auxotrophic diploid that was heteroallelic for the mutations *trp5-18/trp5-21* in the tryptophan synthetase gene. The slope of the curve is 5.5 prototrophs per 10^8 survivors per roentgen. In the absence of prototrophs of spontaneous origin in the culture, it would be possible to detect the effect of exposure to less than 1r of X-rays or an equivalent exposure to other mutagenic agents. In practice, the sensitivity is limited by the background of spontaneous prototrophs that accumulate during the growth of the culture. The spontaneous background can be minimized by using very small cultures grown from small inocula. The background can be further reduced, for ultimate senstivity, by selectively killing the prototrophs with Nystatin before treating with the mutagen (Strömnaes and Mortimer, 1968).

C. Genetic and Molecular Mechanisms

The genetic mechanisms of mitotic segregation have been relatively well defined. The corresponding molecular mechanisms, however, are still a subject for speculation. Genetically, two types of mitotic segregation processes can be distinguished—reciprocal and nonreciprocal. Reciprocal crossing-over between homologous chromosomes at the four-strand stage followed by a normal mitotic disjunction accounts for the majority of the sectored colonies that result from irradiation of a heterozygous diploid with UV or X-rays. However, a significant proportion of the mitotic segregants produced by radiation and most of those produced by certain chemical mutagens (e.g., EMS) cannot be explained on the basis of reciprocal crossing-over. This nonreciprocal process is generally termed *gene conversion*.

Except for their lower frequencies and the differences that are the consequences of mitotic chromosome mechanics, these reciprocal and nonreciprocal processes are comparable genetically to meiotic crossing-over and meiotic gene conversion, respectively.

In terms of genetic information, the mechanisms of these meiotic processes are generally understood. Reciprocal crossing-over between homologous chromosomes occurs at the four-strand stage and results in reciprocal interchange of long segments of chromosomes. It has been estimated that in yeast there are, on the average, from one to four such exchanges in each chromosome arm of each cell undergoing meiosis (Mortimer and Hawthorne, 1966b). At mitosis the frequency is probably lower by a factor of 10,000. The simplest explanation of the nonreciprocal process, gene conversion, is that it results in the replacement of an informational segment in one homolog with information identical to that contained in the corresponding section of a nonsister homolog (Fogel and Mortimer, 1969). The length of the informational segment transferred corresponds to approximately 100 nucleotide pairs. It has been estimated that approximately 2% of a haploid genome is replaced in meiosis by this process. The consequences of gene conversion are thereby different from both crossing-over and mutation.

Recent advances in our understanding of the enzymatic mechanisms of replication and repair of DNA in *Escherichia coli* have inspired several authors to propose molecular models for crossing-over and gene conversion in fungi (Whitehouse and Hastings, 1965; Holliday, 1964; Stahl, 1969). The elements of these models have been compared by Stahl (1969). They all assume the involvement of two processes: (1) breakage and rejoining of DNA duplexes, and (2) local DNA synthesis of one kind or another. In this latter feature they bear a certain resemblance to current models for mutagenesis (Drake, 1969). Accordingly, we may anticipate that the mode of action of mutagenic agents in the stimulation of mitotic segregation may, in some cases, correspond to the mode by which they produce mutations.

D. Experimental Systems for Studying Mitotic Segregation

1. Segregation of Homozygosity

The essential requirement for detection of mitotic segregation in a heterozygote is the ability to recognize the expression of the recessive phenotype. Some of the most frequently used types of marker combinations are illustrated in Fig. 3. The strain whose genotype is illustrated would normally accumulate a red pigment because of the homozygosity for *ade2*. Segregation of homozygosity at the *ade5* locus, however, blocks pigment production and results in a white phenotype (Roman, 1956a). Colonies of this strain grown on yeast-extract medium (YEPD) for several days are red but eventually develop white papillations as a result of spontaneous mitotic segregation of *ade5*. When irradiated cells are plated on this medium, some red/white sectored colonies develop as a result of mitotic segregation in one of the first several divisions. The above strain, which is also heterozygous for the cycloheximide (actidione) resistance marker *cyh2* (formerly designated *ac₂ʳ*), is inhibited by 1 μg/ml of cycloheximide, the allele for resistance being recessive to the allele for sensitivity (Wilkie and Lee, 1965). Mitotic segregants that are homozygous for the recessive resistance allele, however, can be selected either directly or indirectly (by replica plating). The latter method allows isolation of the sensitive sector of the colony as well. Finally, homozygosity for any recessive auxotrophic markers can be detected by replica plating to suitable omission media or directly on media containing the dye Magdala Red (Horn and Wilkie, 1966; Esposito, 1968; Strömnaes, 1968). On complete medium containing this dye, colonies of respiratory-competent prototrophic cells stain a pale pink but respiratory-competent auxotrophs stain intensely red. Petite colonies also accumulate the dye.

Table 3 lists a number of specific marker systems that have been used in conjunction with these methods to detect mitotic segregation in heterozygotes.

2. Segregation of Prototrophy

Selection and measurement of prototroph frequencies in cultures of noncomplementing heteroallelic diploids are the same as measuring reversion frequencies of haploids or homoallelic diploids except that the frequencies are much greater. Special precautions must be taken to avoid cultures that have a high background of spontaneous prototrophs. This is easily accomplished by growing small cultures (no more than 5 ml) from small inocula (less than 100 cells). Qualitative or semiquanitative spot tests for heteroallelic reversion can be achieved by replica plating on solid medium (Fink, 1966). Prototrophs are detected as small colonies appearing on the replica print after 2 or 3 days of incubation. The sensitivity of the test can be controlled by varying the number of cells transferred.

Many physical, chemical, physiological, and genetic factors affecting the frequency of heteroallelic reversion have been studied. Table 4 outlines some examples of the studies that have been reported.

3. Systems for Distinguishing Between Reciprocal and Nonreciprocal Segregation

When reciprocal recombination is responsible for mitotic segregation of homozygosity, the two reciprocal products are found in separate clones in a sectored colony (assuming both sectors are viable). To characterize a segregation as reciprocal or nonreciprocal therefore requires that both sectors be isolated and analyzed. Sectored colonies are readily detected by any of the indirect selection methods mentioned above. As a general approach, the sector with the dominant phenotype can be sporulated and analyzed by either tetrad or random spore analysis. This approach, however, is laborious and therefore not suitable for quantitative studies or as a screening method. However, with marker systems in which the recessive phenotype can be selected, heterozygous dominant sectors can be distinguished from homozygous dominant sectors by much simpler tests (Roman, 1967, and personal communication). The principle of this method is illustrated in Fig. 3 for recessive resistance to cycloheximide and for the red/white adenine color system. The distinction is based on the fact that heterozygotes are relatively unstable and continuously segregate recessive clones. In practice, treated cells are plated and allowed to form colonies on nonselective medium. In the case of *cyh2*, the colonies are replica-plated onto selective medium to detect sectors. The dominant phenotype sectors (cycloheximide sensitive) are then incubated on selective medium for several days more. Heterozygous sectors are distinguished from homozygous ones by the appearance of small clones of resistant segregants. The use of the adenine color selection system is similar. White homozygous *ade5* segregants are selected in favor of the red parent phenotype, forming white papillations after prolonged incubation (Roman, 1956*b*).

E. Relationship Between Reciprocal and Nonreciprocal Mitotic Segregation

1. Observed Patterns of Mitotic Segregation

Gene conversion and reciprocal crossing-over may reflect different aspects of the same molecular events. The recombination models proposed by Holliday (1964), Whitehouse and Hastings (1965), and Stahl (1969) explain conversion as a consequence of the molecular events in the region of exchange. Only when a heterozygous site is within the exchange region is conversion possible. Reciprocal crossing-over occurs when the markers are sufficiently removed from the exchange region to not be involved in the conversional events. The different models all account for the observations that only about half of the gene conversion events result in recombination of outside markers.

The same general relationships between meiotic reciprocal and non-reciprocal recombination can be observed in mitotic segregations. This is illustrated by the data in Table 5, which were obtained with the selective systems diagrammed in Fig. 3. Diploid cells were treated with ethyl methanesulfonate (3% solution for 2 hr at 30°C; Roman, personal communication)

TABLE 5. Genotypes of Sectored Colonies Resulting from EMS-Induced Mitotic Segregation

Sensitive sector	Resistant sector			
	$\dfrac{cyh2 \quad ade5}{cyh2 \quad +}$		$\dfrac{cyh2 \quad ade5}{cyh2 \quad ade5}$	
$\dfrac{+}{+}$	0		5	
$\dfrac{cyh2}{+}$	18		7	

TABLE 6. Relation Between Mitotic Sectoring and Centromere Distance

Locus	Centromere distance[a] (C.M.)	Percentage frequency of sectoring		
		X-ray[a]	UV[a]	EMS[b]
leu1	2.5	0.14	0.26	0.48
trp5	21	0.23	0.41	0.40
met13	86	1.3	1.1	0.56
tyr3	104	1.5	1.2	0.56
lys5	112	1.5	1.3	0.44
ade5,7	167	2.3	1.8	1.13
Percentage sectored colonies:		2.9	2.4	2.9

[a] Data from Nakai and Mortimer (1969).
[b] Data from MacKay (unpublished).

and plated on complete medium. After colonies had developed, they were replica-plated to cycloheximide-containing medium. Colonies sectored for resistance to cycloheximide were picked and further characterized with respect to heterozygosity or homozygosity of the cycloheximide-sensitive sector and of the distal ade5 marker. Of the 30 colonies analyzed, five were reciprocally recombinant at the cyh2 locus and all five were recombinant for the distal marker. The other 25 were nonreciprocal at cyh2, and of these only seven were recombinant for the distal marker. The relative frequencies of induced reciprocal and nonreciprocal segregation are strikingly different for different mutagenic agents. The majority of the sectored colonies induced by UV or X-rays are reciprocal, whereas most of those induced by EMS are nonreciprocal (Roman, 1967; Nakai and Mortimer, 1969; see also Table 5). Furthermore, the relationship between the frequency of mitotic sectoring at a locus and its distance from its centromere depends on the inducing agent. The data summarized in Table 6 illustrate the correlation between sectoring frequency and centromere distance observed with UV or X-rays and the absence of such a correlation with EMS. This is evidently a reflection of the differences in relative frequencies of reciprocal and nonreciprocal events, the former, but not the latter, being dependent on the centromere distance.

Mitotic segregation of prototrophs from heteroallelic diploids is nearly always nonreciprocal, and exhibits the expected relationship to recombination of outside markers (see references in Table 2).

2. Implications for Evaluation of Mutagens

Mitotic crossing-over and mitotic gene conversion provide two empirically different responses to mutagenic agents. Although the molecular basis of these differences are not understood, the observations provide a promising approach to the analysis of the genetic consequences of mutagenic action. The observations also provide further criteria for characterizing mutagenic agents.

Mitotic segregation of prototrophs from heteroallelic diploids provides a simple, highly senstive, and relatively specific method for measuring induced gene conversion (Fig. 4). Technically, this method is more convenient than measuring nonreciprocal segregation of homozygosity.

The quantitation of induced mitotic crossing-over, on the other hand, presents a more difficult problem. Direct analysis of the two sectors of each colony, by the simplified mitotic methods discussed, is tedious, but is the only reliable method available that permits analysis of sufficiently large samples. Concomitant sectoring of a distal marker is not a completely reliable criterion for reciprocal exchange, as illustrated by the data in Table 5. The differences are sufficient, however, to provide a convenient semiquantitative method for characterizing inducing agents. With the marker system diagrammed in Fig. 3, cycloheximide-resistant colonies can be selected directly on YEPD containing 1 μg/ml of cycloheximide. The ratio of the number of red colonies to the number of white colonies will then be an index of the relative frequencies of reciprocal and nonreciprocal segregations induced.

V. ACKNOWLEDGMENTS

We thank Miss Vivian MacKay for permission to use her unpublished data.

VI. REFERENCES

Abbondandolo, A., and Loprieno, N. (1967), *Mutation Res. 4*, 31–36.
da Cunha, M. F. (1969), Thesis, Universidade do Rio de Janeiro.
de Serres, F. J. (1964), *J. Cell. Comp. Physiol. 64* (Suppl. 1), 1–18.
Drake, J. W. (1969), *Ann. Rev. Genet. 3*, 247–268.
Esposito, R. E. (1968), *Genetics 59*, 191–210.
Fink, G. R. (1966), *Genetics 53*, 445–459.
Fink, G. R., and Lowenstein, R. (1969), *J. Bacteriol. 100*, 1126–1127.
Fogel, S., and Hurst, D. D. (1963), *Genetics 48*, 321–328.
Fogel, S., and Mortimer, R. K. (1969), *Proc. Nat. Acad. Sci. 62*, 96–103.
Friis, J., and Roman, H. (1968), *Genetics 59*, 33–36.
Gilmore, R. A. (1967), *Genetics 56*, 641–658.
Gilmore, R. A., Stewart, J. W., and Sherman, F. (1968), *Biochim. Biophys. Acta 161*, 270–272.
Gits, J. J., and Grenson, M. (1967), *Biochim. Biophys. Acta 135*, 507–516.
Grenson, M., Mousset, M., Wiame, J. M., and Bechet, J. (1966), *Biochim. Biophys. Acta 127*, 325–338.

Guglielminetti, R., Bonatti, S., Loprieno, N., and Abbondandolo, A. (1967), *Mutation Res. 4*, 441–447.

Hawthorne, D. C. (1969), *Mutation Res. 7*, 187–197.

Hawthorne, D. C., and Mortimer, R. K. (1963), *Genetics 48*, 617–620.

Hawthorne, D. C., and Mortimer, R. K. (1968), *Genetics 60*, 735–742.

Holliday, R. (1964), *Genet. Res. Camb. 5*, 282–304.

Holliday, R. (1969), *Genetics 50*, 323–335.

Horn, P., and Wilkie, D. (1966), *J. Bacteriol. 91*, 1388.

Hurst, D. D., and Fogel, S. (1964), *Genetics 50*, 435–458.

James, A. P. (1955), *Genetics 40*, 204–213.

James, A. P., and Lee-Whiting, B. (1955), *Genetics 40*, 826–831.

Kakar, S. N. (1963), *Genetics 48*, 957–966.

Lacroute, F. (1963), *Compt. Rend. Acad. Sci. 257*, 4213–4216.

Leupold, U. (1955), *Arch. Jul. Klaus-Stiftung 30*, 506–516.

Leupold, U. (1958), *Cold Spring Harbor Symp. Quant. Biol. 23*, 161–170.

Lindegren, G., Hwang, Y. L., Oshima, Y., and Lindegren, C. C. (1965), *Can. J. Genet. Cytol. 7*, 491–499.

Lingens, F., and Oltmanns, O. (1964), *Z. Naturforsch. 19b*, 1058–1065.

Lingens, F. and Oltmanns, O. (1966) *Z. Naturforsch. 21b*, 660–663.

Loprieno, N. (1966), *Mutation Res. 3*, 486–493.

Loprieno, N., and Clarke, C. H. (1965), *Mutation Res. 2*, 312–319.

Loprieno, N., Abbondandolo, A., Bonatti, S., and Guglielminetti, R. (1968), *Genet. Res. Camb. 12*, 45–54.

Loprieno, N., Gugielminetti, R., Bonatti, S., and Abbondandolo, A. (1969a), *Mutation Res. 8*, 65–71.

Loprieno, N., Bonatti, S., Abbondandolo, A., and Guglielminetti, R. (1969b), *Mol. Gen. Genet. 104*, 40–50.

Mackler, B., Douglas, H. C., Will, S., Hawthorne, D. C., and Mahler, H. R. (1965), *Biochemistry 4*, 2016–2020.

Magni, G. E., and Puglisi, P. (1966), *Cold Spring Harbor Symp. Quant. Biol. 31*, 699–704.

Magni, G. E., and von Borstel, R. C. (1962), *Genetics 47*, 1097–1108.

Magni, G. E., von Borstel, R. C., and Sora, S. (1964), *Mutation Res. 1*, 227–330.

Manney, T. R. (1964), *Genetics 50*, 109–121.

Manney, T. R., and Mortimer, R. K. (1964), *Science 143*, 581–582.

Manney, T. R., Duntze, W., Janosko, N., and Salazar, J. (1969), *J. Bacteriol. 99*, 590–596.

Mortimer, R. K. (1959), *Radiation Res. Suppl. 1*, 394–402.

Mortimer, R. K., and Hawthorne, D. C. (1966a), *Ann. Rev. Microbiol. 20*, 151–168.

Mortimer, R. K., and Hawthorne, D. C. (1966b), *Genetics 53*, 165–173.

Mortimer, R. K., and Hawthorne, D. C. *in* "The Yeasts" (A. H. Rose and J. S. Harrison, eds.) pp. 385–460, Academic Press, London.

Mortimer, R. K., Brustad, T., and Cormack, D. V. (1965), *Radiation Res. 26*, 465–482.

Moustacchi, E., and Marcovich, H. (1963), *Compt. Rend. Acad. Sci. 256*, 5646–5648.

Nagai, S. (1963), *J. Bacteriol. 86*, 299–302.

Nakai, S., and Mortimer, R. (1967), *Radiation Res. Suppl. 7*, 172–181.

Nakai, S., and Mortimer, R. (1969), *Mol. Gen. Genet. 103*, 329–338.

Nashed, N., and Jabbur, G. (1966), *Z. Vererb. 98*, 106–110.

Nasim, A., and Clarke, C. H. (1965), *Mutation Res. 2*, 395–402.

Parker, J. H., and Sherman, F. (1969), *Genetics 62*, 9–22.

Parry, J. M., and Cox, B. S. (1965), *J. Gen. Microbiol. 40*, 235–241.

Parry, J. M., and Cox, B. S. (1968a), *Mutation Res. 5*, 373–384.

Parry, J. M., and Cox, B. S. (1968b), *Genet. Res. 12*, 187–198.

Puglisi, P. P. (1968), *Mol. Gen. Genet. 103*, 248–252.

Raypulis, E. P., and Kozhin, S. A. (1966), *Tr. Moscow Soc. Natural. 22*, 135–139.

Resnick, M. A. (1969), *Genetics 62*, 519–531.

Roman, H. (1956a), *Compt. Rend. Trav. Lab. Carlsberg 26*, 299–314.

Roman, H. (1956b), *Cold Spring Harbor Symp. Quant. Biol. 21*, 175–185.

Roman, H. (1963), *in* "Methodology in Basic Genetics" (W. J. Burdette, ed.) pp. 209–227, Holden-Day San Francisco.

Roman, H. (1967), *J. Cell. Biol. 70* (Suppl. 1), 116–118.

Roman, H., and Jacob, F. (1957), *Compt. Rend. Acad. Sci. 245*, 1032–1034.

Roman, H., and Jacob, F. (1958), *Cold Spring Harbor Symp. Quant. Biol. 23*, 155–160.

Schwaier, R. (1965), *Z. Vererb. 97*, 55–67.

Schwaier, R., Nashed, N., and Zimmermann, F. K. (1968), *Mol. Gen. Genet. 102*, 290–300.

Sherman, F. (1963), *Genetics 48*, 375–385.

Sherman, F., Stewart, J. W., Margoliash, E., Parker, J., and Campbell, W. (1966), *Proc. Nat. Acad. Sci. 55*, 1498–1504.

Slonimski, P. P., Perrodin, G., and Croft, J. H. (1968), *Biochem. Biophys. Res. Commun. 30*, 232–239.

Snow, R. (1968), *Mutation Res. 6*, 409–418.

Stahl, F. (1969), *Genetics 61* (Suppl. 1), 1–13.

Strömnaes, O. (1968), *Hereditas 59*, 197–220.

Strömnaes, O., and Mortimer, R. K. (1968), *J. Bacteriol. 95*, 197–200.

Surdin, Y., Sly, W., Sire, J., Bordes, A. M., and de Robichon-Szulmajster, H. (1965), *Biochim. Biophys. Acta 107*, 546–566.

Tavlitski, J. (1949), *Ann. Inst. Pasteur 76*, 497–509.

Whitehouse, H. L. K., and Hastings, P. J. (1965), *Genet. Res. Camb. 6*, 27–92.

Wilkie, D., and Lee, B. K. (1965), *Genet. Res. Camb. 6*, 130–138.

Wilkie, D., and Lewis, D. (1963), *Genetics 48*, 1701–1716.

Winge, O., and Roberts, C. (1958), *in* "Chemistry and Biology of Yeast" (A. H. Cook, ed.) pp. 123–156, Academic Press, New York.

Woods, R. A., and Bevan, E. A. (1966), *Heredity 21*, 121–130.

Zimmermann, F. K. (1968a), *Mol. Gen. Genet. 101*, 171–184.

Zimmermann, F. K. (1968b), *Mol. Gen. Genet. 103*, 11–20.

Zimmermann, F. K. (1969), *Z. Krebsforsch. 72*, 65–71.

Zimmermann, F. K., and Schwaier, R. (1967), *Mol. Gen. Genet. 100*, 63–76.

Zimmermann, F. K., and von Laer, U. (1967), *Mutation Res. 4*, 377–379.

Zimmermann, F. K., Schwaier, R., and von Laer, U. (1966a), *Z. Vererb. 98*, 152–166.

Zimmermann, F. K., Schwaier, R., and von Laer, U. (1966b), *Z. Vererb. 98*, 230–246.

Author Index

Subject Index